The Physiology of Bioelectricity in Development, Tissue Regeneration, and Cancer

T0225548

Biological Effects of Electromagnetics Series

Series Editors

Frank Barnes
University of Colorado
Boulder, Colorado, U.S.A.

Ben Greenebaum
University of Wisconsin–Parkside
Somers, Wisconsin, U.S.A.

Advanced Electroporation Techniques in Biology and Medicine,
edited by Andrei G. Pakhomov, Damijan Miklavčič, and Marko S. Markov

The Physiology of Bioelectricity in Development, Tissue Regeneration, and
Cancer, *edited by Christine E. Pullar*

The Physiology of Bioelectricity in Development, Tissue Regeneration, and Cancer

Edited by
Christine E. Pullar

CRC Press
Taylor & Francis Group
Boca Raton London New York

CRC Press is an imprint of the
Taylor & Francis Group, an **informa** business

CRC Press
Taylor & Francis Group
6000 Broken Sound Parkway NW, Suite 300
Boca Raton, FL 33487-2742

First issued in paperback 2017

© 2011 by Taylor and Francis Group, LLC
CRC Press is an imprint of Taylor & Francis Group, an Informa business

No claim to original U.S. Government works

ISBN-13: 978-1-4398-3723-8 (hbk)
ISBN-13: 978-1-138-07783-6 (pbk)

This book contains information obtained from authentic and highly regarded sources. Reason-able efforts have been made to publish reliable data and information, but the author and publisher cannot assume responsibility for the validity of all materials or the consequences of their use. The authors and publishers have attempted to trace the copyright holders of all material reproduced in this publication and apologize to copyright holders if permission to publish in this form has not been obtained. If any copyright material has not been acknowledged please write and let us know so we may rectify in any future reprint.

Except as permitted under U.S. Copyright Law, no part of this book may be reprinted, reproduced, transmitted, or utilized in any form by any electronic, mechanical, or other means, now known or hereafter invented, including photocopying, microfilming, and recording, or in any information storage or retrieval system, without written permission from the publishers.

For permission to photocopy or use material electronically from this work, please access www.copyright.com (http://www.copyright.com/) or contact the Copyright Clearance Center, Inc. (CCC), 222 Rosewood Drive, Danvers, MA 01923, 978-750-8400. CCC is a not-for-profit organiza-tion that provides licenses and registration for a variety of users. For organizations that have been granted a photocopy license by the CCC, a separate system of payment has been arranged.

Trademark Notice: Product or corporate names may be trademarks or registered trademarks, and are used only for identification and explanation without intent to infringe.

Visit the Taylor & Francis Web site at
http://www.taylorandfrancis.com

and the CRC Press Web site at
http://www.crcpress.com

Contents

Foreword

It is our pleasure to introduce this volume in the series of books on the effects of electric and magnetic fields on biological systems. This series of books is intended to update and extend material that was reviewed in the third edition of the *Handbook of Biological Effects of Electric and Magnetic Fields*. The activity in this field includes both issues associated with low levels of exposures, including concerns about the possible health effects of extended exposures, and applications of electric and magnetic fields for therapeutic applications.

Our hope is that this volume will be valuable as a guide to the current state of knowledge and techniques for those who wish to understand the physiology of bioelectricity in tissue regeneration in the clinical treatment of human diseases. Future volumes will have the same goals for other aspects of the biological effects of electric and magnetic fields.

Frank Barnes
Ben Greenebaum

Preface

Bioelectricity was discovered by Galvani in the late 1700s. During a public experiment in Bologna, Italy, in 1794, he showed that the cut end of a frog sciatic nerve from one leg could stimulate contractions when it touched the muscles of the opposite leg, a demonstration of the injury potential. Over two hundred years later, we have strong evidence that direct current, electric field gradients (bioelectricity) exist in all developing and regenerating animal tissues and have, more recently, been measured around tumors and sites of inflammation. Ion flow appears to be fundamental in all aspects of bioelectricity. Indeed, we have evolved to use about half of our cells' energy, in the form of ATP, maintaining ion concentration gradients and electric fields, highlighting their importance to life. Moreover, ionic current flow and electric field gradients are interdependent and inseparable vectors with both magnitude and direction, providing cues for the spatial organization of developing/regenerating multicellular tissues.

In recent years, advances in technology have facilitated the accurate measurement of endogenous electric fields around sites of tissue disruption. Meanwhile, state-of-the-art molecular approaches have demonstrated that bioelectricity can control the directionality of cell migration, the speed of migration, proliferation, apoptosis, differentiation, and orientation. This level of physiological control can initiate and coordinate complex regenerative responses in development and wound repair and may play a role in cancer progression and metastasis.

Here we bring together, for the first time, current research in the area of bioelectricity.

Methods for detecting endogenous electric field gradients and studying applied electric fields in the lab are provided. The roles of bioelectricity in guiding cell behavior during morphogenesis and orchestrating higher-order patterning are described. The response of stem cells to applied electric fields reveals bioelectricity as an exciting new player in tissue engineering and regenerative medicine. How electric signals control corneal wound repair and skin reepithelialization, angiogenesis, and inflammation is explored in depth, and the bioelectric responses of cells derived from the musculoskeletal system are described. The bioelectrical guidance of neurons and the beneficial application of voltage gradients to promote regeneration in the spinal cord are also explored in depth. Finally, we reveal that bioelectricity can play a significant role in the progression of cancer. This relatively new field has the potential of revealing novel cancer biomarkers, providing a new method for early detection of cancer and producing bioelectricity-based therapies to target both the tumor and metastatic cancer cells.

This book is a multidisciplinary compilation intended for biologists, biomedical scientists, engineers, and clinicians, but would be beneficial to anyone with an interest in development, regeneration, cancer, and tissue engineering. It would also make an ideal textbook for students in biology, medicine, biophysics, and biomedical engineering.

Finally, I thank all the authors who contributed their time and effort to make this book a reality and hopefully a success!

Christine E. Pullar

About the Editor

Christine E. Pullar is a lecturer at the University of Leicester in the UK. She received her PhD in immune cell signal transduction from the University of Sheffield, UK. She moved to the University of California, Davis, in 2000 to work as a postdoctoral scientist with Richard Nuccitelli and Rivkah Isseroff on a project studying how keratinocytes are able to sense and migrate directionally in response to applied physiological strength electric fields in the context of wound repair. After two years she obtained her own funding in the form of a five-year NIH Career Award to continue her studies on cell migration and wound repair.

In 2006, she moved to the Department of Cell Physiology and Pharmacology at the University of Leicester to establish a laboratory studying the physiological processes in wound repair, with a focus on both the role of the beta-adrenergic receptor family and endogenous electric fields. Her lab is currently funded by the Wellcome Trust, the Medical Research Council, and the British Skin Foundation. Her work has a strong translational flair, and projects aim to promote healing in chronic wounds and reduce wound scarring; several patents are held in this area. She is a member of a number of cell biology and dermatology societies, was a board member for the Society for the Physical Regulation of Biology in Medicine (SPRBM) from 2006 to 2009, and is active in the Bioelectromagnetics Society (BEMS), in addition to chairing the Gordon Conference on Bioelectrochemistry in 2012. She has delivered invited lectures at more than twenty international meetings and is active in mentoring young scientists within the research community.

Contributors

Huai Bai
Unit of Laboratory Medicine
West China Second Hospital of
    Sichuan University
Chengdu, People's Republic of China

Richard Ben Borgens
Center for Paralysis Research
Department of Basic Medical Sciences
School of Veterinary Medicine
Purdue University
West Lafayette, Indiana

J. Chlöe Bulinski
Department of Biological Sciences
Columbia University
New York, New York

Michael Cho
Department of Bioengineering
University of Illinois
Chicago, Illinois

Mustafa B. A. Djamgoz
Division of Cell and Molecular Biology
Neuroscience Solutions to Cancer
    Research Group
Imperial College London
London, United Kingdom

Najmuddin J. Gunja
Department of Biomedical
    Engineering
Columbia University
New York, New York

Francis X. Hart
Department of Physics
University of the South
Sewanee, Tennessee

Clark T. Hung
Department of Biomedical
    Engineering
Columbia University
New York, New York

Michael Levin
Biology Department
and
Tufts Center for Regenerative and
    Developmental Biology
Tufts University
Medford, Massachusetts

Francis Lin
Department of Physics and Astronomy
University of Manitoba
Winnipeg, Manitoba

Colin D. McCaig
School of Medical Sciences
Institute of Medical Sciences
University of Aberdeen
Aberdeen, Scotland

Richard Nuccitelli
BioElectroMed Corp.
Burlingame, California

Christine E. Pullar
Department of Cell Physiology and
    Pharmacology
University of Leicester
Leicester, United Kingdom

Ann M. Rajnicek
School of Medical Sciences
Institute of Medical Sciences
University of Aberdeen
Aberdeen, Scotland

Vidya Rao
Department of Bioengineering
University of Illinois
Chicago, Illinois

Brian Reid
Department of Dermatology
School of Medicine
Center for Neurosciences
University of California, Davis
Davis, California

Shan Sun
Department of Bioengineering
University of Illinois
Chicago, Illinois

Igor Titushkin
Department of Bioengineering
University of Illinois
Chicago, Illinois

Entong Wang
Department of Otolaryngology-Head
    and Neck Surgery
General Hospital of Air Force
Beijing, People's Republic of China

Yili Yin
Wellcome Trust Centre for Gene
    Regulation and Expression
College of Life Sciences
University of Dundee
Dundee, Scotland

Min Zhao
Department of Dermatology
School of Medicine
Center for Neurosciences
University of California, Davis
Davis, California

Zhiqiang Zhao
Institute of Medical Sciences
University of Aberdeen
Aberdeen, Scotland

1 Measuring Endogenous Electric Fields

Richard Nuccitelli
BioElectroMed Corp.
Burlingame, California

CONTENTS

INTRODUCTION

We are electric beings. Every one of our senses utilizes ionic currents and electric fields at the heart of its transduction mechanism. You are able to read these words because a steady "dark current" is flowing through the rods in your retina, and this current is being modulated by light absorbed by rhodopsin (Figure 1.1). Light-activated rhodopsin triggers an amplifying signal transduction cascade that leads to the reduction in the cGMP gating of the Na^+ channel, and this results in a reduction in the dark current running through that rod. This change in current modifies the rod's membrane potential and influences signaling to other cells in the visual cortex that eventually forms an image in your brain. You are able to hear because compressions of air impinging on your eardrum move hairs or cilia on the cells of your inner ear. These cilia exhibit a "gating spring" mechanism that opens ion channels at their tips when they vibrate to transduce the vibration into membrane potential changes in the hair cell. It is these electrical signals that the brain perceives as noise. Similarly, the senses of

1

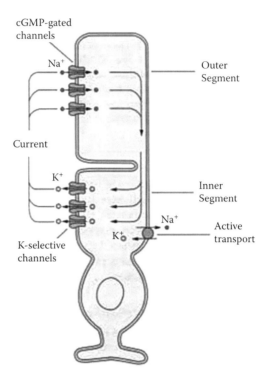

FIGURE 1.1 **See color insert.** Diagram of a single retinal rod cell illustrating the segregation of ion channels that leads to the generation of a dark current. Na^+ channels found only in the outer segment are gated by cGMP and pass the positive inward current there. K^+ channels are localized in the inner segment and pass the outward current. Photon absorbance by rhodopsin in the outer segment triggers a transduction reaction that results in the reduction of cGMP and leads to the reduction of the inward Na^+ current.

touch, taste, and smell all have receptor systems that generate electrical signals sent to the brain.

Our dependence on electric fields (EF) extends beyond our senses. Every cell in our body generates a voltage difference across the plasma membrane called a membrane potential that is about 70 mV, inside negative. Every organ is surrounded by a monolayer of cells called an epithelium that generates a potential difference across itself of 30 to 100 mV, inside positive. These membrane potentials and transepithelial potentials are generated by ion concentration gradients or ion flow across membranes or cell layers. So it should not surprise you that about half of the energy coming from our mitochondria in the form of ATP is used by our cells and organs to generate these ion concentration gradients (Clausen et al., 1991). The fact that we have evolved to expend half of our energy in generating ion concentration gradients and electric fields within our cells and tissues illustrates the importance of these fields for our life process. This book brings together many examples of the use of bioelectric signaling in development, regeneration, and cancer, and it is only natural to begin with a discussion of how such endogenous electric fields are detected.

THE INJURY POTENTIAL/ENDOGENOUS WOUND ELECTRIC FIELD

The injury potential was discovered by Galvani in the late 1700s (Piccolino, 1997, 2000). He demonstrated that the severed end of a frog sciatic nerve in one leg could stimulate contractions when it touched the muscles in the other leg. An injury potential is a steady, long-lived, direct current (DC)–induced voltage gradient within extracellular spaces that is generated by current flowing out of a tissue wound (McCaig et al., 2005). Injury currents normally flow out of wounds because the epithelium surrounding all organs is polarized and generates a voltage difference across itself, inside positive. When a low-resistance pathway is generated across the epithelium, this transepithelial potential will drive current out through that pathway.

Our skin is composed of the stratum corneum, epidermis, and dermis. The epidermis has multiple layers of polarized epithelial cells or keratinocytes. These keratinocytes are highly polarized, with the majority of Na^+ channels located in the apical membrane and the majority of K^+ channels found within the basolateral membranes along with the Na^+/K^+ ATPase. This polarized distribution of channels and pumps leads to a net movement of Na^+ across the epidermis from the apical side to the basal side, generating a transepidermal ion flow that returns through the paracellular route between the epidermal cells. This flow between the epidermal cells generates a voltage difference across the epidermis that is referred to as the transepidermal potential (TEP) (Figure 1.2) (Nuccitelli, 2003). The transepithelial potential is proportional to the resistance of the paracellular pathway, but typically will be between 15 and 60 mV, inside positive across human epidermis (Nuccitelli, 2003). Wherever the resistance across the epidermis is compromised, i.e., at a wound or where tight junction resistance is reduced (e.g., in developing chick embryos at the primitive streak (Jaffe and Stern, 1979), the posterior intestinal portal (Hotary and Robinson, 1990), or at the forming limb bud (Altizer et al., 2001; Robinson, 1983)), current will flow through the low-resistance pathway. This "leakage" current that flows back under the stratum corneum generates a lateral electric field gradient that will be proportional to the resistivity in that narrow space. The wound site becomes more negative than the surrounding tissue and is therefore the cathode of the wound EF (Nuccitelli, 2003; Ojingwa and Isseroff, 2003) (Figure 1.2). It is essential to point out here that moisture is required in order to carry the ionic flow, perhaps underlying the importance of a moist healing environment to facilitate wound healing (Ojingwa and Isseroff, 2003).

The existence of wound-induced DC EFs has been demonstrated experimentally in bovine cornea, human skin, and other multilayered epithelia (see review by McCaig et al., 2005). As the collapse of the TEP at the wound occurs immediately upon injury, it is the earliest guidance cue present within the wound, and

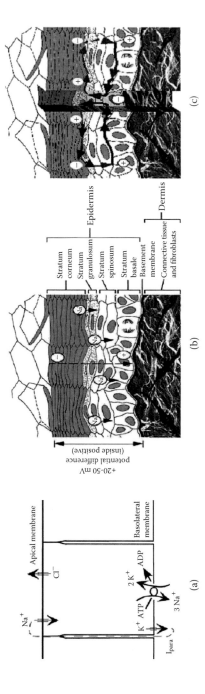

FIGURE 1.2   Generation of skin wound electric fields. (a) Diagram of a typical epithelial cell in a monolayer with Na$^+$ and Cl$^-$ channels localized on the apical plasma membrane and K$^+$ channels localized on the basolateral membranes along with the Na$^+$/K$^+$-ATPase. This asymmetric distribution of ion channels generates a transcellular flow of positive current that must flow back between the cells through the paracellular pathway (I$_{para}$). This current flow generates a transepithelial potential that is positive on the basolateral side of the monolayer. (b) Unbroken skin maintains this "skin battery" or transepidermal potential of 20 to 50 mV. (c) When wounded this potential drives current flow through the newly formed low-resistance pathway, generating a lateral electric field whose negative vector points toward the wound center at the lower portion of the epidermis and away from the wound on the upper portion just beneath the stratum corneum. (Redrawn from Nuccitelli, *Curr. Top. Dev. Biol.*, 58, 1–26, 2003.)

all cell "wound healing" behaviors within approximately 1 mm of a wound edge take place under the influence of this steady electric field.

ALTERING THE INJURY POTENTIAL CAN CAUSE DEVELOPMENTAL DEFECTS AND ALTER THE RATE OF WOUND HEALING

Modulating this endogenous current can cause serious developmental defects and can alter the rate of wound repair. Axolotl embryos were placed in physiological EFs during either gastrulation or neurulation (Metcalf and Borgens, 1994). In the absence of an externally applied EF, abnormalities occurred in 17% of embryos, while 62 and 43% developed abnormally in applied fields of 75 and 50 mV/mm, respectively. In a second series of experiments *Xenopus* embryos were impaled at the blastopore stage with glass microelectrodes and current was injected into the embryos to reverse the current normally exiting the blastopore at stages 14 to 16. Eighty-seven percent of embryos developed gross developmental abnormalities (Hotary and Robinson, 1991). In addition, ionic currents can be measured exiting stage 15 to 22 chick embryos via the posterior intestinal portal (PIP) (Hotary and Robinson, 1990). Inserting hollow capillaries under the dorsal skin to shunt approximately 30% of the ionic current away from the PIP resulted in 92% of these embryos developing gross abnormalities. Finally, enhancing and reducing the transcorneal potential difference can significantly enhance or delay the healing of rat corneal wounds (Reid et al., 2005). All of these observations suggest that imposed electric fields can have strong effects on development and wound healing, and this leads us to suspect that the endogenous field plays an important role in our normal development and physiology. How have such endogenous electric fields been detected?

MEASURING BIOELECTRIC FIELDS

INVASIVE APPROACHES

Bioelectric fields are present whenever there is a potential difference between two regions in an organism. They are most directly measured using an invasive approach that inserts a microelectrode into each region and measures the two voltage values there. Introducing a microelectrode into precise locations in the body is very challenging and can even perturb the field one is trying to measure. Nevertheless, some successful measurements have been made in developing embryos of the chick (Hotary and Robinson, 1990, 1992), frog (Hotary and Robinson, 1994), and newt (Metcalf and Borgens, 1994; Metcalf et al., 1994; Shi and Borgens, 1995). Since this work has already been reviewed (Nuccitelli, 2007), I will not discuss it here except to say that the range of electric fields measured within regions with an active current flow is 20 to 40 mV/mm. In nearly all of these examples, by "shorting" the electric field or imposing a different field, the electric field was found to play a role in the development of the organism.

NONINVASIVE APPROACHES

The **self-referencing vibrating probe** detects the flow of transcellular or transembryonic ionic currents outside of cells and tissues by vibrating a small platinum sphere in the extracellular medium between two points 10 to 30 μm apart (Jaffe and Nuccitelli, 1974; Hotary et al., 1992; Reid et al., 2007). The small voltage gradient generated by the flow of current through the medium is detected by signal averaging and noise filtering with a lock-in amplifier. This technique has been used to study more than thirty cell types and has revealed that most cells have an asymmetrical distribution of ion channels that naturally leads to a transcellular ionic current density on the order of 1 to 10 μA/cm^2 (Nuccitelli, 1988, 1990). Analogously, most epithelia that have been studied exhibit extracellular current densities on the order of 10 to 100 μA/cm^2 that are flowing through the encapsulated organ or embryo. Since this approach can only measure current flowing **outside** of the cell or tissue, the electric field that is generated by this current as it flows inside the cell or tissue can only be estimated based on the resistivity of the cytoplasm or tissue.

The **BioKelvin Probe** (Baikie and Estrup, 1998; Baikie et al., 1999) or **Dermacorder**® (Nuccitelli et al., 2008) can be used to detect electric fields on conductive surfaces. We have applied this relatively new approach to study the surface potential distribution on the epidermis of mammalian skin near lesions such as wounds, skin tumors, and bacterial infections. This technique measures the surface potential of the epidermis noninvasively by vibrating a small metal disk or probe in the air just above the skin. This probe forms one plate of a parallel plate capacitor, and the first conductive layer encountered below it becomes the other plate. This is usually the outer layer of epidermis since the stratum corneum is nonconductive. Therefore, the Dermacorder is limited to detecting epidermal surface potentials only and cannot detect electric fields that might exist below the outer conductive surface.

HOW DOES THE DERMACORDER WORK?

The effective charge on the capacitor formed by the probe and the epidermal surface is proportional to the voltage difference between these two surfaces, so the charge movement goes to zero when the voltage is the same on both plates. Our approach is to apply a biasing voltage to the skin and determine the value of the biasing voltage that is equal and opposite to the epidermal surface potential there. We do this by holding the probe plate at 2.5 V and applying alternating values of –7.5 V and +12.5 V to the surface of the skin. We then measure the peak-to-peak voltage (V_{ptp}) generated at the output for each of these biasing voltages (V_b) and plot these two values of V_{ptp} and V_b (applied voltage). When a line is drawn between these points, the V_b value for which $V_{ptp} = 0$ indicates the unknown surface potential of the epidermis and the slope of the line provides a measure of distance from the surface, which can be shown as follows: The output of the

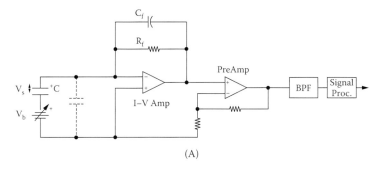

(B)

FIGURE 1.3 Dermacorder. (A) Circuit diagram of the head stage of Dermacorder probe. V_s represents the surface potential of the skin that forms a capacitor with the probe tip. V_b represents the backing potential applied to the skin. Since the probe is vibrating, the capacitor, C, will oscillate. The charge movement on and off of the capacitor will be equal to zero when V_b is equal and opposite to the unknown surface potential, V_s. (B) Bench-mounted Dermacorder scanning over a mouse wound. The probe is recessed within the housing and stepper motors are used to scan it over a 1 cm long region of skin. A second stepper motor is used to maintain a constant distance between the probe and the skin.

Dermacorder circuit (Figure 1.3A) when the probe is vibrated above a surface is given by

$$V_o = (V_s + V_b)GRC\omega(d/d_o)\sin(\omega t + \phi)$$

where d is the oscillation amplitude, d_o is the average distance between the sample and probe tip, G is the amplifier gain, ω is frequency, and R is the input resistance of the current-to-voltage converter. From this, one can see that $V_{ptp} = mV_b + c$, where $m = 2GRC\omega(d/d_o)$.

Thus, if V_{ptp} is plotted vs. V_b, we get a straight line whose slope is inversely proportional to the distance between probe and sample. This provides a very sensitive method for maintaining a constant spacing between probe and sample. This positional information is used to provide feedback to a z axis motor to hold this distance constant for every subsequent measurement as the probe scans along the surface.

What Have We Learned from This Technique So Far?

While this approach was first introduced to biology sixty years ago, there are not many published papers that apply this technique to biological systems. The first application to biological systems was developed by Bluh and Scott (1950) and was "used in preliminary experiments for the mapping of surface potential of various plant materials" (p. 868) without presenting any data. Carl Hertz was the next to use this approach to measure the geoelectric effect in coleoptiles (Hertz, 1960; Grahm and Hertz, 1962, 1964). He observed a 50 mV difference in surface potential between the top and bottom surfaces of coleoptiles beginning fifteen minutes after they were rotated 90° from a vertical to horizontal orientation. The amplitude of this surface potential difference increased with the time the plants were horizontally exposed in a log-linear manner. More recently, Iain Baikie developed a multitip BioKelvin probe that he used to measure the photoelectric response of coleoptiles (Baikie et al., 1999) while working at the BioCurrents Research Center at the Marine Biological Labs in Woods Hole, Massachusetts. He reported that an increase in surface potential averaging 65 mV in the plane of stimulus drifts down the shoot following blue light exposure at a rate of 2.4 cm/h, which is twice as fast as reported previously (Grahm, 1964).

I also worked at the BioCurrents Center during 2002, where we adapted this technique to be used on mammalian skin and called the benchtop prototype the Bioelectric Field Imager (Nuccitelli et al., 2008). About that same time Pamela Nuccitelli and I started BioElectroMed Corp., and during the past several years BioElectroMed has developed a portable BioKelvin probe called the Dermacorder (Figure 1.3B). We have been using this new instrument to detect the electric field near skin wounds, skin tumors, and bacterial infections.

Electric Fields Near Mammalian Skin Wounds

We first used the Bioelectric Field Imager to map the electric field near linear full thickness wounds in mouse skin (Nuccitelli et al., 2008). The epidermis of mammalian skin is a multilayered and polarized epithelium that generates a voltage difference across itself, inside positive (Figure 1.2). This transepidermal potential difference will drive current out of any low-resistance pathway in the skin that is formed by wounding. This "wound current" must complete the circuit by flowing back between the epidermis and stratum corneum and then into the intact epidermal cells surrounding the wounded region. It is this wound current flowing between the epidermis and stratum corneum that generates the lateral electric field that is detected by the Bioelectric Field Imager and Dermacorder. When using the Dermacorder to scan wounds, it is important to cover the skin surface with a thin plastic wrap such as Saran Wrap to avoid liquid work function artifacts. As can be visualized in Figure 1.4A, current flowing out of a wound that is a few millimeters wide makes the wounded region appear negative and the edge of the wound relatively positive. For a pinprick wound (Figure 1.4B), the detected current flow is all moving away from the wound, so the surface potential will appear positive there. A typical linear wound field is illustrated in Figure 1.4C–E, in which Figure 1.4D shows the surface potential distribution and Figure 1.4E shows a cross section through the middle of Figure 1.4D. These wound fields can be decreased by applying the Na^+ channel blocker amiloride to the skin (Figure 1.4F,G). This supports the hypothesis that these wound fields are generated by current flow between the epidermis and stratum corneum. A second example of the field near a wider wound is shown in Figure 1.4H–J, and Figure 1.4K,L shows the field measured near normal skin as a control for comparison to the other figures.

One good application of the Dermacorder is to monitor wound closure noninvasively. Since the lateral electric field is generated by current flow out of the wound, once the wound closes, the outward current should be greatly reduced along with the lateral field. We have examined this and present some data that correlate the histology with the lateral electric field at the wound site (Figure 1.5). We find that there is an excellent correlation between the histology of the epidermal regeneration and the electric field detected by the Dermacorder. By day 6 after wounding, the epidermis has returned to normal and the electric field has also returned to control levels. While the histological analysis captures a single time point and destroys the wounded region in order to fix, embed, and section the tissue to prepare the images in Figure 1.5, the Dermacorder data were obtained with a single one-minute long scan of the wound without disturbing the skin at all. That makes possible the continuous, noninvasive monitoring of wound closure.

ELECTRIC FIELDS NEAR SKIN TUMORS

We have also investigated the electric field of mouse skin near malignant lesions. We have found that both malignant melanomas and squamous cell carcinomas, created by injecting cultured tumor cells just beneath the skin, generate electrical signals (Figure 1.6A–F). The surface potential of the epidermis over the lesion appears more negative than the surrounding skin about 70% of the time, but about 20% of our measurements indicate a positive surface potential. This observation was puzzling because the histology showed that these tumors developed below the epidermis, and it was not clear why they would perturb the surface potential. These experiments were conducted using hairless, immune-competent SKH-1 mice. However, when we began using athymic Nu/Nu mice, this change in epidermal surface potential was reduced. That led us to suspect the involvement of an immune response in this signal generation.

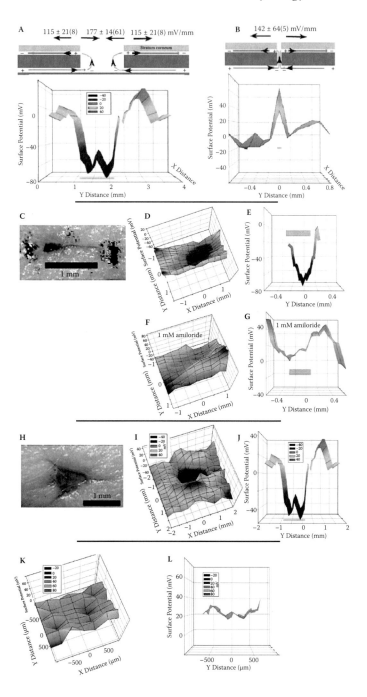

FIGURE 1.4 **See color insert.** Summary of results observed on mouse skin wounds. Pink bars mark the wound location on the scan. (A) Common field profile for wounds with a significant break in the epidermis. *(continued on next page)*

ELECTRIC FIELD NEAR BACTERIAL INFECTIONS

We tested this hypothesis by injecting bacteria into the skin in the same numbers as we were using for tumor generation. We observed a negative surface potential over the epidermis near the injection site beginning about two days following injection (Figure 1.6G–I). This pattern was quite similar to that observed near skin tumors. We suspect that the activated lymphocytes release molecules such as interleukins that can influence the permeability of the epidermis. This could result in the flow of ionic currents between the epidermis and stratum corneum that could generate the observed surface potential change. Much more work is needed to test this hypothesis.

SUMMARY

We have only just begun to use the Dermacorder to detect surface potentials of mammalian skin. The initial results are interesting and raise many questions that must be addressed before we can understand how these signals are generated. The Dermacorder excels as a noninvasive detector of wound healing and detects interesting surface potential changes near sites of skin tumors and bacterial infections. Thus, it may turn out to be a good detection system for inflammation in the skin. Much work is needed to improve our understanding of these changes in epidermal surface potential.

FIGURE 1.4 *(continued)*   (B) Common field profile for a wound with very little separation of the epidermis. (C) Photomicrograph of linear wound scanned in D to G. (D) Two-dimensional scan of voltage profile over wound. (E) Cross section of D along the y axis at x = 50. (F,G). Bioelectric Field Imager (BFI) scans made after the application of 1 mM amiloride to the wound. (H) Photomicrograph of a larger, nonlinear wound. (I) Two-dimensional BFI scan of wound in H. (J) Cross section of I along y axis at 50×. (K,L): Control scans of a 3 mm^2 region of unwounded mouse skin with a layer of polyvinyl film (Saran Wrap) adhering closely to it. (K) Two-dimensional profile of surface potential indicates all regions fall within a range of 60 mV. (L) Cross section view of a 200 mm wide strip from K along the y axis at x = 50. (From Nuccitelli et al., *Wound Repair Regen.*, 16, 432–41, 2008.)

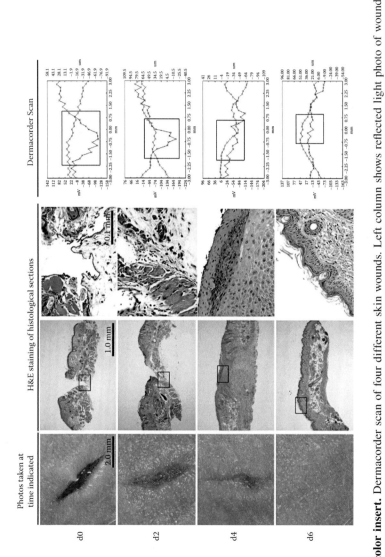

FIGURE 1.5   See color insert. Dermacorder scan of four different skin wounds. Left column shows reflected light photo of wound on the day indicated to the left of the photo. Two center columns are stained histological sections taken at 40× and 400× magnification. The rectangular outline indicates the region of the image that is further magnified on the right center column. The far right shows the Dermacorder scan of the wound on the left. The blue line represents the surface potential and the red line represents the surface topology. The rectangle in the far right column outlines the region of the scan taken when the probe was over the wound.

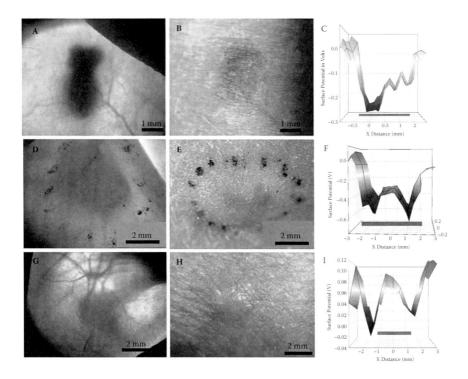

FIGURE 1.6   See color insert. Skin tumors and bacterial infections influence the sur-face potential on the epidermis above them. (A) Transillumination photo of a melanoma. (B) Reflected light microscopy of the melanoma shown in A. (C) Bioelectric Field Imager scan of the melanoma shown in A and B. Location of the melanoma on the scan is indicated by the red bar. (D) Transillumination photo of a squamous cell carcinoma. (E) Reflected light microscopy of the same squamous cell carcinoma shown in D. (F) Bioelectric Field Imager scan of the squamous cell carcinoma shown in D and E. (G) Transillumination photo of bacterial infection three days after injecting 10^6 *Staphylococus epidermidis*. (H) Reflected light photo of bacterial injection region shown in G. (I) Dermacorder scan of the bacterial infection on day 3.

REFERENCES

Altizer, A. M., Moriarty, L. J., Bell, S. M., Schreiner, C. M., Scott, W. J., and Borgens, R. B. 2001. Endogenous electric current is associated with normal development of the vertebrate limb. *Dev. Dyn.* 221:391–401.

Baikie, I. D., and Estrup, P. J. 1998. Low cost PC based scanning Kelvin probe. *Rev. Sci. Instrum.* V69:3902–7.

Baikie, I. D., Smith, P. J. S., Porterfield, D. M., and Estrup, P. J. 1999. Multitip scanning Bio-Kelvin probe. *Rev. Sci. Instrum.* 70:1842–50.

Bluh, O., and Scott, B. I. H. 1950. Vibrating probe electrometer for the measurement of bioelectric potentials. *Rev. Sci. Instrum.* 21:867–68.

Clausen, T., Van, H. C., and Everts, M. E. 1991. Significance of cation transport in control of energy metabolism and thermogenesis. *Physiol. Rev.* 71:733–74.

Grahm, L. 1964. Measurements of geoelectric and auxin-induced potentials in coleoptiles with a refined vibrating electrode technique. *Physiol. Plant.* 17:231–61.

Grahm, L., and Hertz, C. H. 1962. Measurements of the geoelectric effect in coleoptiles by a new technique. *Physiol. Plant.* 15:96–114.

Grahm, L., and Hertz, C. H. 1964. Measurements of the geoelectric effect in coleoptiles. *Physiol. Plant.* 17:186–201.

Hertz, C. H. 1960. Electrostatic measurement of the geoelectric effect in coleoptiles. *Nature* 187:320–21.

Hotary, K. B., Nuccitelli, R., and Robinson, K. R. 1992. A computerized 2-dimensional vibrating probe for mapping extracellular current patterns. *J. Neurosci. Methods* 43:55–67.

Hotary, K. B., and Robinson, K. R. 1990. Endogenous electrical currents and the resultant voltage gradients in the chick embryo. *Dev. Biol.* 140:149–60.

Hotary, K. B., and Robinson, K. R. 1991. The neural tube of the *Xenopus* embryo maintains a potential difference across itself. *Dev. Brain Res.* 59:65–73.

Hotary, K. B., and Robinson, K. R. 1992. Evidence of a role for endogenous electrical fields in chick embryo development. *Development* 114:985–96.

Hotary, K. B., and Robinson, K. R. 1994. Endogenous electrical currents and voltage gradients in *Xenopus* embryos and the consequences of their disruption. *Dev. Biol.* 166:789–800.

Jaffe, L. F., and Nuccitelli, R. 1974. An ultrasensitive vibrating probe for measuring extracellular currents. *J. Cell Biol.* 63:614–28.

Jaffe, L. F., and Stern, C. D. 1979. Strong electrical currents leave the primitive streak of chick embryos. *Science* (Washington, DC) 206:569–71.

McCaig, C. D., Rajnicek, A. M., Song, B., and Zhao, M. 2005. Controlling cell behavior electrically: Current views and future potential. *Physiol Rev.* 85:943–78.

Metcalf, M. E. M., and Borgens, R. B. 1994. Weak applied voltages interfere with amphibian morphogenesis and pattern. *J. Exp. Zool.* 268:323–38.

Metcalf, M. E. M., Shi, R. Y., and Borgens, R. B. 1994. Endogenous ionic currents and voltages in amphibian embryos. *J. Exp. Zool.* 268:307–22.

Nuccitelli, R. 1988. Physiological electric fields can influence cell motility, growth, and polarity. *Adv. Cell Biol.* 2:213–33.

Nuccitelli, R. 1990. The vibrating probe technique for studies of ion transport. In *Noninvasive techniques in cell biology*, ed. J. K. Foskett and S. Grinstein, 273–310. New York: Wiley-Liss.

Nuccitelli, R. 2003. A role for endogenous electric fields in wound healing. *Curr. Top. Dev. Biol.* 58:1–26.

Nuccitelli, R. 2007. Endogenous electric fields in animals. In *Bioengineering and biophysical aspects of electromagnetic fields*, ed. F. S. Barnes and B. Greenebaum, 35–50. Boca Raton, FL: CRC Press.

Nuccitelli, R., Nuccitelli, P., Ramlatchan, S., Sanger, R., and Smith, P. J. 2008. Imaging the electric field associated with mouse and human skin wounds. *Wound Repair Regen.* 16:432–41.

Ojingwa, J. C., and Isseroff, R. R. 2003. Electrical stimulation of wound healing. *J. Invest. Dermatol.* 121:1–12.

Piccolino, M. 1997. Luigi Galvani and animal electricity: Two centuries after the foundation of electrophysiology. *Trends Neurosci.* 20:443–48.

Piccolino, M. 2000. The bicentennial of the Voltaic battery (1800–2000): The artificial electric organ. *Trends Neurosci.* 23:147–51.

Reid, B., Nuccitelli, R., and Zhao, M. 2007. Non-invasive measurement of bioelectric currents with a vibrating probe. *Nat. Protoc.* 2:661–69.

Reid, B., Song, B., McCaig, C. D., and Zhao, M. 2005. Wound healing in rat cornea: The role of electric currents. *FASEB J.* 19:379–86.

Robinson, K. R. 1983. Endogenous electrical current leaves the limb and prelimb region of the *Xenopus* embryo. *Dev. Biol.* 97:203–11.

Shi, R., and Borgens, R. B. 1995. Three-dimensional gradients of voltage during development of the nervous system as invisible coordinates for the establishment of embryonic pattern. *Dev. Dyn.* 202:101–14.

2 Investigation Systems to Study the Biological Effects of Weak Physiological Electric Fields

Francis X. Hart
Department of Physics
University of the South
Sewanee, Tennessee

CONTENTS

INTRODUCTION

This volume features the results of experiments on cell systems exposed to electric fields (Chapters 3–12). Such experiments require a broad set of skills for their successful execution and interpretation because of their cross-disciplinary nature. The independent variable, the electric field or a related parameter to which the cell is exposed, must be well characterized. A variety of confounding factors that can reduce or even completely eliminate the action of the field must be identified and eliminated.

In this chapter considerable attention will be given to the complicated relationships that determine how the electrical parameters to which the cell is actually exposed depend on the applied field and various properties of the cells in their exposure chamber. The basic electrical parameters of interest in determining the actual exposure of cells to the field are the electric field at the cell surface, Es, and the change in transmembrane potential difference, ΔV. These parameters are not constant, but vary with position along the surface of the cell.

One of the major issues in this area of research is how the fields are detected or transduced by the cells. One possibility is through the opening of voltage-gated channels that depends on ΔV (Cho et al., 2002). Two other possibilities, electrodiffusion/osmosis (Jaffe, 1977; Jaffe and Nuccitelli, 1977; McLaughlin and Poo, 1981) and electromechanical transduction (Hart, 2008, 2010), depend on Es. Nuccitelli (1988) has reviewed the basic features as well as limitations of the voltage-gated channel and electrodiffusion/osmosis models. Hence, how ΔV and Es vary around the cell is key for the interpretation of experimental results. As will be seen, nearly all papers in this area calculate ΔV and ignore Es. This chapter will provide more information regarding the determination of Es.

In addition, a variety of potentially confounding factors are identified in this chapter and suggestions are offered for their elimination. A successful experiment, with conclusions properly drawn from carefully measured data, requires attention to these details. An example of a well-conceived experimental design is presented at the end of the chapter.

DOSIMETRY

This section describes the relationship between the applied electric field and the electric field to which the cell is actually exposed. The electric fields are those produced by a direct current (DC) source with electrodes connected to the medium containing the cells, although the same considerations would apply to low-frequency alternating current (AC) fields. Induced electric fields produced by changing magnetic fields are not considered here. Discussion of the dosimetry for such fields may be found elsewhere (Stuchly and Xi, 1994; Hart, 1996). It is important that both electrodes are ultimately connected to the medium. If one electrode is in air and the other in the medium, then the electric field is almost entirely in the air gap and the cells are not exposed to a significant field.

Experimenters commonly report that the field that a cell experiences is equal to the applied voltage, Vo, divided by the electrode separation, L. This section will demonstrate that the field that a cell actually experiences varies considerably over its surface and differs from Eo = Vo/L. Eo will be designated as the applied field. Authors use a variety of units to describe the field strength. A simple conversion gives 100 mV/mm = 100 V/m = 1 V/cm.

Although cell membranes are relatively good insulators, a small, but measurable (Jaffe and Nuccitelli, 1974) electric current passes through them when an external electric field is applied. However, the resulting cytoplasmic electric field is much smaller than the applied field. When placed in a medium of conductivity, g, to which an electric field is applied, the cells to a good approximation act as barriers to the passage of current, which must flow around them. It will be assumed for the remainder of this chapter that the cytoplasmic field is negligibly small compared to the external field. The membrane also acts as a capacitor with the lipid bilayer serving as the dielectric. For frequencies on the order of 1 MHz and higher, electric fields can also couple capacitively across the membrane so that much larger currents are produced inside the cell. The discussion here is confined to lower frequencies.

At the instant an electric field is applied, charges begin to flow in the cell interior to the inner surface of the membrane and produce a polarization field in the opposite direction. The net electric field in the cell interior is thus reduced to a relatively small value on a timescale of microseconds. The magnitude of this induced charge on the membrane will be estimated later.

DETERMINATION OF EO

Conceptually, the simplest way to determine Eo is to measure Vo and L. It is then essential to place the voltage-measuring probes as close to the chamber inlets as possible so that voltage drops across agar bridges and other components are not included. Some authors, however, use the relationship Eo = I/gA, where I is the measured current flowing through the system and A is the cross-sectional area for its passage through the cell chamber. Accurate determination of A requires the careful measurement of the height and width of the fluid in the chamber, which will vary from experiment to experiment as chambers are continually being rebuilt. The medium in the chamber is generally not connected directly to the metal electrodes providing the current, as will be discussed later, but via an opening of area A* to a medium reservoir. Use of the relationship with A* then yields the field at the chamber entrance. Only if A* = A will it yield the applied field in the region of the cells.

A further complication can arise if the cells are suspended throughout the medium, rather than plated on a substrate. If the concentration of cells in suspension is not relatively sparse, then the obstructing cross section they present to the current cannot be neglected. The conductivity to be used in the relationship for Eo is then not g, the conductivity of the medium alone, but the conductivity of the cell-medium mixture, which must be obtained by computation, as discussed later. The simplest procedure is just to use Eo = Vo/L.

CELL SHAPE AND ORIENTATION

It must be emphasized that the field to which a cell is actually exposed, Es, is not simply the applied potential difference, Vo, divided by the electrode separation, L. Es depends on the shapes and relative distribution of the cells. An analogy with fluid flow may help in visualizing the field distribution. The cells, with their insulating membranes, form barriers to the flow of current just as rocks form barriers to the flow of water in a stream. The direction of the electric field is determined by the direction of current flow, and thus the shape of obstacles in its path. Consider a rock shaped like an ellipsoid, a figure that has the cross section of an ellipse in the xy, yz, and zx planes. The semiaxes of the ellipsoid are designated a, b, and c. If two are equal, then the object is a spheroid. If all three semiaxes are equal, then the object is a sphere. Many plated cells, such as keratinocytes, can be conveniently represented in shape as hemiellipsoids, that is, the top half of a full ellipsoid.

Ellipsoidal Cells: General Considerations

Suppose a > b > c. If the long axis, a, of the ellipsoidal rock is parallel to the flow of water, it presents a relatively small profile to the flow, which moves smoothly around it. Conversely, if the long axis is perpendicular to the flow, it presents a larger profile and the water is diverted more in order to flow around it. The same principle applies to the flow of current around an ellipsoidal cell. The electric field lines flow more smoothly around a cell with its long axis parallel to the field than for a cell with the long axis perpendicular to the field. As the water must move around the rock, the flow is minimal at the front and back ends of the rock and maximal along the sides. Similarly, the electric field at the cell surface is minimal on the ends facing the electrodes and maximum along the sides for either orientation of the cell.

Mathematical solutions for the electric field exterior to the cell and within the membrane are obtained from Laplace's equation (Reitz et al., 1992). This second-order, partial differential equation describes the spatial variation of the electric potential for regions in which there is no net charge density. Closed-form analytical solutions can be found only for shapes with high symmetry, whereas numerical solutions are required for irregularly shaped cells. Numerical solutions presented here were obtained using the software package COMSOL Multiphysics™. Recall that the electric field within the cell is essentially zero.

Of particular interest here is the field at the cell surface, Es. One of the main conditions imposed on the solution for the external field is that the component of the field perpendicular to the plane of the surface, the "normal" component, must be nearly zero at the surface. If it were not, then there would be a current flow into the membrane comparable to that flowing around the cell. Such flow is impossible because the membrane is to a good approximation an insulator. Hence, at the surface the electric field is "transverse" or parallel to the surface. As will be noted later, this restriction is somewhat relaxed for media with a relatively low conductivity.

One of the features of the solutions of Laplace's equation is that the external electric field depends only on the shape of the cell, but not on its absolute size. For commonly used media the field is also independent of the conductivity of the surrounding medium. For example, the maximum electric field at the surface of a spherical cell is 1.5 Eo, independent of the radius of the cell (Hart and Marino, 1986). That field is transverse at 90° to the direction of the applied field, that is, at the cell's "equator."

As the electric field is the change of potential with respect to position, the external potential must vary slowly toward the surface, but can vary more rapidly along the surface. Inside the cell the electric potential is essentially constant because the field is almost zero there. It is approximately equal to the average of the potential over the cell surface.

Consider an ellipsoidal cell with its long axis parallel to the field. There will be a potential difference between its two ends. Because there is essentially no difference in potential in the cell interior, that potential difference appears across the two membranes. The longer the cell, the greater is the difference in potential between its ends and the greater is the transmembrane potential difference, ΔV. Because the membrane thickness, d, is small compared to the dimensions of the cell, the membrane may be regarded locally as a parallel plate capacitor. The field within the membrane, Em, is then Em = ΔV/d. The larger the cell, the larger is Em.

For spherical cells

$$\Delta V = 1.5 \text{ Eo R } \cos\theta \qquad (2.1)$$

where R is the cell radius and θ indicates the angular position on the cell surface relative to the direction of the field (Grosse and Schwan, 1992). θ is 0 in the direction of the field. ΔV is greatest in magnitude at the two poles (θ = 0 and 180°) and is 0 at 90°. A depolarization of the membrane is produced on the side of the cell facing the cathode (negative electrode) and a hyperpolarization on the anode-facing (positive electrode) side.

The surface charge density, σ, at the interface of two media is the difference in the normal components of the electric displacement vectors in the two media (Reitz et al., 1992). At the cytoplasm-membrane interface such a surface charge is produced when the external field is applied. In that case σ = (KmEm – KcEc)ε_o, where Km and Kc are, respectively, the dielectric constants of the membrane and cytoplasm, Ec is the electric field in the cytoplasm and ε_o is the permittivity of free space (8.85 × 10^{-11} Farads/m). Since Em >> Ec, σ ~ Km ΔV ε_o/d.

For an applied electric field of 100 V/m and a cell radius of 20 μm, ΔV = 3 mV at θ = 0. For a membrane thickness of 5 nm, Em = 6 × 10^5 V/m there. A membrane dielectric constant of 5 yields σ = 2.66 × 10^{-4} C/m^2. If monovalent ions with a charge of 1.6 × 10^{-19} C are assumed to be the source of the charge, then the induced surface ion density is about 1,700 ions/μm^2. Similar, but more complicated expressions apply to spheroidal and ellipsoidal cells. ΔV is maximum in magnitude at the two apices and 0 halfway around if an axis of the cell is aligned with the field.

Ellipsoidal Cells: Numerical Modeling

To obtain an overall perspective on the variation of both Es and ΔV with cell shape and orientation a numerical calculation is required. As will be noted later, the analytical solutions published in the literature do not generally report the electric field distribution, but only changes in transmembrane potential. Because two of the three proposed mechanisms for the detection of fields by cells involve Es, it is important that its variation be understood as well. The calculations for which the results are shown in Figures 2.1 to 2.4 were performed using COMSOL Multiphysics. Because of its negligible size relative to the dimensions of the cell, meshing of the membrane must be avoided. The use of two modes, one for the cell and one for the medium, and coupling via normal current continuity avoids this problem. The details for setting up such a model are described elsewhere (Pucihar et al., 2006).

Figure 2.1a–c illustrate the field distributions for a hemiellipsoidal cell with its long axis parallel to the field in a, perpendicular to the field in b, and with a hemispherical shape in c. The dimensions of the hemiellipsoidal cell are 90, 40, and 15 μm. An electric field of 100 V/m is applied in the x direction. The z axis is perpendicular to the plane of the substrate (vertically up); the 15 μm dimension represents the height of the plated cell. The y axis is in the plane of the substrate and perpendicular to the applied field. The hemispherical shape was used to represent the occasional balling up of a cell into a "fried egg" appearance. The volume of the hemisphere was set equal to the volume of the hemiellipsoid so that its radius was 23.8 μm. Recall that the field values are independent of the size of the cell. In the field distribution plots of Figures 2.1A–C, 2.3, and 2.4, the cells are shown as white areas, as they do not belong to the exterior mode of the calculation. The total electric field, Et, is given by $Et^2 = Ex^2 + Ey^2 + Ez^2$, where Ex, Ey, and Ez are the components of the field in the x, y, and z directions, respectively. The fields were evaluated at a height of 2 μm above the substrate to eliminate minor surface irregularities introduced in the numerical meshing process.

In each case at distances significantly greater than the cell dimensions the electric field is equal to Eo. As the cell is approached along the x axis, Et decreases and approaches zero at the cell surface. The electric field is maximum at the cell surface at the y axis, that is, at 90° to the applied field direction. The maximum Es, however, is greatest for case b, with the long axis perpendicular to the field, and least for case a, with the long axis parallel to the field.

Figure 2.2 illustrates in more detail the variation of Et along the x, y, and z axes for these three cells. All distances are measured from the center of each cell. The z axis values are calculated directly over the center of the cell. As the x axis is in the direction of the field, the field along that axis is essentially directed radially

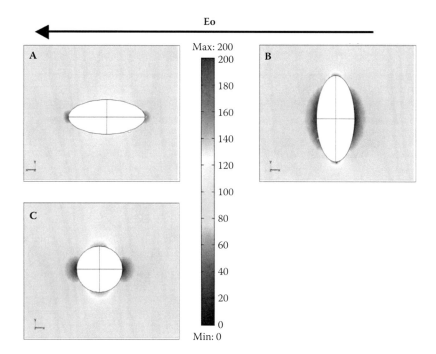

FIGURE 2.1   See color insert. The total electric field distribution around a hemiellipsoidal cell placed in a uniform, 100 V/m electric field directed along the x axis. The field is evaluated in the xy plane at a height z = 2 μm above the substrate. The dimensions of the cell are 90, 40, and 15 μm (the height of the plated cell). The long axis of the cell is parallel to the field in (A) and perpendicular to the field in (B). In (C) the distribution around a hemispherical cell of the same volume is illustrated. The color range is from 0 (deep blue) to 200 V/m (red) for an applied field of 100 V/m (green).

inward toward the cell. Because the y and z axes are perpendicular to the x axis, the field components along the y and z axes are tangent to the cell surface.

Note that because the projections along the axes of the three cells are different due to their orientations, the cell with the parallel orientation extends farther out in the x direction than does the perpendicular orientation with the hemispherical cell in between. Similarly, the cell with the perpendicular orientation extends farther out from the center in the y direction than do the other two. By symmetry the y and z tangential fields are identical for the hemispherical cell. Recall also that these field results depend only on the shape of the cell and are independent of its size.

For distances from the cell comparable to the cell dimensions and greater, the electric field is equal to the applied field. As one approaches the cell along the x axis, the field decreases until reaching essentially zero at the cell surface. Because the cell with its long axis parallel to the field extends farther out, this decrease must occur more rapidly. The decrease is slowest for the cell with its

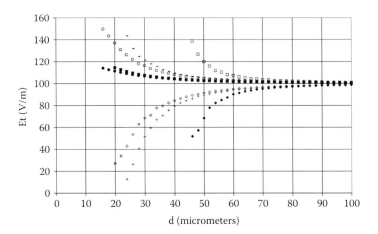

FIGURE 2.2   See color insert. The total electric field at various distances from the center of the cell described in Figure 2.1. The solid symbols represent the field components for the long axis parallel to the field, and the open symbols represent the field components for the long axis perpendicular to the field. The blue diamonds represent the field values along the x axis; the red squares, along the y axis; the green circles, along the z axis (distance above the substrate). The + signs represent the values along the x axis for the hemispherical cell; the − signs, along either the y or z directions for that cell.

long axis perpendicular to the field as that cell extends the least in the direction of the applied field.

Approaching the cell along the y axis, one observes the tangential field steadily increasing and reaching its maximum value at the cell surface. That value is greatest for the cell with its long axis perpendicular to the field and least for the cell with its axis parallel to the field. The surface field is largest in the former case because the current is more diverted in its flow around the hemielliptical cell, so the surfaces of constant potential have a sharper curvature there. The values for the hemispherical cell are intermediate. The maximum tangential field in that case approaches 1.5 Eo, as expected from the analytical solution (Hart and Marino, 1986).

As one approaches the cell from above along the z axis, the values appear to show that the field is greatest for the hemispherical cell. The reason is that the hemispherical cell extends farther above the surface than do either of the ellipsoidal orientations, so that at d ~ 25 μm one is closer to the hemispherical cell. However, at the top of each cell the tangential surface field is 164 V/m for the hemiellipsoidal cell perpendicular to the field, 151 V/m for the hemispherical cell, and 115 V/m for the hemiellipsoidal cell parallel to the field. The tangential field at the top of the cell reflects the curvature of the surface there.

Taking the difference in the inside and outside potentials at the tip of the cell along the x axis yields ΔV-max. For the long axis of the cell parallel to the field ΔV-max is 5.1 mV, and it is 3.1 mV for the long axis perpendicular to the field. For the hemispherical cell ΔV-max is 3.6 mV, in agreement with Equation 2.1. These results confirm the direct relationship between the length of the cell in the direction of the field and the change in transmembrane potential.

In summary, the field at the cell surface close to the plane of the substrate is essentially zero facing either electrode and maximum at 90°. That field is greatest for the long axis of the cell perpendicular to the field. In that orientation, however, ΔV-max is smallest. The converse holds for the long axis parallel to the field with the greatest ΔV-max, but the smallest surface field. The field at the top of the cell is largest for the cell with the long axis perpendicular to the field.

Ellipsoidal Cells: Analytical Modeling

In this section it is assumed that the medium and the cytoplasm are good conductors, with conductivities on the order of 1 S/m. If the medium conductivity is on the order of 0.01 S/m or lower, a more complicated analysis is required, as discussed later. It is also assumed that the field is applied parallel to one of the axes of the ellipsoid. If that is not the case, then the applied field can be decomposed into components along each axis and the resulting solutions combined at the end of the calculation. The applied field is assumed to be uniform so that the cell must be much smaller in size than the dimensions of the chamber and not located near a wall or electrode.

There is an important exception to the previous statement. The analytical solutions are obtained for an isolated ellipsoid suspended in the interior of a medium. If a cell is plated on the substrate of the chamber, however, its bottom surface is a plane of symmetry. Hence, the solution for a plated, hemiellipsoidal cell is identical to that for the suspended, full ellipsoid.

Some of the references for calculations of transmembrane potential are concerned primarily with high fields, but the resulting equations are also applicable to the low-field experiments described in this volume.

The simplest analytical model for a cell is a layered sphere (Hart and Marino, 1986). Although the emphasis of that paper is on fields in tissue, its results also apply to an individual cell. The interior of the cell corresponds to a sphere of radius a, and the membrane is a concentric shell of inner radius a and outer radius b. The solution to Laplace's equation provides the electrical potential, which is sufficient to determine the transmembrane potential. To determine the electric field one must take the gradient of the potential. For coordinate systems more complicated than spherical, that procedure is much more difficult, so that only transmembrane potential values are generally discussed. The full expressions for the fields are presented in the Hart and Marino paper for spheres. (It should be noted that several of the equations in that paper were printed with errors but used correctly in the calculations. A corrected version is available from this author at fhart@sewanee.edu.)

Spheroids are divided into two categories: prolate, which has an elongated shape with axes a > b = c, and oblate, which has a flattened shape with axes a < b = c. The first solutions for the transmembrane potential difference for spheroidal cells with tables to assist in its calculation were reported in the early 1970s (Bernhardt and Pauly, 1973). Because the solutions for spheroidal and ellipsoidal systems involve the evaluation of elliptic integrals, and this early work was done prior to the general use of desktop computers, values are provided only for "standard" cells of 10 microns in size. The full expressions for the potential, as well as the transmembrane potential, for both prolate and oblate spheroids are now available in the literature (Kotnik and Miklavcic, 2000). Moreover, an expression for the transmembrane potential produced on a spheroidal cell with an arbitrary orientation relative to the applied field and a useful, general table to assist with its evaluation are also available (Valic et al., 2003).

Such analysis has been extended (Gimsa and Wachner, 2001) to obtain an expression for the transmembrane potential of a general ellipsoidal cell at an arbitrary orientation with respect to the applied field. The actual ellipsoid plus shell is replaced by a Maxwell equivalent body having the same geometry as the original ellipsoid, but with electrical properties calculated from those of the cell and medium. They also analyze a cylindrical cell as the limiting case of an infinitely long spheroid.

Ellipsoidal Cells: Poorly Conducting Media

Because higher media conductivities lead to greater heat dissipation, as will be discussed later, some researchers use low-conductivity media for the cells in order to minimize possible thermal side effects, particularly for large electric fields. Typical media conductivities are on the order of the conductivity of the cytoplasm, about 1 S/m. For media conductivities down to about 0.01 S/m there is little significant difference in the expressions for Es and ΔV. Changing the conductivity of the medium changes the magnitude of the current, but not its spatial distribution. Care must be taken, however, lest the reduced ionic content of the medium lead to ionic diffusion from the cells.

However, for media conductivities below about 0.01 S/m the expression for ΔV becomes more complicated than Equation 2.1 (Grosse and Schwan, 1992; Kotnik et al., 1997). The transmembrane conductivity may no longer be negligible compared to that of the medium. In the glycocalyx coions and counterions may provide a transverse conductivity along the cell as well. For such low-resistivity media there can be significant transfer of current into the cell interior, so that a field is now present there, the normal component of the field is no longer zero at the surface, and the change in transmembrane potential is reduced from its level for conductive media. A sensitivity analysis (Kotnik et al., 1997) details how ΔV depends on cell size, membrane conductivity, and cytoplasmic conductivity in the case of low-conductivity media.

Cells of Arbitrary Shape

If the cells are not ellipsoidal in shape, then a numerical solution is required. A combination of the finite element and boundary element methods has been used (Liu et al., 2003) to compute ΔV for "biconcave" erythrocytes. They found that ΔV was somewhat smaller for their model erythrocyte than for a spherical cell of comparable size. Because their method required an analytical expression for the surface of the cell, it would not be applicable to cells of arbitrary shape for which a full numerical method must be used. Their approach has been extended (San Martin et al., 2006) to asymmetric cells, such as stromatocytes, to find the variation of ΔV along the cell surface. ΔV tends to be higher for unusually shaped cells than for symmetric erythrocytes.

For cells with a more irregular shape a full numerical treatment is necessary. A finite element model has been used (Pucihar et al., 2006) to calculate ΔV in Chinese hamster ovary (CHO) cells. The geometry of their model was constructed from a sequence of microscopic cross-sectional images. The results compared well with analytical solutions for spherical and spheroidal cells and also with potentiometric fluorescence measurements of ΔV. In contrast, the ΔV values obtained by the analytical solution for an approximation of the CHO cell shape by a hemiellipsoid introduced considerable error. It is clear that finite element modeling must be used to obtain realistic values for ΔV and Es for cells of irregular shape.

The COMSOL Multiphysics method described earlier for hemiellipsoidal cells was used to calculate the field distribution around a stylized model of a neurite, as shown in Figure 2.3. That model was developed by splicing thin hemiellipsoidal segments to a hemiellipsoid base to form a composite object. The segments were added in the x direction (parallel to the field), in the y direction (perpendicular

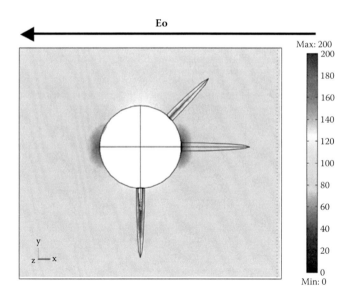

FIGURE 2.3 See color insert. The total electric field distribution around a stylized model of a neurite placed in a uniform, 100 V/m electric field directed along the x axis. The field is evaluated in the xy plane at a height z = 3 μm above the substrate. The color range is from 0 (deep blue) to 200 V/m (red) for an applied field of 100 V/m (green).

to the field), and at 45° to the field. Ellipsoids were used rather than cylinders to avoid physically unrealistic sharp edges for the extensions.

The results for Et are shown for the xy plane at a height of 3 μm above the substrate with an applied field of 100 V/m. Because of the sloping shape of the extensions, the field pattern varies considerably with the height above the substrate for small distances. As previously, Et is smallest in the direction of the field and largest at 90°. The field is particularly enhanced over the tops of the y axis and 45° extensions. The height of 3 μm was chosen to highlight this enhancement. The small blank area in the y axis projection indicates a field greater than the scale limit, which is a convenient 200 V/m in all the Et plots for ready comparison among them. The field in that part of the projection reaches 227 V/m. Local surface fields, which are always tangential to the surface, can be much greater than the applied field for cells with long extensions, depending on their direction.

DENSE CELL SYSTEMS

If the separation of cells in a suspension is large compared to the cell dimensions, then the expressions for the transmembrane potential for individual cells are still applicable and the effective conductivity of the system is essentially that of the medium. As the cell density increases, however, a point is reached at which the transmembrane potential and effective conductivity decrease. The transmembrane potential must then be expressed in terms of f, the volume fraction of the cells. For spherical cells the maximum change in transmembrane potential can be expressed as $\Delta Vmax = 1.5\ Eo\ R\ (1 - 0.38\ f)$ (Ramos et al., 2006). Thus, for $f \sim 0.1$, $\Delta Vmax$ is reduced by about 5%. Higher cell densities produce even more significant changes.

Analytical and numerical calculations indicate that for spherical cells the decrease in ΔV depends also on the actual arrangement of the cells in the system (Pavlin and Miklavcic, 2009). However, the effective conductivity depends only on the volume fraction of the cells, not on their actual arrangement, and can be represented well by either the Maxwell or Tobias effective medium theories.

For nonspherical cells only a numerical approach can be used—particularly for Es. The COMSOL Multiphysics method described earlier was used to calculate the field distribution within a closely spaced group of hemiellipsoidal cells exposed to a 100 V/m field, as shown in Figure 2.4. The surface field distribution is quite complicated, as expected. At the front and rear (x axis) ends of the group the field is lower than the average, while it is higher at the sides (y axis) of the group, as would be the case for an individual cell. However, the cell at the top experiences a much higher than average field over its surface, whereas the cell below it experiences a much lower than average surface field. Similar large variations are apparent throughout the group. Care should then be taken in the interpretation of results for closely spaced cells, as they typically experience a variety of actual field exposures in addition to the problem of chemical signaling, which is discussed later.

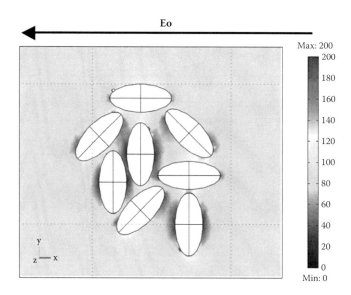

FIGURE 2.4   See color insert. The total electric field distribution within a closely spaced group of hemiellipsoidal cells placed in a uniform, 100 V/m electric field directed along the x axis. The field is evaluated in the xy plane at a height z = 2 μm above the substrate. The color range is from 0 (deep blue) to 200 V/m (red) for an applied field of 100 V/m (green).

PULSED FIELDS

In some cases the electric field is not constant in time, but is pulsed. For unipolar pulses the electric field is either on or off, but when on, it always has the same direction. For bipolar pulses the direction of the electric field is reversed at various times. Pulsed fields cannot be treated as simple, constant, direct current fields. Instead, the variation of the field with time must be expressed as a combination (Fourier series) of sinusoidal signals (Hart, 1987). How many of these component signals must be used depends on the complexity of the waveform. Although the emphasis of that paper is on fields in tissue, its results also apply to an individual cell. For a cell, unless the component frequencies approach the MHz range, each component may be analyzed separately, as above, and there is little distortion of the pulse shape by the cell itself.

CONFOUNDING FACTORS

The preceding sections have emphasized that the actual field to which a cell is exposed is not simply Vo/L, but is considerably more complicated. Other, unwanted factors, however, may be present that provide additional independent variables in the measuring process. Their presence could modify, or even eliminate completely, the effects produced by the electric field. These variables can appear in four different forms: electrical, thermal, chemical, and mechanical.

ELECTRICAL

The power supply providing the DC current, the leads in the circuit, and the cell apparatus must form a closed loop for current to flow. As a closed loop, however, it is also a receiving antenna for other electromagnetic signals in the environment, such as from a nearby incubator or computer.

In order to ensure that current flow is being maintained during the experiment, it is a wise idea to provide a microammeter or some other current-measuring device in the supply circuit. Such a precaution would detect if an agar bridge (to be described later) happens to fail during the course of an experiment. A particularly useful idea is to insert a resistor in series with the apparatus and to measure the voltage drop across the resistor to indicate the current flow. The resistor should be small in comparison to the resistance of the cell chamber itself. The voltage drop can be monitored by a computer data acquisition system for a permanent record or by an oscilloscope. An advantage of this method is that if the power supply is set to zero output, but still connected to the circuit (so that the antenna loop is still intact), any significant signals induced in the circuit by surrounding equipment can be detected. Inexpensive DC power supplies may contain a small AC signal (ripple) superimposed on the desired constant voltage. This ripple can also be readily measured by this method.

Cells placed in a nonuniform electric field experience an attractive force, dielectrophoresis, toward the region of stronger field that could be mistaken for galvanotaxis. As described previously, cells placed in an electric field become electrically polarized. In a nonuniform field the charge in the stronger field experiences a greater attractive force. It is also possible that such field gradients in themselves may produce biological effects in cells that would not be observed in uniform fields. For these reasons the use of electrodes that produce strong field gradients, such as needle electrodes, should be made with care.

THERMAL

It is well known that cells respond to changes in their thermal environment. Thus, it is important that the temperature of the system is maintained at the desired level during the course of the experiment. This goal can be achieved by either monitoring the temperature during the course of the experiment itself or by measuring the temperature during the course of a preliminary experiment. Monitoring during the course of an actual experiment could be problematic because the presence of the probe might distort the electric field. A preliminary experiment in which the temperature is mapped throughout the cell chamber would detect the presence of thermal gradients to which the cells would be exposed. Problems can arise because of the heating of the medium due to the passage of current. That problem can be reduced by the use of a less conductive medium, but then diffusion problems can arise as noted previously. On the other hand, if the cells are maintained at physiological temperatures (e.g., 37°C), then there may be heat loss to the room temperature surroundings. A careful, preliminary temperature mapping is required to determine whether such problems are present.

The possible temperature rise can be estimated by relating the energy dissipated by the passage of the current to the energy gained by the fluid in the chamber. Suppose, as previously, that the chamber has a length L with cross-sectional area A and is filled with a fluid of conductivity σ. The electrical resistance offered by the fluid is then $R = L/A\sigma$. In a time Δt the energy dissipated is $V^2\Delta t/R$. The energy gained by the fluid during that time is $Mc\Delta T$, where M is the fluid mass, c is its specific heat, and ΔT is the temperature increase. M can be expressed as ρAL, where ρ is the fluid density. Equating the energy dissipated to the energy gained, one obtains $\Delta T = (V/L)^2(\sigma/\rho c)\Delta t$. Recall that the electric field is $E = V/L$, so that the temperature rise depends quadratically on the applied field. Increasing the conductivity of the fluid also increases the temperature rise. The longer the time the field is applied, the greater is the temperature increase. It may appear surprising that the increase appears to be independent of the chamber area A. Although a greater A means that more energy is absorbed, it also provides a greater amount of fluid to absorb that increase and the effects cancel. However, it should be noted that the chamber is not thermally isolated from its surroundings. A considerable amount of the heat dissipated will flow to the surrounding materials by thermal conduction so that the above equation will overestimate ΔT.

The temperature rise can alternatively be related to the electric current, I, for a given chamber fluid and dimensions. The energy dissipated can also be represented as $I^2R\Delta t$. For a given R, doubling I doubles V, and thus quadruples ΔT, as seen in the above equation. In practice one should attempt to minimize I if the experiment is to be conducted for a long time. For example, the chamber described in the "Experimental Design" section has a width of ~6 mm, a depth of ~0.2 mm, and a length of ~ 25 mm; σ ~ 1 S/m R ~ 2×10^4 Ω. An applied electric field of 100 mV/mm then requires V = 2.5 V and I = V/R ~ 10^{-4} A, a small current.

CHEMICAL

Cells are certainly known to be affected by the presence of chemicals. Electrochemical reactions at the electrodes can generate contaminants that could enter the experimental chamber and modify the results. Choice of electrode material is critical. Reactive electrodes should be avoided. Chlorided silver electrodes are preferred. There are three methods for dealing with this problem:

1. Make the chamber very long and keep the experiment brief. In this way there may not be enough time for the contaminants to move to the central area where the cells are present. This method is not recommended.
2. Provide a gentle, transverse flow of medium across the chamber to remove any introduced contaminants. This method makes the design of the experimental system more complicated. The temperature of the fluid

must be the same as that in the chamber in order to eliminate temperature gradients. Moreover, fluid shear is also known to produce physiological effects on cell systems. Preliminary experiments should be performed to ensure that the cross-flow itself is not producing competing effects.

3. Introduce agar bridges between the electrodes and the cell chamber. These bridges can be hollow plastic tubes filled with an agar solution in the medium. The result should be a gel that electrically connects a small bath into which an electrode is inserted to a second bath in direct contact with the cell chamber. The bridge provides a long, viscous path that increases the time for small molecules to diffuse from the electrodes into the chamber to values beyond the time frame of the experiment.

Perhaps the most important chemicals to which the cells are exposed are contained in the medium. A certain cell line may respond differently in different media. Even different batches of a medium, commercially obtained or lab produced, may have sufficient variation to produce different effects. A precaution is to preserve for some time a small sample of medium used in the experiments so that it can be checked later for unusual contents if unusual results are obtained.

Cells themselves excrete chemicals into their surroundings. Individual cells may be attracted to or repelled by the chemicals emitted by a cell cluster, with a resulting modification of galvanotactic behavior. It would be wise to check whether isolated cells are responding differently than cell clusters in a particular experiment.

MECHANICAL

Cells attach to substrates via the binding of integrins to proteins on the substrate material. A variety of substrate proteins can attach to different integrins, which are connected, in turn, to the intracellular structure and cell signaling processes. Mechanical factors that affect that binding can, by themselves, produce physiological effects.

The texture of the substrate can provide signals to the cell. Even on a nanometer scale, substrate topography affects the spreading of osteoblasts (Kunzler et al., 2007) and the upregulation of a variety of genes in fibroblasts (Dalby et al., 2002). The effects of electric field and nanoscale topography can even be combined to influence the alignment of epithelial cells (Rajnicek et al., 2008).

The nature of the substrate is important as well. The migration of macrophages on glass in an applied electric field has a random component in addition to the field-induced component, and microfilaments are rearranged too (Cho et al., 2000). In contrast, there is only a field component to the migration, with no microfilament reorganization for either laminin or fibronectin substrates. The chemically stimulated migration of monocytes was enhanced on a laminin substrate compared to fibronectin or collagen types I and IV (Penberthy et al., 1995), whereas neutrophil migration was the same for the first three of these substrates, but less on collagen IV.

It is clear that both the nature and the texture of the substrate are important. If cell migration is the object of study, investigators should make sure that the substrate being used supports migration of the cell type they are investigating. Furthermore, the substrate should be applied as smoothly as possible. Nanoscale grooves produced by brushing with coarse bristles could influence the migration pattern.

Applied mechanical forces are known to produce physiological effects. For example, mechanosensitive ion channels play a role in the mechanically stimulated increase of Ca^{2+} in neurons (Sanchez et al., 2007), and cyclic pressure changes decreased the proliferation of keratinocytes (Nasca et al., 2007). Fluid shear forces are also known to produce a wide variety of effects on cells (Tarbell et al., 2005; Makino et al., 2007; Kwon and Jacobs, 2007). Such forces might be of concern if a transverse fluid flow is being used to remove electrode by-products from the cell chamber. Prior to the electric field experiments tests should be performed to ensure that the transverse flow is not, itself, producing any effect.

Mechanical vibrations can also produce physiological effects in cells, such as the proliferation of chondrocytes (Kaupp and Waldman, 2008), as well as the release of NO and PGE_2, and COX-2 mRNA expression related to oscillations of the cell nucleus (Bacabac et al., 2006). Care should be taken that the cell chamber is mechanically isolated from significant sources of vibraion, such as incubators or coils.

OTHER POTENTIAL CONFOUNDERS

Cells lines are generally obtained from commercial suppliers or the European (Invitrogen) or American Tissue Culture Collection (ATCC) and can be stored in sterile cryovials in liquid nitrogen until required. Primary cells may be isolated directly from tissue under the appropriate government licensing approval procedures, which vary from country to country, or directly from a commercial supplier (e.g., Invitrogen). Before final placement in the experimental chamber the cells are subjected to a variety of processes, such as repeated washings with a particular serum, trypsinizing to remove them from a dish, spinning in a centrifuge to form a pellet, counting and diluting, etc. Standard procedures for these processes have been developed for particular cell lines. These procedures should be used and reported. Having the same person always perform these tasks leads to improved consistency in the results.

Depending on the cell line, the cells are responsive only for a certain number of subculturings into a new vessel (passages) (see ATCC, 2007). An initial couple passages may be required for the cells to respond to the imposed electric field, and they may become senescent after too many passages. It is wise to perform some preliminary tests to determine the optimal range of passage numbers over which the cells are behaving properly, and then to record the passage number of the cells used in case an experiment produces unusual results. To show an effect is robust, several replications are necessary. When using primary cells that have been isolated from a single donor, it is recommended that experiments be repeated with a second or third donor to ensure that an effect is not specific to one donor.

As indicated, there are many potential confounders that can modify or eliminate the proposed effect. It is wise to have a positive control for the system to test for the presence of confounders. One way to achieve this is to replicate a well-known effect each time an experiment is performed. If the cells do not respond as expected, then there is something wrong with the setup or the cells are not healthy for some reason. In that case, the experiment must be aborted. Only if the cells are clearly healthy and responding as expected can the results of a new experiment be trusted.

EXPERIMENTAL DESIGN

Figure 2.5a,b illustrates the apparatus used in a typical galvanotaxis experiment (Pullar, 2009). A DC power supply provides the voltage used to guide the cells. Leads from the supply's terminals are connected via chlorided silver electrodes, shaped as coils to increase the surface area for passing current, to two outer reservoirs that contain medium. A small amount of putty or tape attached to the lead end of each coil can be used to fasten it to the chamber edge so that it does not slip out of the reservoir during the course of the experiment. Agar bridges approximately 6 cm in length connect each outer reservoir to an inner reservoir that is in immediate contact with the chamber. An ammeter (or a small series resistor) should be connected in series between one pole of the power supply and one electrode to monitor the current. A loss of current would indicate that an agar bridge has slipped out of a reservoir, or that the level of medium in the reservoir has fallen significantly. In either case the field in the chamber would have vanished due to the break in the circuit.

The custom chamber is mounted in a custom-heated stage, which is not shown. Heat is transferred to the chamber by thermal conduction from the stage. The temperature of the stage should be kept slightly higher than the desired value for the cells because of heat loss to the surroundings. Preliminary tests should be carried out to determine how much higher the stage temperature should be.

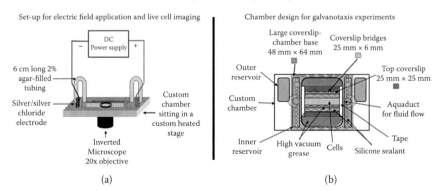

FIGURE 2.5   See color insert. Apparatus used in a typical galvanotaxis experiment. The electrical connections used to apply the field are illustrated in (a). The design of the experimental cell chamber is shown in (b). (From Pullar, *J. Wound Technol. Iss.*, 6, 20–24, 2009.)

There is a large, rectangular opening, approximately 38 × 25 mm, in the center of the apparatus. A large glass coverslip, 64 × 48 mm, is fastened to the bottom of the apparatus by silicone sealant to provide a base for the cell chamber. The cells will ultimately be plated on that base along a strip 25 mm long and about 6 mm wide along its centerline. On either side of that strip coverslip bridges 25 mm long and about 6 mm wide are cut from a 25 × 25 mm coverslip with a diamond pencil and are fastened to the base by silicone sealant. The lengths of these bridges are such that they just reach the plastic, inner sides of the apparatus. In this way, a very shallow, long groove is formed for the cells. Each end of this central groove meets an opening (aquaduct) in the plastic that is connected to an inner reservoir. This aquaduct is about 6 mm wide and 1 mm deep at the bottom of the plastic.

The central reservoir can now be coated with substrate, e.g., collagen. A sterile solution of collagen 1 is applied to the glass on the bottom of the reservoir and the chamber is incubated at 37°C for 30 minutes. After aspirating the collagen solution from the reservoir, the collagen-coated glass is washed briefly with a sterile physiological solution and the cells can be plated onto it in the appropriate cell media and replaced at 37°C to allow the cells to settle and adhere to the collagen substrate for usually 2 to 3 hours. At this point the custom chamber is ready for assembly. The media is removed and a small volume (150 μl) of experimental media is added on top of the cells. Another coverslip, 25 × 25 mm, is fastened, with vacuum grease on top of the groove and side coverslips, ensuring that no bubbles are present in the central reservoir. Tape is fastened along the aquaduct ends and vacuum grease is applied liberally around the sides to eliminate the possibility of fluid leakage, and therefore current leakage, during the experiment. Medium is now added to the two outer reservoirs and one of the inner reservoirs. Light suction is applied to the other inner reservoir to draw medium from the first reservoir through the groove and over the cells to ensure that continuous flow can occur through the central viaduct. Once flow continuity has been achieved, the filling of the inner reservoirs can be completed and the custom chamber can be placed inside the customized heating unit on the microscope stage.

The cells should be viewed through the microscope to ensure that they are healthy and plated in a suitable manner. The DC power supply can now be turned on and current applied. Because there may be measurable voltage drops across the agar bridges, fine wire electrodes should be placed in the two inner reservoirs close to the aquaducts to measure the actual voltage drop, which is Vo. Because the groove containing the cells has a very small cross section and a relatively long length, L, the applied electric field is uniform and equal to Vo/L. After any necessary adjustment to the DC output voltage is made, the wires should be removed. A software package, e.g., Velocity or Openlab (Improvision, Invitrogen), is then used to take pictures of the cells at predetermined time points and allow manual tracking of each cell to calculate the speed and directionality of migration of each cell within an experiment; see Chapter 6 for more details.

Song et al. (2007) provide additional information regarding the application of electric fields to three-dimensional cultures, the design of cross-flow systems to minimize chemical gradients, and troubleshooting procedures.

CONCLUSIONS

A successful experiment requires the proper characterization of the independent variable, such as the surface electric field or change in transmembrane potential, as well as the elimination of confounding factors that could lead to erroneous conclusions. To decide whether the coupling mechanism of the cell to the field depends on Es or ΔV requires a numerical modeling of a realistically shaped cell. Es values can be readily obtained otherwise only for spherical cells. Cell preparation procedures should follow well-established guidelines and be carried out in a consistent manner. These procedures should be carefully documented and reported. Just prior to the start of an experiment the condition of the cells should be checked. Positive control experiments should be run as a measure of cell viability. Progress in this area requires both careful modeling and laboratory technique.

REFERENCES

ATCC. 2007. Passage number effects in cell lines. *ATCC Technical Bulletin* 7. American Type Culture Collection.

Bacabac, R. G., T. H. Smit, J. J. W. A. Van Loon, B. Z. Doulabi, M. Helder, and J. Klein-Nulend. 2006. Bone cell responses to high-frequency vibration stress: Does the nucleus oscillate within the cytoplasm? *FASEB J* 20:858–64.

Bernhardt, J., and H. Pauly. 1973. On the generation of potential differences across the membranes of ellipsoidal cells in an alternating electric field. *Biophysik* 10:89–98.

Cho, M. R., J. Marler, H. Thatte, and D. Golan. 2002. Control of calcium entry in human fibroblasts by frequency-dependent electrical stimulation. *Front Biosci* 7:a1–a8.

Cho, M. R., H. S. Thatte, R. C. Lee, and D. E. Golan. 2000. Integrin-dependent human macrophage migration induced by oscillatory electrical stimulation. *Ann Biomed Eng* 28:234–43.

Dalby, M. J., S. J. Yarwood, M. O. Riehle, H. J. H. Johnstone, S. Affrossman, and A. S. G. Curtis. 2002. Increasing fibroblast response to materials using nanotopography: Morphological and genetic measurements of cell response to 13-nm-high polymer demixed islands. *Exp Cell Res* 276:1–9.

Gimsa, J., and D. Wachner. 2001. Analytical description of the transmembrane voltage induced on arbitrarily oriented ellipsoidal and cylindrical cells. *Biophys J* 81:1888–96.

Grosse, C., and H. P. Schwan. 1992. Cellular membrane potentials induced by alternating fields. *Biophys J* 63:1632–42.

Hart, F. X. 1987. Pulse shape distortion by tissue. *J Bioelec* 6:93–107.

Hart, F. X. 1996. Cell culture dosimetry for low-frequency magnetic fields. *Bioelectromagnetics* 17:48–57.

Hart, F. X. 2008. The mechanical transduction of physiological strength electric fields. *Bioelectromagnetics* 29:447–55.

Hart, F. X. 2010. Cytoskeletal forces produced by extremely low-frequency electric fields acting on extracellular glycoproteins. *Bioelectromagnetics* 31:77–84.

Hart, F. X., and A. A. Marino. 1986. Penetration of electric fields into a concentric-sphere model of biological tissue. *Med Biol Eng Comput* 24:105–8.

Jaffe, L. F. 1977. Electrophoresis along cell membranes. *Nature* 265:600–2.

Jaffe, L. F., and R. Nuccitelli. 1974. An ultrasensitive vibrating probe for measuring steady extracellulsr currents. *J Cell Biol* 63:614–28.

Jaffe, L. F., and R. Nuccitelli. 1977. Electrical controls of development. *Ann Rev Biophys Bioeng* 6:445–76.

Kaupp, J. A., and S. D. Waldman. 2008. Mechanical vibrations increase the proliferation of articular chondrocytes in high-density culture. *Proc IMechE* 222(Part H):695–703.

Kotnik, T., F. Bobanovic, and D. Miklavcic. 1997. Sensitivity of transmembrane voltage induced by applied electric fields—A theoretical analysis. *Bioelectrochem Bioenerg* 43:285–91.

Kotnik, T., and D. Miklavcic. 2000. Analytical description of transmembrane voltage induced by electric fields on spheroidal cells. *Biophys J* 79:670–79.

Kwon, R. Y., and C. R. Jacobs. 2007. Time-dependent deformations in bone cells exposed to fluid flow *in vitro*: Investigating the role of cellular deformation in fluid flow-induced signaling. *J Biomechanics* 40:3162–68.

Kunzler, T. P., C. Huwiler, T. Drobek, J. Voros, and N. D. Spencer. 2007. Systematic study of osteoblast response to nanotopography by means of nanoparticle-density gradients. *Biomaterials* 28:5000–6.

Liu, C., D. Sheen, and K. Huang. 2003. A hybrid numerical method to compute erythrocyte TMP in low-frequency electric fields. *IEEE Trans Nanobiosci* 2:104–9.

Makino, A., H. Y. Shin, Y. Komai, S. Fukuda, M. Coughlin, M. Sugihara-Seki, and G. W. Schmid-Schonbein. 2007. Mechanotransduction in leukocyte activation: A review. *Biorheology* 44:221–49.

McLaughlin, S., and M. Poo. 1981. The role of electro-osmosis in the electric-field-induced movement of charged macromolecules on the surface of cells. *Biophys J* 34:85–93.

Nasca, M. R., A. T. Shih, D. P. West, W. M. Martinez, G. Micali, and A. S. Landsman. 2007. Intermittent pressure decreases human keratinocyte proliferation *in vitro*. *Skin Pharmacol Physiol* 20:305–12.

Nuccitelli, R. 1988. Physiological electric fields can influence cell motility, growth, and polarity. *Adv Cell Biol* 2:213–33.

Pavlin, M., and D. Miklavcic. 2009. The effective conductivity and the induced transmembrane potential in dense cell system exposed to DC and AC electric fields. *IEEE Trans Plasma Sci* 37:99–106.

Penberthy, T. W., Y. Jiang, F. W. Luscinskas, and D. T. Graves. 1995. MCP-1-stimulated monocytes preferentially utilize b_2-integrins to migrate on laminin and fibronectin. *Am J Physiol* 269:C60–68.

Pucihar, G., T. Kotnik, B. Valic, and D. Miklavcic. 2006. Numerical determination of transmembrane voltage induced on irregularly shaped cells. *Ann Biomed Eng* 34:642–52.

Pullar, CE. 2009. The biological basis for electric stimulation as a therapy to heal chronic wounds. *J Wound Technol Iss* 6:20–24.

Rajnicek, A. M., L. E. Foubister, and C. D. McCaig. 2008. Alignment of corneal and lens epithelial cells by co-operative effects of substratum topographies and DC electric fields. *Biomaterials* 29:2082–95.

Ramos, A., D. O. H. Suzuki, and J. L. B. Marques. 2006. Numerical study of the electrical conductivity and polarization in a suspension of spherical cells. *Bioelectrochemistry* 68:213–17.

Reitz, J. R., F. J. Milford, and R. W. Christy. 1992. *Foundations of electromagnetic theory*. 4th ed. Reading, MA: Addison-Wesley.

Sanchez, D., U. Anand, J. Gorelik, C. D. Benham, C. Bountra, M. Lab, D. Klenerman, R. Birche, P. Anand, and Y. Korchev. 2007. Localized and non-contact mechanical stimulation of dorsal root ganglion sensory neurons using scanning ion conductance microscopy. *J Neurosci Methods* 159:26–34.

San Martin, S. M., J. L. Sebastian, M. Sancho, and G. Alvarez. 2006. Modeling normal and altered human erythrocyte shapes by a new parametric equation: Application to the calculation of induced transmembrane potentials. *Bioelectromagnetics* 27:521–27.

Song, B., Y. Gu, J. Pu, B. Reid, Z. Zhao, and M. Zhao. 2007. Application of direct current electric fields to cells and tissues *in vitro* and modulation of wound electric field *in vivo*. *Nature Protocols* 2:1479–89.

Stuchly, M. A., and W. Xi. 1994. Modelling induced currents in biological cells exposed to low-frequency magnetic fields. *Phys Med Biol* 39:1319–30.

Tarbell, J. M., S. Weinbaum, and R. D. Kamm. 2005. Cellular fluid mechanics and mechanotransduction. *Ann Biomed Eng* 33:1719–23.

Valic, B., M. Golzio, M. Pavlin, A. Schatz, C. Faurie, B. Gabriel, J. Teissie, M.-P. Rols, and D. Miklavcic. 2003. Effect of electric field induced transmembrane potential on spheroidal cells: Theory and experiment. *Eur Biophys J* 32:519–28.

3 Endogenous Bioelectric Signals as Morphogenetic Controls of Development, Regeneration, and Neoplasm

Michael Levin
Biology Department and Tufts Center for
Regenerative and Developmental Biology
Tufts University
Medford, Massachusetts

CONTENTS

INTRODUCTION

Embryonic development, regeneration, and prevention of cancer during adult remodeling all require the generation and maintenance of complex order on several scales of size and organization. Endogenous electric fields, ion flows, and transmembrane potential gradients are a powerful system underlying crucial aspects of morphogenetic regulation. Bioelectrical signals produced by ion channels and pumps are increasingly recognized as determinants of cell proliferation, apoptosis, differentiation, shape, orientation, and migration. Moreover, exciting recent data using state-of-the-art molecular approaches *in vivo* have demonstrated that rational changes in ion flow can initiate complex regenerative responses and modulate patterning in a variety of cell types. Thus, bioelectrical signals are not only a profoundly interesting area of investigation for cell, developmental, and evolutionary biology, but also an exciting target for the development of biomedical strategies in regenerative medicine.

Modern efforts to understand cell regulation and pattern formation, as well as mainstream approaches to address biomedical problems such as cancer and injury, focus largely on biochemical signals. Thus, a majority of pathway discovery and tool development has been based on understanding transcriptional control networks, and gradients of protein ligands of various receptors. However, alongside chemical signals functions a fascinating and rich system of bioelectrical signals mediated by the steady-state electrical properties of cells and tissues. Despite much classical data on the roles of endogenous bioelectric signals in limb and spinal cord regeneration (Borgens, 1986; Borgens et al., 1986, 1990), cell and embryonic polarity (Bentrup et al., 1967; Novák and Bentrup, 1972; Novak and Sirnoval, 1975), growth control (Cone and Tongier, 1971, 1973), and migration guidance of numerous cell types (McCaig et al., 2005), this field is unfamiliar to modern cell and developmental biologists (although some well-known processes, such as the fast, electrical polyspermy block (Grey et al., 1982; Jaffe and Schlichter, 1985), are in fact good examples of such signaling).

Functional experiments throughout the last decades showed that some bioelectric events were not merely physiological correlates of housekeeping processes, but rather provided specific instructive signals regulating cell behavior during embryonic development and regenerative repair (Borgens et al., 1989; Jaffe, 1982; Levin, 2007b). This chapter discusses the roles of ion-based physiological processes in guiding cell behavior during morphogenesis, outlining controls of individual cell behavior and the unique properties of electrical processes that may underlie the orchestration of higher-order patterning.

Bioelectric signals are generated by specific ion channels and pumps within cell membranes. The segregation of charges achieved by ion fluxes through such transporter proteins gives rise to a transmembrane voltage potential (usually on the order of -50 to -70 mV, inside negative). Ion channels and pumps are localized to distinct regions of some cell types; in particular, the differences of ion flux through the apical and basal membranes of epithelial cells result in a parallel arrangement of battery-like cells, which in turn give rise to a transepithelial potential (McCaig et al., 2005; Robinson, 1989) (see Figure 1.2). Thus, all cells— not just excitable neurons and muscle—generate and receive steady-state bioelectrical signals. All of these bioelectrical phenomena—transmembrane potentials, electric fields through tissue and surrounding fluids, isoelectric and iso-pH cell groups established by gap junctions (Fitzharris and Baltz, 2006), and fluxes of individual ions—carry information to the source cell as well as to its neighbors, and in some cases (Zimmermann et al., 2009) to distant locations.

Early discoveries of "animal electricity" can be traced to Luigi Galvani in the late 1700s, and as early as 1903, it was discovered that hydroids have a specific electrical polarity (Mathews, 1903). However, the majority of the literature in this rich field has come from several subsequent major waves of research effort. E. J. Lund, through the 1920s and 1930s, focused on currents and showed that polarity of numerous developing and regenerating plant and animal systems was predicted by, and in some cases controlled by, the bioelectric polarity of ion flows *in vivo* (Lund, 1947). H. S. Burr (1930s and 1940s) focused on measuring and correlating voltage gradients with future developmental pattern in a wide range of species and organs (Burr, 1944; Burr and Northrop, 1935); the measurements suggested that the voltage gradients are quantitatively predictive of morphology, and it was hypothesized that the measured fields carried patterning information (an example of the "second anatomy" underlying patterning decisions (Slack, 1982)).

Some of the best early functional results were obtained by Marsh and Beams (1947a, 1949, 1957), who were able to specifically control anterior-posterior polarity in planarian regeneration by applying exogenous bioelectrical fields to cut fragments of the flatworms. Enormously influential for the field was the subsequent work of Lionel Jaffe and coworkers, including Richard Nuccitelli, Ken Robinson, and Richard Borgens (Borgens et al., 1989; Borgens, 1982, 1983; Hotary and Robinson, 1994; Jaffe, 1981, 1982; Jaffe and Nuccitelli, 1974, 1977; Nuccitelli, 1980, 1992; Robinson, 1989; Robinson and Messerli, 1996), who demonstrated that electrical properties of individual cells, epithelia, neural structures, and entire limbs were instructive for growth, pattern, and anatomical polarity. Indeed, bioelectrical signals are likely to be highly conserved, playing roles in patterning of diverse phyla spanning from pollen tubes (Certal et al., 2008; Feijo et al., 1999, 2001; Michard et al., 2008) to mammals (Borgens et al., 1989).

The rise of molecular genetics has drawn attention away from a huge literature of not only descriptive, but also solid, well-controlled functional work using physiological techniques. However, in the last decade, state-of-the-art experiments have begun to identify proteins responsible for the well-characterized bioelectric signals, the genetic networks that shape them, and the mechanisms that allow cells to transduce the information into changes of behavior. Molecular and cell biology are now being applied to this problem in the areas of wound healing (Chapters 5–9), neural guidance (Chapters 10–11), and cell orientation to physiological electric fields (Chapters 5–10) (Pullar et al., 2006; Pullar and Isseroff, 2005; Rajnicek et al., 2006, 2007; Zhao et al., 1999a, 1999b, 2006), as well as the role of specific ion transporter activity in tail regeneration, left-right patterning, control of adult stem cells (Chapter 4) and regenerative polarity, and the switch between embryonic stem cell and neoplastic phenotypes (Adams et al., 2006, 2007; Levin et al., 2002; Morokuma et al., 2008a; Nogi and Levin, 2005; Oviedo and Levin, 2007; Sundelacruz et al., 2008). Indeed, the potential of harnessing these powerful controls of cell function for applications to human medicine are numerous (Adams, 2008; Funk et al., 2009; Nuccitelli et al., 2009; Shapiro et al., 2005).

Although many modern workers may be unaware of the history and data of the field of bioelectricity as a patterning signal, the connection between molecular genetic pathways and bioelectric signaling is being forged by the data themselves (Forrester et al., 2007). A variety of relevant channelopathies have now been discovered by unbiased approaches (Arcangeli et al., 2009; Stuhmer et al., 2006; Teng et al., 2008; Wang, 2004), though ion transporters are usually de-prioritized for analysis when they show up on comparative microarray or profiling experiments because it is not yet second nature for cell and molecular biologists to think in terms of bioelectrical signaling. It is hoped that by highlighting the techniques and tools now available, and illustrating strategies for integrating bioelectrical signals with mainstream pathways, scientists in multiple subfields will consider that modulation of ion flows, currents, and voltages may be at the root of their favorite patterning or mispatterning problem when ion channels and pumps are identified in genetic screens or subtraction analyses. A superb example of such a convergence is the recent elegant study implicating sodium-hydrogen exchange in the establishment of planar polarity in *Drosophila* (Simons et al., 2009), a relationship that was predicted by bioelectric signals during embryonic left-right patterning (Aw and Levin, 2009).

Several comprehensive recent reviews address the role of electric fields and specific ion transporter proteins in wound healing (Huttenlocher and Horwitz, 2007; Nuccitelli, 2003; Reid et al., 2005; Zhao et al., 2006), neoplastic growth (Arcangeli et al., 2009; Kunzelmann, 2005), and cell cycle (Blackiston et al., 2009; Lang et al., 2005, 2007; McCaig et al., 2002, 2005; Wonderlin and Strobl, 1996). Thus, here I focus on transmembrane potential (V_{mem}), and consider the unique properties of bioelectrical signals as carriers of morphogenetic information in complex organs and tissues. The novel techniques being used in this field and exciting recent data promise significant advances for regenerative biology and biomedicine (Levin, 2007b).

BIOELECTRICAL CONTROLS OF CELLULAR FUNCTION: MOLECULAR PATHWAYS AND MECHANISMS

STATE-OF-THE-ART TOOLS FOR RESEARCH IN BIOELECTRIC SIGNALING

A variety of new reagents and methodologies have been developed for molecular analysis of bioelectric signals in regenerative contexts (Adams, 2008; Adams and Levin, 2006b). Tools for the characterization of bioelectrical events now include highly sensitive, ion-selective, fast-response extracellular electrode probes (Messerli et al., 2009; Reid et al., 2007; Smith et al., 2007); fluorescent reporter dyes (Figure 3.1) enabling the real-time monitoring of pH, membrane voltage, and ion content in any optically accessible tissue (Morley et al., 1996; Moschou and Chaniotakis, 2000; Steinberg et al., 2007; Wolff et al., 2003; Yun et al., 2007); and nanoscale voltage reporters (Tyner et al., 2007). Especially exciting will be the use of multiple physiological dyes in fluorescence activated cell sorting (FACS) experiments to identify subpopulations of stem and other cell sets that differ in key bioelectric properties, as has been observed for HUVEC cells (Yu et al., 2002) and rat hippocampal cells (Maric et al., 1999). Importantly, such experiments on dissociated cells will clearly highlight properties that are cell autonomous vs. those physiological conditions that can only be maintained *in vivo*.

To determine whether ion flow is a causal factor in a particular instance of patterning, and to inexpensively and rapidly implicate specific ion transporter proteins for further molecular validation, an inverse drug screen can be performed (Adams and Levin, 2006a). This is a chemical genetics approach that capitalizes on a tiered (least specific → more specific) tree-based distribution of known blocker compounds that enables an efficient binary search for likely candidates (ion channels and pumps targeted by the inhibitors). In this strategy, nonspecific reagents are most useful, because they have the potential to rule out large regions of the candidate list from further consideration, saving considerable effort. This can be used to probe endogenous bioelectrical mechanisms, and has resulted in the identification of channels and pumps as novel components of left-right patterning in vertebrates and invertebrates (Levin, 2006), anterior-posterior polarity during flatworm regeneration (Nogi and Levin, 2005), and stem cell regulation (Morokuma et al., 2008a; Oviedo and Levin, 2007; Sundelacruz et al., 2008).

Many experiments make use of external applications of electric fields (Robinson, 1989; Song et al., 2007); it is now also possible to use DNA constructs encoding wild-type, mutant, and dominant-negative channel and pump proteins in gain- and loss-of-function approaches to modulate different aspects of ion flux with molecular resolution. Specific cells can be induced to depolarize or hyperpolarize the membrane, or pump individual ions across it at a defined rate (Miller, 2006; Wyart et al., 2009). The molecular-genetic control of specific ion fluxes in cells using well-characterized constructs also enables loss-of-function and rescue experiments, allowing elegant demonstrations of necessity and sufficiency for bioelectrical signals (Figure 3.2). For example, a channel can be blocked by a pharmacological inhibitor, and the signaling function restored by overexpression of a drug-insensitive mutant (Kosari et al., 2006). This strategy has been used to induce tail regeneration (Adams et al., 2007) at nonregenerative stages and drastically alter the positioning and proliferation of neural crest derivatives (Morokuma et al., 2008a). The work of neurobiologists and kidney physiologists has resulted in the availability of a large number of expression constructs encoding ion transporters that can be used as molecular tools for rationally altering the electrical activity of cells and tissues. In cases where maternal proteins

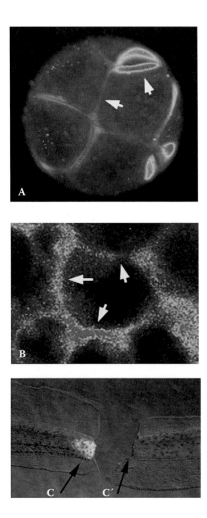

FIGURE 3.1   See color insert. Ascertaining transmembrane potential *in vivo* using volt-age-sensitive dyes. Fluorescence of the voltage-sensitive dyes DiBAC and DiSBAC (Krotz et al., 2004; Wolff et al., 2003) reveals transmembrane potential in early frog embryo blastomeres (A) as well as COS cells in monolayer culture (B). Both kinds of cells exhibit significant variations of membrane voltage level around the cell surface, indicating that a single V_{mem} number for a given cell drastically underestimates the amount of information that can be encoded in the plasma membrane's physiological state and potentially communicated to neighboring cells. Blastemas of regenerating (C) and nonregenerating (C') tadpole tails differ significantly in their membrane voltage. (Images in A and B courtesy of D. S. Adams.)

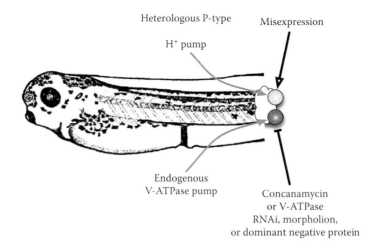

FIGURE 3.2 **See color insert.** Strategy for molecular rescue experiment to test role of ion transporter in patterning. In tadpoles, the V-ATPase H⁺ pump complex is required to regenerate the tail after amputation (Adams et al., 2007). A rescue experiment, in which endogenous V-ATPase is inhibited by concanamycin (pharmacological blocker) or mor-pholino/RNAi (genetically), can be performed by misexpressing a heterologous (yeast) proton pump, PMA1 (Bowman et al., 1997; Masuda and Montero-Lomeli, 2000), which bears no sequence or structure similarity to the V-ATPase. The resulting restoration of regenerative ability (Adams et al., 2007) proves that it is the bioelectrical signal (proton pumping), not some cryptic other role of V-ATPase proteins, that is responsible for the induction of tail regeneration. Different ion transporter proteins with similar physiological functions can be used to test loss and gain of function for individual ion fluxes in many contexts, with molecular specificity.

are not involved, morpholino- and RNAi-mediated knockdown can also be used (Oviedo and Levin, 2007).

Indeed, analysis of the patterning phenotypes induced by such constructs can be used to dissect the mechanism of action, by distinguishing among different aspects of bioelectrical signals. For example, rescue via electroneutral transport-ers can differentiate between the importance of voltage changes and that of flux of specific ions (Adams et al., 2006). Pore mutants can be used to distinguish between ion conductance roles (true bioelectric signaling) and possible func-tions of channels/pumps as scaffolds or binding partners for other proteins. For example, in the Na⁺/H⁺ exchanger, both ion-dependent and ion-independent func-tions control cell directionality and Golgi apparatus localization to wound edge (Denker and Barber, 2002a); similarly, the ability of the NKCC1 transporter to induce a primary axis in frog embryos is not dependent on its current-passing function (Walters et al., 2009). Gating channel mutants and pumps with altered kinetics can, respectively, be used to reveal upstream signals controlling the bio-electric events, and the temporal properties of the signal. Misexpression of gap junction proteins can alter boundaries of isopotential cell fields, testing for the role of multicellular voltage domains *in vivo*. Together, these tools can now be used to integrate bioelectrical signals with canonical downstream and upstream pathways, identifying transduction mechanisms leading from ion flow to patterning decisions.

CELLULAR-LEVEL PROCESSES CONTROLLED BY BIOELECTRIC SIGNALS

A functionally useful regenerative response requires integration of proliferation, cell movement, and differentiation into needed cell types to restore large amounts of organized tissue. Endogenous bioelectrical signals participate in the control of all key aspects of those cell functions:

1. *Cell movement and positioning* is an important component of regeneration (Han et al., 2007); movement of progenitor cells toward wounds is observed in planaria (Salo and Baguna, 1985), zebrafish brain (Zupanc, 2006), and mammalian stem cell homing (Chute, 2006). One of the earliest observed effects of electric fields was growth (extension of processes) or migration (toward the anode or cathode) of slime mold and vertebrate as well as invertebrate cells (Anderson, 1951; Hyman and Bellamy, 1922). Modern protocols avoid artifacts created by polarization of substratum molecules or by release of electrode products into medium (Song et al., 2007). Despite controversy (Robinson and Cormie, 2008) over which cell types respond to physiological strength electric fields (usually on the order of 50 mV/mm, and as high as 500 mV/mm within the neural tube (Borgens et al., 1994; Hotary and Robinson, 1991)), it is clear that a large variety of embryonic and somatic cells exhibit directional movement (galvanotaxis) in electric fields of the magnitude often found *in vivo* (Nishimura et al., 1996; Pullar and Isseroff, 2005; Stump and Robinson, 1983; Zhao et al., 1997). In embryos, it has been suggested that patterns of voltage gradients form coordinates guiding cell movement during complex morphogenetic processes (Shi and Borgens, 1995). Electric guidance also occurs in several types of tumor cells (Yan et al., 2009); recently, voltage-gated sodium channels have been strongly implicated in this phenomenon (Fraser et al., 2005; Mycielska and Djamgoz, 2004) (see Chapter 12), suggesting that endogenous bioelectric states may be a factor in metastatic invasion. It is also now known that bioelectric events are important not only for the generation of guidance signals, but also for cell-autonomous responses to electric fields during migration (Schwab, 2001), where channels such as $K_{Ca}3.1$ (KCNN4) and K_{atp} (van Lunteren et al., 2002) are required for cells to interpret instructive signals for the direction of cell movement (Schwab et al., 1995).
2. *Differentiation* into numerous and appropriate cell types is required for regeneration. Early links between ion flow and differentiation were observed by Barth and coworkers, who showed that ventral ectoderm explants could be differentiated into a variety of different cell types

by careful modulation of extracellular medium ion content (Barth and Barth, 1974a, 1974b). Bioelectric signals apply not only to embryonic cell differentiation, but also to stem cells, which have unique profiles with respect to ion channel expression and physiological state (Balana et al., 2006; Biagiotti et al., 2005; Chafai et al., 2006; Flanagan et al., 2008; Park et al., 2007; Ravens, 2006; Wang et al., 2005; Wenisch et al., 2006). Moreover, it has been recently shown by functional experiments that membrane voltage controls human mesenchymal stem cell (MSC) differentiation *in vitro* (Sundelacruz et al., 2008). Much remains to be learned about this process, but it is known, for example, that Kir2.1 (KCNJ2) channel-mediated hyperpolarization controls differentiation in human myoblasts via a calcineurin pathway (Konig et al., 2006). Oligondendrocyte progenitor cell lineage progression is regulated by a glutamate-receptor-mediated potassium channel blockade (Gallo et al., 1996). Importantly, a degree of de-differentiation can be induced by ionic modulation (Cone and Cone, 1976; Harrington and Becker, 1973), and even mature neurons can be coaxed to reenter the cell cycle by long-term depolarization. This raises the possibility that a degree of stem-cell-like plasticity could be induced in terminally differentiated somatic cells by bioelectric signals (Cone and Cone, 1976; Cone and Tongier, 1971; Filek et al., 2005; Stillwell et al., 1973).

3. *Mitotic rate* is also controlled by transmembrane potential (Blackiston et al., 2009; Sundelacruz et al., 2009). A comparative analysis of membrane voltage potential (Figure 3.3) of various kinds of cells revealed a striking relationship between depolarization and control of differentiation and proliferation (Binggeli and Weinstein, 1986). Numerous studies have implicated K^+ currents as protagonists of proliferation and cell cycle progression (MacFarlane and Sontheimer, 2000; Rouzaire-Dubois et al., 1993), reviewed in MacFarlane and Sontheimer (2000) and Wonderlin and Strobl (1996). Cell proliferation appears to be controlled mostly by membrane potential (Arcangeli et al., 1993; Cone, 1974), although the effect is not always cell autonomous: depolarized cells can induce distant neural crest derivatives (frog embryo melanocytes) to overproliferate (Morokuma et al., 2008a).

A considerable literature now exists on the role of specific ion transporters, including a sodium-hydrogen exchanger (NHE) and a variety of K^+ and Cl^- channels, in the functional control of cell cycle progression, although many questions remain about mechanistic details (Boutillier et al., 1999; MacFarlane and Sontheimer, 2000; Ouadid-Ahidouch and Ahidouch, 2008; Putney and Barber, 2003; Valenzuela et al., 2000). In the zebrafish eye, the V-ATPase is required for retinoblast proliferation (Nuckels et al., 2009). Thus, because of its many patterning roles spanning from the elongation of the tadpole tail (Adams et al., 2007) to that of pollen tubes (Certal et al., 2008), as well as in neural stem cells in the regenerating fish brain (Zupanc et al., 2006), H^+ efflux is a widely relevant signal for efforts to augment regenerative growth. Knockdown or inhibition of two K^+ channels ($K_V 1.3$/KCNA3 and $K_V 1.5$/KCNA5) resulted in cell cycle arrest at G1 in rat oligodendrocyte precursors (Chittajallu et al., 2002; Ghiani et al., 1999); this effect was characterized by accumulation of p27 and p21. Overexpression of the hyperpolarizing Kir4.1 channel shifts a significant number of glioma cells from the G_2/M phase into the quiescent G_0/G_1 stage of the cell cycle (Higashimori and Sontheimer, 2007). Blockade of other K^+ channels results in similar signaling cascades, suggesting convergent mechanisms downstream of the activity of many diverse channels (Ghiani et al., 1999).

4. *Shape change, orientation, and polarization* of a wide variety of cells by electric signals have long been recognized (Marsh and Beams, 1945, 1946, 1947b). The pioneering studies of Lionel Jaffe in the alga *Fucus* showed how self-electrophoresis could be used in determination of cell polarity and outgrowth in single cells (Bentrup et al., 1967; Bentrup and Jaffe, 1968; Jaffe, 1966, 1968). Indeed, several groups have used the orientation response of cells in external electrical fields to dissect the molecular mechanisms of sensation of, and response to, non-cell-autonomous electrical signals, implicating microtubules, microfilaments, camp, integrins, and Rac1 (Erskine and McCaig, 1995a, 1995b; Erskine et al., 1995; McCaig and Dover, 1993; Moriarty and Borgens, 2001; Pullar et al., 2006; Pullar and Isseroff, 2005; Rajnicek et al., 2006; Stewart et al., 1996; Wang et al., 2000; Zhao et al., 1996, 1997, 1999b). More recently, it has been shown that depolarization of cells *in vivo*, by misexpression of an inhibitory potassium channel subunit, increased dendricity of neural crest derivatives (melanocytes), radically changing their cellular morphology (Morokuma et al., 2008a), although melanocytes are not galvanotactic (Grahn et al., 2003).

FIGURE 3.3 (*see facing page*)   **See color insert.** Transmembrane potential as a determinant of proliferation and plasticity. Membrane voltage is correlated to, and indeed controlled by, proliferative potential and differentiation state. This sample of data (from Binggeli and Weinstein, 1986) illustrates the observation that tumor and embryonic (proliferative) cells tend to be highly depolarized; in contrast, terminally differentiated quiescent cells tend to be strongly polarized. Data suggest that this relationship is functional and not merely an epiphenomenon. The position of the liver (highlighted in yellow) near the proliferative component of the scale is consistent with the relationship between membrane voltage and cellular plasticity, in light of its unique (high) regeneration potential. The properties of stem cells along this scale remain to be investigated, although early indications are that at least human mesenchymal stem cells follow the same trend during differentiation (Sundelacruz et al., 2008). (From Levin, *Trends Cell. Biol.*, 17, 262–71, 2007b.)

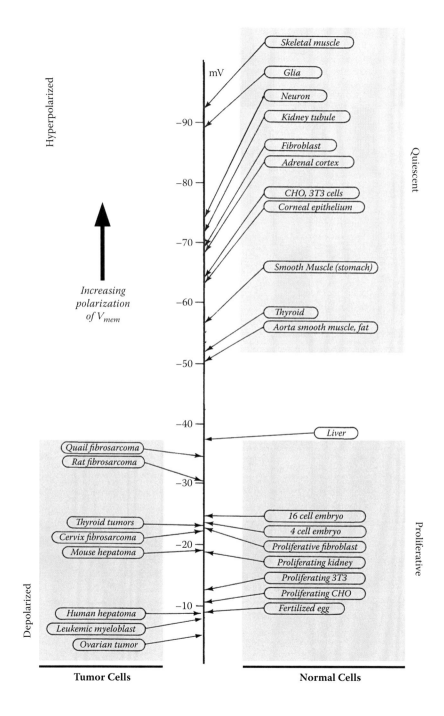

5. *Cell elimination through programmed cell death* is known to be a part of regeneration in a variety of systems utilizing stem cells (Chera et al., 2009; Hwang et al., 2004), tissue renewal (Nadal-Ginard et al., 2003; Tseng et al., 2007), and transdifferentiation (Mescher et al., 2000). Apoptosis is regulated by hyperpolarization via a set of K^+ and other channels (Gilbert et al., 1996; Lang et al., 2005; Wang et al., 1999, 2002); for example, inhibition of K^+ channels can promote apoptosis (Chin et al., 1997; Miki et al., 1997, 2001), while activation of K^+ channels can inhibit it (Lauritzen et al., 1997, 2003). Surprisingly, programmed cell death has recently been shown to be required for regeneration in vertebrates (Tseng et al., 2007) and invertebrates (Chera et al., 2009), suggesting that tight control over programmed cell death (by bioelectric means as well as chemical) may need to be an important consideration for design of regenerative interventions.

Thus, the data point to transmembrane potential and channel-mediated ion flux as a broadly conserved aspect of orchestrating the proliferation, elimination, differentiation, and movement of cells. This is of particular relevance for bioengineers (see Chapter 4) and those seeking to transition findings in regenerative biology into therapeutics (see Chapter 11): bioelectric events are a powerful, largely untapped set of cellular control knobs (Adams, 2008). Gaining the ability to modulate cell number, position, and identity provides the opportunity to manage the alteration or generation of any desired shape.

HIGHER-LEVEL INTEGRATION: THE ROLES OF BIOELECTRIC SIGNALS IN MORPHOGENESIS

Roles for endogenous currents and fields have been found in numerous developmental systems, and in several cases, spatially instructive signaling was demonstrated (Dimmitt and Marsh, 1952; Marsh and Beams, 1950, 1952, 1957; Rose, 1972, 1974). Examples include guidance of gastrulation, neurulation, and organogenesis in chick, frog, and axolotl embryos (Borgens and Shi, 1995; Hotary and Robinson, 1992; Shi and Borgens, 1995; Stern, 1982), growth control and size determination in segmented worms (Kurtz and Schrank, 1955), control of anterior-posterior polarity during worm regeneration (Marsh and Beams, 1947a, 1949, 1950, 1952, 1957), and neural differentiation and patterning in zebrafish and *Xenopus* embryos (Jones and Ribera, 1994; Pineda et al., 2006; Svoboda et al., 2001; Uzman et al., 1998).

Use of ion-based signals in higher-order patterning necessitates coupling groups of cells with respect to electrical signals (Palacios-Prado and Bukauskas, 2009). This often occurs through gap junctions (Levin, 2007a; Nicholson, 2003). This coupling not only augment cells' ability to sense extracellular electric fields (Cooper, 1984), but gap junctional communication is a common mechanism for organizing cells into functional domains, for example, when delimiting regions of neurogenic precursors in the spinal cord (Russo et al., 2008), or mediating the communication between adult stem cells and the cells of their niche during planarian regeneration (Oviedo and Levin, 2007).

The simplest examples of the roles of ionic signals in multicellular systems involve healing epithelial layers, where the fields resulting from disruption of the integrity of the polarized layer provide guidance cues for growth of migratory cells that repair the wound; much molecular data are now available about the alveolar epithelium (Trinh et al., 2007) and the cornea in particular, where not only electric fields (Chao et al., 2007; Huttenlocher and Horwitz, 2007; Raja et al., 2007; Zhao et al., 2006) but also cell-autonomous changes in transmembrane potential (Chifflet et al., 2003, 2005) are involved. In the alveolar epithelium, for example, the epidermal growth factor (EGF)–stimulated repair process is dependent on the activity of K_{atp} and KCNQ1 channels (Trinh et al., 2008). Other tissues where bioelectric cues contribute to repair include the spinal cord (Borgens et al., 1981, 1987, 1999), and indeed this modality is now in human clinical trials with paralyzed patients (Shapiro et al., 2005) (see Chapter 11).

A more complex example of morphogenetic control by bioelectric cues is revealed by the role of currents during appendage regeneration. Reviews of the early work of bioelectric effects on regeneration of amphibian limbs (augmentation of innervation, control of polarity, and alteration of differentiation) are given (Borgens et al., 1989; Borgens, 1984; Smith, 1970). Amputated amphibian limbs maintain a current of injury—a direct current signal that is very different in regenerating and nonregenerating animals. In the latter, the current decreases slowly as the limb heals, while the former exhibits first a positive polarity (similar to the nonregenerative organism), and then a sharp switch to negative polarity, the peak voltage of which occurs at the time of maximum cell proliferation. For example, in salamanders and newt limbs, which have superb regenerative ability, several hours after amputation the density of stump current density reaches 10 to 100 mA/cm^2 and the electric field is on the order of 50 mV/mm (Borgens et al., 1984). Currents leave the end of the stump and reenter the skin around the limb. The relevant currents can be measured for weeks—much longer than the time needed for the damaged cells to either recover or die, refuting the simple model that the fields reflect passive ion leaks from damaged cells. The studies that correlated changes in voltage and currents were followed by functional experiments. Interfering with the required regeneration gradients via electrical isolation, shunting, ion channel blockers, or exogenous reversal of the gradient inhibited regeneration in several systems (Borgens, 1982; Hotary and Robinson, 1992; Jenkins et al., 1996; Novak and Sirnoval, 1975), demonstrating that these biophysical events were necessary factors regulating regeneration.

Another set of crucial experiments demonstrated sufficiency of the electrical signals for inducing or augmenting regeneration (Sisken, 1992; Sisken et al., 1993). Guided by measurements of field density, voltage gradient, and direction in endogenous regenerating systems, several labs showed that application of exogenous fields (with physiological parameters) can induce limb regeneration in species that normally do not regenerate, including amphibia (Borgens et al., 1977; Sharma and Niazi, 1990; Smith, 1967, 1981), aves (Sisken and Fowler, 1981), and possibly even mammals (Becker, 1972; Sisken et al., 1984), although the rodent data have not been widely reproduced. For example, when a 0.1 mA DC current was artificially pulled out of the stumps of amputated adult *Xenopus* and *Rana* forelimbs, treated animals (but not controls) formed broad bifurcated structures (Borgens et al., 1977) containing nerve trunks within the cartilage core and mature epidermal papillae. Cathodal current initiated partial regeneration (including extension of severed ulna, and production of muscle, ligament, and isolated partially segmented cartilage). Implantation of sham electrodes (carrying no current) produced no deviations from the normal response.

Recently, molecular details have been uncovered about the guidance of regenerative events in vertebrate appendages. The tail of *Xenopus* tadpoles contains spinal cord, muscle, vasculature, skeletal, and epidermal components (Beck et al., 2009; Tseng and Levin, 2008). A combination of pharmacological and molecular-genetic analyses using dominant-negative and constitutively active ion transporters implicated H^+ pumping from the wound as an instructive factor in regeneration (Adams et al., 2007), controlling the appearance of proliferative cells and being required for the correct pattern of innervation. Thus, tadpoles normally rely on the V-ATPase hydrogen pump and other currents (Reid et al., 2009) to drive regeneration during early stages. More importantly, during later stages, when tadpoles cannot regenerate, the entire regenerative cascade can be reproduced (Adams et al., 2007) by artificially driving H^+ efflux by misexpression of a heterologous (yeast) pump PMA-1 (Masuda and Montero-Lomeli, 2000). The details of the bioelectrical circuits in the regenerating tail, in *Xenopus* and axolotl models, are beginning to be worked out using ion-selective self-referencing probe technology (Reid et al., 2009) and fluorescent physiological reporter dyes (Ozkucur et al., 2010).

How Are Changes in Membrane Voltage Transduced to Canonical Pathways?

Bioelectric signals function both upstream and downstream of biochemical and genetic elements (Figure 3.4). Ion flows are produced by channels and pumps (which are regulated by transcriptional, translational, and gating mechanisms). In turn, flows control the expression of other genes and the function of physiological mechanisms at the cell surface and in the cytoplasm. Biophysical processes can often achieve considerable patterning in the absence of changes in transcription or even translation, due to the rich regulation of ion transporter activity and the redistribution of macromolecules by electric fields. For example, the stimulation of the sodium-hydrogen exchanger in tumor cell lines results from increased affinity of the internal H^+ binding site, not from changes in expression (Reshkin et al., 2000); likewise, the electrophoretic mechanisms underlying early left-right patterning in frog embryos occur during the first few cleavages, when the zygotic genome is not transcribed (Adams et al., 2006; Fukumoto et al., 2005b). Interestingly, a bimodal distribution of V_{mem} in cortical astrocytes (−68 and −41 mV) becomes unimodal after knockdown of Kir4.1 (−45 mV) (Kucheryavykh et al., 2007), illustrating how expression of the same protein (Kir4.1) can result in two different cell populations, distinguished physiologically.

Despite the many signals that can be generated and processed by bioelectrical pathways without change of mRNA or protein levels, eventually these processes feed into subsequent pathways that alter gene expression. A commonly occurring example of feedback in the interplay between bioelectrical and molecular-genetic mechanisms is the determination of a cell's anatomical (apical-basal) polarity by the electric fields, in turn produced by specifically localized ion transporters (Bentrup et al., 1967; Denker and Barber, 2002a, 2002b; Jaffe, 1968).

Specialized sensory cells can distinguish signals as weak as 5 nV/cm (Kalmijn, 1971, 1982); moreover, these mechanisms can exhibit window effects (Gruler and Nuccitelli, 1991), where a stronger applied signal does not necessarily induce the same effects as a more physiological one. The most common mechanism linking membrane voltage change and downstream events is calcium influx (voltage-sensitive Ca^{++} channels) (Sasaki et al., 2000). However, in some instances of K^+-dependent signaling, Ca^{++} fluxes were not affected by K^+ channel activity, showing that not all effects on cell behavior depend on Ca^{++} (Malhi et al., 2000) (Figure 3.4C'). Additional mechanisms that transduce electrical signals into second messenger cascades (Levin, 2009) include modulation of the activity of voltage-sensitive small-molecule transporters (e.g., the serotonin transporter, which converts membrane voltage into the influx of specific chemical signals) (Fukumoto et al., 2005a); redistribution of charged receptors along the cell surface (Lin-Liu et al., 1984; Poo and Robinson, 1977a; Stewart et al., 1996); directional electrophoresis of morphogens through cytoplasmic spaces (Brooks and Woodruff, 2004; Levin et al., 2006; Woodruff and Telfer, 1980; Woodruff, 2005); and activation of integrin signaling by conformational changes in membrane proteins (Arcangeli and Becchetti, 2006; Cherubini et al., 2005; Hegle et al., 2006). These elements can be capitalized upon, for the design of bioelectrical intervention in regenerative processes.

Several more unfamiliar possibilities may be fertile areas for future work. First, it is now clear that the nuclear membrane possesses its own complement of ion transporters, the activity of which expands the relevance of bioelectricity past cell surface events (Mazzanti et al., 2001) and opens the possibility of specific gene regulation by the membrane potential across nearby nuclear envelope regions (Bustamante, 1994; Bustamante et al., 1994, 1995). Second, direct changes of specific transcriptional element activity by intracellular potassium ion concentration might mediate ion-specific events independent of membrane voltage per se; this mechanism can involve the DNA-binding activity of such important signaling molecules as p53, *forkhead*, and CREB (cAMP response element-binding protein) (Tao et al., 2006). Third, depolarization has recently

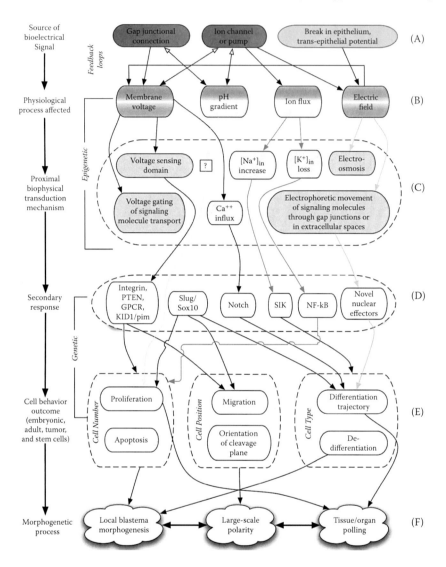

FIGURE 3.4 See color insert. Integration of bioelectrical events into molecular signaling cascades. (**A**) Expression of ion channels or pumps, gap junctional connections,
and epithelial damage all give rise to bioelectric signals. (**B**) These signals manifest as
changes in transmembrane potential, pH gradients, specific ion flows, or electric fields.
In the first two rows, pink shading indicates non-cell-autonomous signals, while purple
indicates cell-autonomous cues. Some nodes are both. (**C**) These processes are transduced via a variety of proximal epigenetic mechanisms, including voltage-sensing
domains on proteins, electro-osmosis, gating of morphogen transporters, and movement of specific ions like calcium. Green indicates a true electrical effect, while yellow
indicates a biochemical effect due to ion identity. (**C'**) Illustration of different possible
mechanisms by means of which membrane voltage levels and intracellular pH and K⁺
concentrations are transduced to transcriptional responses. *(continued on next page)*

been shown to lead to subcellular translocation of NRF-2 transcription factor, providing a mechanistic link between membrane voltage and transcriptional targets (Yang et al., 2004).

An exciting recent discovery involves VSP—a phosphoinositide phosphatase that converts $PI(3,4,5)P_3$ to $PI(4,5)P_2$ in a manner regulated by a voltage sensor domain (Murata et al., 2005). Local levels of $PI(4,5)P_2$ control the cytoskeleton and nuclear effectors. The identification of a protein able to transduce membrane voltage into all of the potential downstream pathways controlled by this powerful second messenger system (Li et al., 2002) provides a plethora of testable hypotheses of how membrane depolarization functions in a variety of patterning systems involving migration, apoptosis, and proliferation. Crucially, it was shown that wound healing control by endogenous electric fields is mediated by PTEN (Zhao et al., 2006), adding weight to the possibility that PTEN could be a widely conserved and important means of integrating cell-autonomous ion flows into second messenger and transcriptional responses.

UNIQUE FEATURES OF BIOELECTRICAL SIGNALING PROCESSES

Paradigms for understanding developmental patterning, as well as strategies for modulating morphogenesis in regenerative medicine, must take into account the properties of key signaling pathways. Bioelectric signals have several interesting properties that distinguish them from the dynamics of biochemical cascades (morphogen gradients).

Bioelectric networks are essentially recursive, with feedback that modulates the processes that created the ion flows. The effects can be direct, as when changes in membrane voltage gradient affect the gating of voltage-sensitive ion channels, which in turn alters membrane potentials further. The loop can also exist at the level of transcription, as ion channel activity modulates ion channel gene expression (Ribera, 1998). Likewise, gap junctions shape electrical properties of cell

FIGURE 3.4 *(continued)*   Some can impinge directly on transcriptional machinery, while others (schematized by connections to nucleus as a whole) need to be transduced by canonical mechanisms. Clockwise from the top, these include: entry of small molecule signals through voltage-gated gap junctions, voltage-dependent conformation changes in integrin-associated proteins, Ca^{++} entry through voltage-gated Ca^{++} channels, entry of small signaling molecules such as neurotransmitters through voltage-powered transporters, voltage-sensitive phosphatase (VSP) activity, and changes in cytoplasmic content of H^+, K^+, and other ions. 5-HT = serotonin, an example of a small molecule signal that could traverse gap junctional (GJC) paths under an electrophoretic force. Blue lightning bolts represent membrane voltage potentials. (D) These processes feed into several known genetic signaling pathways, including NF-kB, Notch, PTEN, Slug/Sox10, and integrins. (E) Downstream of these signaling molecules are changes in cell cycle, apoptosis, position, orientation, and differentiation. (F) The ultimate result of orchestrated changes in cell behavior is morphogenetic processes, including patterning of blastemas and embryonic fields, polarity decisions on several scales, and polling of remote tissues that enable wounds to decide what already exists and what must be recreated. The arrows indicate sample cases where the whole pathway has been traced for bioelectrical control of patterning. (From Levin, *Semin. Cell. Dev. Biol.*, 20, 548, 2009.)

groups and are themselves sensitive to changes in transmembrane potential and pH. This offers very rich opportunities for biological systems to use ion flows to implement both positive and negative feedback mechanisms. An example of the former, hydrogen-potassium exchanger regulation via potassium-sensitive NF-kB (Zhang and Kone, 2002) can be used to amplify small physiological signals, while the latter, such as the depolarization-induced activation of the hyperpolarizing V-ATPase pump (Jouhou et al., 2007), can be used to ensure robustness of patterning against perturbations. For example, consistent left-right patterning is driven by bioelectric cues in early embryogenesis (Adams et al., 2006; Aw et al., 2008; Levin et al., 2002; Morokuma et al., 2008b; Shimeld and Levin, 2006), despite the significant differences in actual ion content among normal embryos (Gillespie, 1983), likely because the physiological networks can buffer against such genetic and environmental variability.

Bioelectrical signals span several orders of magnitude in scale and in levels of organization, controlling the distribution of subcellular organelles (Fang et al., 1999; Poo and Robinson, 1977b), the behavior of single-cell organisms such as pollen tube extension and *Acetabularia* cellular specialization/regeneration (Certal et al., 2008; Feijo et al., 1999, 2001; Messerli and Robinson, 1997, 1998, 2003; Messerli et al., 1999, 2000, 2008; Novák and Bentrup, 1972; Novak and Sirnoval, 1975; Rathore et al., 1991; Robinson and Messerli, 2002), and the structures of epithelia, appendages, and entire embryos (Borgens, 1984; Borgens and Shi, 1995; Hotary and Robinson, 1990, 1992). While the penetration of endogenous electrical fields into distant tissue is a function of the complex resistivity, and thus often hard to quantify in practice, bioelectric events can exert influence far beyond the local microenvironment. For example, in left-right patterning of the early frog embryo, a pump-driven battery in ventral blastomeres appears to distribute small molecule morphogens across the entire early frog embryo through long-range gap junction paths under an electrophoretic force (Esser et al., 2006; Fukumoto et al., 2005b). Intriguingly, transplanted tumors can induce large-scale changes in voltage potentials detectable at considerable distances from the primary site (Burr, 1941a).

Bioelectrical signals behave according to the intrinsic properties governing electric fields, and are an epigenetic mechanism because physiological networks can regulate and generate order in the absence of changes in DNA, RNA, or protein levels. For example, electrophoresis of signaling molecules through gap junctions (Adams et al., 2006; Brooks and Woodruff, 2004; Esser et al., 2006; Fukumoto et al., 2005b; Levin et al., 2006; Woodruff, 2005; Zhang and Levin, 2009) or through cytoplasmic spaces (Bohrmann and Gutzeit, 1987; Telfer et al., 1981; Woodruff et al., 1988; Woodruff and Telfer, 1980) can redistribute gradients of signaling molecules, and channels/pumps can control each other through post-translational gating/kinetics changes, without directly involving transcription or translation changes (and are thus invisible to the currently popular transcriptome network analyses). They are likely to be an evolutionarily ancient example of living systems capitalizing upon "order for free" (Kauffman, 1993), derived from basic physics, which ensures that injury automatically provides cells with a vector cue indicating the position of the damage. It seems reasonable that the basic mechanism of electric vector pointing toward regions of membrane break could be extended to multicellular structures such as epithelia (Borgens, 1984; Robinson and Messerli, 1996). This could have later incorporated the induced expression of wound-specific transporters (in addition to the passive transepithelial potential leaks) such as the V-ATPase at the tadpole tail blastema (Adams et al., 2007).

As may be expected with any signal that carries important information about the topology of a host organism, the evolutionary arms race has enabled parasites to use electric properties of tissues to home in on and exploit sites of damage. This is seen in metastatic cancer (Djamgoz et al., 2001; Mycielska and Djamgoz, 2004; Yan et al., 2009) and parasitic fungi (Gow and Morris, 1995; McGillivray and Gow, 1986; Morris et al., 1992; Rajnicek et al., 1994; van West et al., 2002).

An interesting and important consequence of multiscale control by bioelectrical signals is their ability to act as "master regulators": to activate coherent downstream morphogenetic cascades via a relatively simple initial signal. It has already been shown in physiological experiments that localized interference with signals such as reversal of potential across the neural tube, or shunting specific currents at various anatomical sites, had broad and global effects on patterning of the entire organism (Borgens and Shi, 1995; Hotary and Robinson, 1992, 1994). With the advent of molecular tools, it is becoming easier to capitalize on this property for augmenting regeneration by providing specific signals. Recent examples of "master control" include the induction of a complete and normal head in posterior blastemas of planaria by junctional isolation (Nogi and Levin, 2005), the demonstration that electric fields in wound healing are an overriding factor with respect to known biochemical pathways (Zhao, 2009), and the induction of the regeneration of a complete and normal tail in *Xenopus* larvae by forced proton pumping (Adams et al., 2007). A single event—the continuous pumping of H^+ at the wound—initiates the complete, normal regeneration of the tail. Its patterning and size are correct and its growth is appropriately halted when it catches up with the size of the tails of uncut controls. Two other illustrations are shown in Figure 3.5. The ability of relatively simple bioelectrical signals to trigger orchestrated morphogenetic subroutines is a very attractive property to consider for regenerative medicine applications: modulation of physical cues can exploit the patterning capacity of the host's genetic programs without needing to micromanage the details of the regenerative process.

One of the key aspects of understanding signaling in morphogenesis is to ask what information is being carried by a given physiological process and what information capacity the signaling system has. For example, a single-membrane voltage value carries limited information; it is likely that the true richness of bioelectrical signaling can only be fully appreciated by considering the microdomains of transporter activity distributed across the entire two-dimensional surface of a cell or epithelium (Figure 3.1A,B): these inhomogeneities comprise a field of potential values that, because of their spatial distribution, can encode enormous amounts of developmental information (Davies et al., 2006; Martens et al., 2004; Wallace,

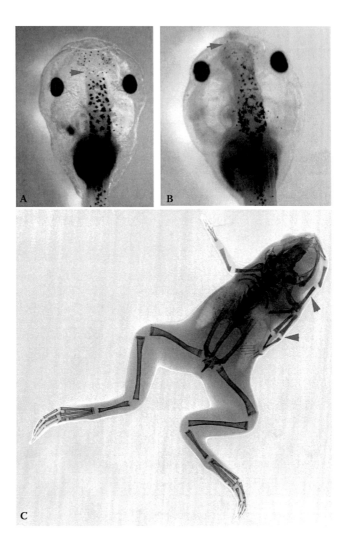

FIGURE 3.5 Sample phenotypes arising from molecular-genetic modulation of bioelectrical cues in *Xenopus* laevis. Unpublished data from our lab showing that misexpression of ion channel constructs during embryogenesis can make coherent changes in pattern. Experiments performed with potassium channels by Sherry Aw result in the normal forebrain (A) being drastically increased (B); red arrow indicates anterior border of forebrain. Similarly, entire limbs can be induced (C), with x-ray imaging revealing the normal skeletal pattern in the ectopic limbs in this adult frog. Image in panel (C) courtesy of Punita Koustubhan and Sherry Aw. (From Levin, *Semin. Cell. Dev. Biol.*, 20, 550, 2009.)

2007). Although it was appreciated as early as 1983 (Kline et al., 1983) that individual cells can have more than a single transmembrane potential value, it is still largely unclear how adjacent domains maintain different voltage values and avoid equalizing short circuits across the underlying cytoplasm. Data now indicate that modulation of potassium channel activity can induce a biphasic (window) effect, and indeed opposite effects on proliferation of adult neural precursor cells can be obtained by inhibition of different potassium (Kir vs. Kv) channels (Yasuda et al., 2008). Thus, it is likely that we still have a very inadequate picture of all of the bioelectrical signals received and generated by cells *in vivo*.

The implication of bioelectrical parameters in regulation suggests the idea of the physiological state space, proposed as a hypothesis for guiding future research in this field. Analyses have shown that generally, plastic, embryonic, stem, and tumor cells tend to be depolarized, whereas quiescent terminally differentiated somatic cells are hyperpolarized (Binggeli and Weinstein, 1986). The use of membrane voltage to control cellular state is a powerful tool (Morokuma et al., 2008a; Perona and Serrano, 1988), but it is likely to be only a primitive approximation to the true richness of bioelectrical control. A more useful idea is that cells can be localized in a multidimensional physiological state space with a number of orthogonal dimensions indicating membrane voltage, intracellular pH, K^+ content, nuclear potential, Cl^- content, surface charge, etc.

One possibility is that cells can be grouped in distinct regions of this state space corresponding to stem cells, tumor cells, somatic cells, and other types of cells that are of interest to regenerative biology (Figure 3.6). This hypothesis implies that in order to make rational changes in cell behavior, (1) data need to be obtained on multiple cell types from different organs and disease conditions, and (2) strategies need to be developed that use pharmacological reagents targeting natively expressed channels/pumps, and misexpression of well-characterized channel/pump constructs, to move cell states into desired regions (e.g., some cells may need to be depolarized by 30 mV and its internal pH acidified in order to induce proliferation). We are currently using quantitative modeling to expand the XYZTG (three-dimensional position, time, and gene expression) space (Megason and Fraser, 2007) to include the systems biology of bioelectrical properties. The end result of the synthesis of experimental and modeling efforts should be the development of targeted channel/pump modulation strategies to achieve desired bioelectrical states of wound tissues for augmentation of regeneration.

One last key aspect of bioelectric signals (Figure 3.7) is due to the fact that the same physiological state can be achieved by the function of many different sets of transporters; at the same time, regulatory (e.g., gating) events can result in the same ion transporter functioning very differently in different cells. The disconnect between a molecular-genetic profile of cells and a bioelectric state is very important: it cannot be assumed that cells expressing the same set of channels and pumps are in the same physiological state. Similarly, comparison of cell types based on microarray or differential expression analysis can be misleading

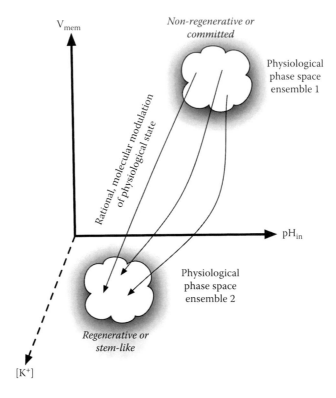

FIGURE 3.6   The hypothesis of the bioelectrical state space. The distinction among cells based on transmembrane potential (Figure 3.3) likely cannot capture the full subtlety of cellular controls by V_{mem} changes. Indeed, cells live in a state space with a number of orthogonal axes corresponding to physiological properties. Here are shown only three (membrane voltage, V_{mem}, internal pH, and K^+ content). A more detailed data set will contain additional semi-independent metrics, such as the content of other ions, nuclear potential, surface charge, etc. One hypothesis is that cell types (e.g., stem cells, cancer cells, nonproliferative cells) will be seen to cluster in different regions of this space. If true, this not only will be a useful diagnostic framework, but also, when coupled with quantitative data and mathematical modeling, can be used for rational modulation of cell behavior. Using well-characterized transporters in gene therapy, and pharmacological reagents targeting endogenous transporters in remaining tissue after injury, bioelectrical properties can be specifically changed to move wound cells from a nonproliferative state toward a more plastic condition that can support large-scale regeneration. (From Levin, *Semin. Cell. Dev. Biol.*, 20, 551, 2009.)

with respect to bioelectric properties. Indeed, knockout of individual channel/pump genes can fail to reveal important aspects of ionic controls because many different transporters can compensate, masking phenotypes. This complexity has a benefit, however: biomedical regeneration efforts could potentially use any convenient channel or pump to achieve the desired change in cell physiology.

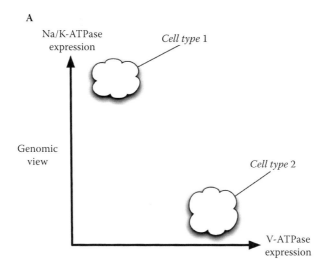

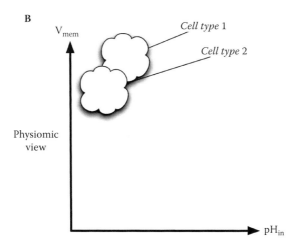

FIGURE 3.7 Traditional molecular-genetic analysis can be misleading when applied to physiological states. (A) Two hypothetical cell types have very different expression patterns. A microarray or differential analysis characterizes them as different, since one cell has low expression of Na,K-ATPase and high expression of V-ATPase, while the other cell is the opposite. (B) However, analysis of pH and membrane potential may reveal that the cells are actually functionally similar because both pumps are hyperpolarizing (although genetically distinct). That is, because of bioelectrical controls, these cells may have the same proliferative potential. (From Levin, *Semin. Cell. Dev. Biol.*, 20, 551, 2009.)

MORPHOGENETIC FIELDS: REGENERATION IN A BROADER CONTEXT

It can be argued that understanding, and learning to control regeneration, requires not only the ability to tweak low-level cell functions, but also a deeper appreciation of how a complex pattern is encoded and emerges from signaling pathways. Indeed, it is clear that the environmental context is a crucial component of cell function; before considering the specific role of bioelectric signals in this multiscale interaction between host and cell, it is useful to consider some little discussed data illustrating patterning control in cancer, regeneration, and development.

Despite high malignancy and euploidy, tumor cells integrated into a wildtype host embryo display normal behavior (Astigiano et al., 2005; Illmensee and Mintz, 1976; Mintz and Illmensee, 1975). Expression of activated Ras potently transforms cultured cells, but *in vivo* results in normal tissue with clonal tumor development only in cells acquiring additional oncogenic modifications (Frame and Balmain, 2000). Chick embryos infected with the v-Src virus have no malignant phenotype, but the same cells in culture undergo massive transformation (Dolberg and Bissell, 1984). Likewise, transduction with a cocktail of factors sufficient to induce an eye from a group of multipotent progenitors only does so in the context of a host, not *in vitro* (Viczian et al., 2009).

One way to view regeneration of complex structures is an example of morphostasis—the maintenance of "target morphology" by an organism (Figure 3.8). This is the shape, defined on multiple scales of size and levels of organization, that a biological system acquires during development, and maintains against cellular turnover (aging), stresses of life (remodeling and wound healing), and major injury (regeneration). This perspective focuses on information processing in cells and tissues, and emphasizes mechanisms common to the patterning events that occur during embryonic development and regeneration, or fail to occur during neoplastic growth (Holcombe and Paton, 1998).

Target morphology can be analyzed via mathematical tools—formalized descriptors allowing comparisons of form and of shape transformation, as well as analyses of complexity (Barkai and Ben-Zvi, 2009; Bryant et al., 1977; French et al., 1976; Slack, 1980, 1982). Its presence is revealed not only through a highly stereotypical outcome of embryonic self-assembly, but also in the morphological remodeling over time, observed in both vertebrate and invertebrate systems where deviations from normal shape are slowly corrected. Examples of patterning driven by nonlocal morphogenetic information include allometric scaling during wholebody remodeling in planaria (Oviedo et al., 2003) and the long-term transformation of a tail into a limb after a tail blastema is grafted at the flank in amphibia (Butler and O'Brien, 1942). Although the origin of blastema cells is local to the site of injury and the initial pattern formation is determined by the original position of the blastema within the donor, the host's morphogenetic fields exert their influence remotely, and slowly transform the ectopic tail into a limb—the structure appropriate to the large-scale global context in which the blastema is placed (Farinella-Ferruzza, 1953, 1956; Guyenot, 1927; Guyenot and Schotte, 1927).

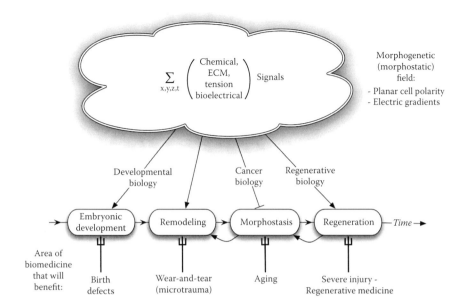

FIGURE 3.8 The morphogenetic field in development, regeneration, and neoplasm and its applications to medicine. The morphogenetic field can be defined as the sum, integrated over three spatial and one temporal dimension, of all nonlocal signals impinging on cells and cell groups in an organism. Functionally, long-range signals (such as planar polarity, standing waves of gene expression and tensile forces, and morphogen gradients) carry information about both the existing and future patterns of the organism. This allows the initial development of a complex form from a single fertilized egg cell, as well as the subsequent maintenance of the form in adulthood against trauma and individual cell loss. Errors in various aspects of the establishment and interpretation of this field result in birth defects, cancer, aging, and failure to regenerate after injury. Thus, almost every area of biomedicine is impacted by our knowledge of how cells interact with and within this set of complex signals. Bioelectrical aspects of the morphogenetic field are crucial, mediating regulation of cell movement, differentiation, and proliferation as part of patterning during development, regeneration, and prevention of neoplasm. ECM = extracellular matrix. (From Levin, *Semin. Cell. Dev. Biol.*, 20, 546, 2009.)

The mediator of pattern formation and remodeling can be viewed as a "morphogenetic field" (Beloussov, 2001; Beloussov et al., 1997; De Robertis et al., 1991; Martinez-Frias et al., 1998; Opitz, 1986)—the sum total of local and long-range patterning signals that impinge upon cells and bear instructive information that orchestrates cell behavior into the maintenance and formation of complex three-dimensional structures (Schnabel et al., 2006). While this is currently studied with respect to gradients of chemical messengers (Lewis, 2008; Schiffmann, 1989, 1991), bioelectric signals are also ideal mediators of distributed, nonlocal field properties in large-scale patterning.

The morphogenetic field, while a classical concept (Child, 1941; Driesch, 1929; Spemann, 1938; Weiss, 1939), has recently been reinvigorated through the discovery and molecular characterization of several long-range patterning systems that use the same genetic components to carry patterning signals in embryonic development and regeneration (Tataria et al., 2006; White and Zon, 2008); it is this same information that may be ignored by cells during neoplasm (Donaldson and Mason, 1975; Lee and Vasioukhin, 2008; Tsonis, 1987). This view is a different perspective on regeneration because, rather than focusing on individual molecules and on the special constraints upon regeneration in adults (e.g., scarring), the goal is to understand, and learn to rationally modulate, large-scale patterning processes and growth control circuits. This is broadly relevant to many other biomedical areas that can be formulated in terms of establishment, maintenance, and deviation of morphology (e.g., aging, birth defects, and cancer). Examples of underlying mechanisms that establish long-range order are planar cell polarity (Klezovitch et al., 2004; Lee and Vasioukhin, 2008; Zallen, 2007; Zhan et al., 2008) and neural signaling. The latter in particular is known to be crucial for regenerative ability (Bryant et al., 1971; Maden, 1978; Yntema, 1959), and involved in the maintenance and organization of multicellular structures in the organism, such as tongue buds (Sollars et al., 2002), which become disorganized when their innervation is perturbed. Further examples of ongoing maintenance of pattern (morphostasis) in adult mammals include gut heterotopias (conversion of regions of the intestinal tract into cells appropriate to a different part of the gut) (Slack, 1985) and the conversion of lymph endothelium to blood endothelium by Prox-1 knockdown (Johnson et al., 2008). This suggests a view somewhat at odds with the usual view that adult body plans are static, terminally differentiated structures.

REGENERATION, CANCER, AND DEVELOPMENT: THREE SIDES OF ONE PUZZLE

A regulative living system must establish pattern during development, recognize damage, and restore missing or damaged tissue. Viewed as a problem of morphogenetic regulation, cancer can be thought of as a disease of geometry: failure of the host to exert control appropriate to the morphogenetic plan, or failure of cells to receive patterning signals appropriate to a multicellular context (Oviedo and Beane, 2009; Rubin, 1985; Rubin, 1990; Soto and Sonnenschein, 2004).

In this view, it is not surprising that regeneration—exertion of morphogenetic cues over rapid new growth—has a fascinating inverse relationship with cancer (Brockes, 1998; Pizzarello and Wolsky, 1966; Tsonis, 1983). Regenerating species rarely exhibit tumors (Polezhaev, 1972). Likewise, many normal cells (in tissues with strong regenerative potential), such as epithelial cells within the small bowel crypt, have very rapid proliferation rates but show an extremely low rate of malignant transformation (Barclay and Schapira, 1983). In the stick insect Dixippus, which has high regenerative ability and a low incidence of cancer, removal of the endocrine gland induces loss of regenerative capacity and spontaneous tumorigenesis (Pflugfelder, 1938, 1939, 1950, 1954).

The complementarity of regeneration and cancer is even more strikingly demonstrated in functional experiments. If limb regeneration is initiated by cutting in the middle of a tumor, the tumor cells can apparently be transformed into normal limb tissue (Brockes, 1998; Donaldson and Mason, 1975; Rose and Wallingford, 1948; Ruben et al., 1966; Tsonis, 1983; Wolsky, 1978). Additional data from the 1960s reveal the ability of regeneration to normalize tumors, turning de-differentiated epithelial cells into normal skin, and perhaps even transdifferentiating them into connective tissue and muscle (Seilern-Aspang and Kratochwill, 1962, 1965). In planaria, anterior regeneration is capable of turning posterior infiltrating tumors into differentiated accessory organs such as the pharynx, which suggests the involvement of regulatory long-range signals (Seilern-Aspang and Kratochwill, 1965) in the interaction between host morphology and tumors.

The more mainstream molecular-level view (focused on tumor suppressor genes, checkpoints, etc.) predicts that proliferative capacity and the plasticity necessary for efficient regeneration would predispose highly regenerative organisms to higher incidence of cancer. The data suggest, in contrast, not only that highly regenerative tissues are resistant to cancer, but that regeneration can bring autonomous growth of tumors under morphogenetic control (Needham, 1936; Rose and Wallingford, 1948; Waddington, 1935). While molecular details remain to be clarified, this can be heuristically interpreted as reassertion of the regional morphogenetic field that overwhelms the mispatterning of a neoplastic cell mass. The taming of tumor cells by the robust patterning controls of normal embryonic development (Mintz and Illmensee, 1975), and the teratomas that develop from implanting embryos (introduction of secondary organizing fields) within a primary organism (Houillon, 1989), have been known for many years, and illustrate another side of the cancer-development-regeneration triad (White and Zon, 2008).

Consistent with a fundamental link between growth control in regeneration and the prevention of neoplasm are little known data indicating a role for innervation (long known to be essential for regeneration) in suppression of cancer. Tumors are chemically induced more easily in denervated rabbit ears than with contralateral controls (Pawlowski and Weddell, 1967). This also occurs in insects (Scharrer, 1945, 1953): in the cockroach, salivary organ and alimentary canal tumors were caused by denervation in 75 to 80% of animals (Scharrer and Lochhead, 1950). This finding is predicted by models in which nervous system components transmit long-range morphogenetic field cues (Becker, 1974; Burr, 1941b, 1944; Burr and Northrop, 1939), but contradicts prediction of local "neural-derived growth factor" models (Cannata et al., 2001; Kumar et al., 2007).

BIOELECTRIC SIGNALS IN CANCER

As predicted by the proposal that ionic mechanisms form a fundamental system of growth control, bioelectric events have been strongly implicated in neoplasm, in addition to their roles in development and regeneration. Indeed, ion transport events are a central component of the shift of the scale of regulation, from a multicellular context (healthy morphostasis) to that of a single cell in the case of cancer. For example, glioma cells do not participate in ion homeostasis, an important altruistic task performed by their nonmalignant counterparts, glia. Instead, gliomas show overgrowth and invasive migration (Olsen and Sontheimer, 2004). Reversion to control loops that favor individual cells over coherent organs, such as that demonstrated by glioma cells, is a hallmark of cancer *in vivo*; changes in the scale of morphogenetic control occur during regeneration (as in the autonomous differentiation of a limb or head blastema) and can be imposed during development, since cancer cells injected into mouse embryos are brought under control and participate in the development of healthy mice (Illmensee and Mintz, 1976; Mintz and Illmensee, 1975), as also occurs when morphogenesis is initiated as part of regeneration (discussed above). Is it possible that a significant component of growth control, and the level of organization from which regulatory signals reach cells, are controlled by bioelectric signals (Popp, 2009).

The proliferation of some tumor cells is dependent on voltage-gated potassium channels (Conti, 2004; Fraser et al., 2000). hERG channels are particularly implicated (Arcangeli, 2005; Bianchi et al., 1998; Lastraioli et al., 2004; Lin et al., 2007; Wang et al., 2002; Wang, 2004), as are two-pore channels such as KCNK9 (Mu et al., 2003). In the case of KCNK9, it is known that its oncogenic potential depends on K^+ transport function, not some other role of the protein (Pei et al., 2003), and in human colorectal cancer tissue KCNK9 K^+ channel expression was shown to be significantly elevated (Kim et al., 2004). A screen of several cervical cancers found the K^+ channel EAG expressed in 100% of the biopsies analyzed, and overexpression of EAG in human cells resulted in more quickly dividing progeny in culture (Farias et al., 2004; Pardo et al., 1999). This result was replicated *in vivo* using mice implanted with human EAG (hEAG) expressing CHO cells. All of the mice receiving CHO hEAG injections formed tumors, while none of CHO controls formed a growth greater than 1mm in size (Pardo et al., 1999). hEAG-1 is a true oncogene since its overexpression drives mammalian cells into uncontrolled proliferation and results in tumor progression in cells injected into immune-suppressed mice (Pardo et al., 1999). Likewise the EAG relative, hERG, is not normally present in most differentiated cells besides the heart, but has been observed in a number of human cancers and neoplastic transformation in prostate epithelium (Klezovitch et al., 2008; Wang et al., 2002). In these cells, hERG appears to recruit tumor necrosis factor receptor (TNFR) to the plasma membrane and cause a subsequent increase in NFκB, a known proliferation control gene.

Manipulation of membrane H^+ flux can confer a neoplastic phenotype upon cells (Perona and Serrano, 1988), and voltage-gated sodium channels potentiate breast cancer metastasis (Fraser et al., 2005). Studies in glioma lines have revealed a role of the ClC-3 chloride efflux channel in cellular division (Habela et al., 2008). Following chloride accumulation, Cl^- efflux is required for mitosis to progress. Inhibition of ClC-3 channels via RNAi resulted in the loss of premitotic condensation and arrest of the cell cycle. These results have also been observed in human prostate cancer lines, and support the role of chloride channels as key regulators of proliferation (Jirsch et al., 1993; Shuba et al., 2000).

Sodium channels have been implicated in mouse cancers *in vivo*. The knockout $APC^{min/+}$ line shares a mutation found in many human colorectal cancers, and subsequently develops multiple intestinal neoplasias (Ousingsawat et al., 2008).

In vivo transepithelial voltage recordings in this line revealed an increase in Na^+ compared to wild-type mice that was the result of an increase in expression of the ENaC Na^+ channel. The downstream targets of Na^+ signaling are not well known, but it was noted that neither Cl^- nor Ca^{2+} absorption was altered in the $APC^{min/+}$ line. Metastatic potential correlates with voltage-gated inward sodium current, and it has been suggested that some sodium channels may be oncofetal genes, encoding signals that are active during the rapid and autonomous growth of tumors and embryos (Brackenbury and Djamgoz, 2006; Diss et al., 2005; Fraser et al., 2005; Onganer and Djamgoz, 2005; Onganer et al., 2005).

CONCLUSION AND FUTURE PROSPECTS

Many properties of biological systems, such as polarity, long-range spatial order, and positional information, are present in the physics of electromagnetic fields. There is now considerable evidence that endogenous DC electric fields, magnetic fields (Asashima et al., 1991; Frohlich, 1950; Paul et al., 1983; Pokorny et al., 1983), and ultra-weak photon emission (Popp et al., 1992) are part of the medium by which information flows in biological systems during dynamic morphostasis and regeneration. Indeed, recent molecular work has revealed, and shown functional roles for, standing patterns of voltage gradients and electric fields *in vivo*. These data have made it plausible that, as suggested decades ago (Becker, 1967; Burr and Northrop, 1935), bioelectric properties form a kind of prepattern in guiding the expression of biochemical morphogenetic cascades; this is seen clearly in the bioelectric properties underlying the future difference between the left and right sides of the vertebrate body plan (Levin, 2006), and the establishment of permanent head organizers in planaria by rapid and transient electrical isolation of blastemas from the rest of the host (Oviedo et al., 2010).

The field faces a number of major challenges. One of the biggest is the lack of sufficient quantitative data. Many measurements of pH, voltage, and ion content are needed from interesting cell types and model systems to flesh out the physiological state space concept, and to compile enough data to develop predictive, quantitative physiological models that encompass the feedback loops and synthesize molecular genetic and bioelectric data (Adams, 2008; Esser et al., 2006; Fischbarg and Diecke, 2005). Issues of information content remain a rich area for discovery: what specific messages are encoded for cells by specific kinetics of individual ion fluxes, discrete ranges of transmembrane and transepithelial voltage, and distinct regions of different potential throughout the membrane of a single cell? Oscillations in membrane voltage on a scale much slower than action potentials (Aeckerle et al., 1985; Dryselius et al., 1994; Grapengiesser et al., 1993; Hulser and Lauterwasser, 1982; Lang et al., 1991; Pandiella et al., 1989; Stokes and Rinzel, 1993) are likely to carry important information and must be incorporated into pathway models. Voltage gradients across nuclear and organelle membranes (Giulian and Diacumakos, 1977) are only beginning to be measured, and their importance for cell function is not yet fully understood (Mazzanti et al., 2001).

At a higher level, there is a significant need for sophisticated modeling. Control networks are rapidly becoming too large to understand "by hand," and data sets are now too complex to allow easy development of consistent models by simple inspection. In particular, there is a major disconnect in the field between the data from groups working out extremely detailed networks of genetic/biochemical interactions, and the need to understand what morphology is generated by any set of regulatory signals. This is not easily gleaned from pathway data, but is necessary if rational changes to shape can be induced in regenerative contexts. Examples of algorithmic, constructive models of the kind that are sorely needed in this field include those made to explain planarian regeneration (Slack, 1980) and size regulation in frog development (Barkai and Ben-Zvi, 2009). Conceptual tools that have been proposed for the understanding of multiscale control of morphogenesis include chaos and dynamical systems theory (Furusawa and Kaneko, 1998a, 1998b, 2000a, 2000b, 2001, 2002, 2003; Kaneko and Yomo, 1997, 1999; Takagi et al., 2000; Yoshida and Kaneko, 2009) as well as evolutionary computation approaches (Andersen et al., 2009; Basanta et al., 2008); the inclusion of bioelectrical cues into such models will be an exciting development and is sure to spur novel experiments.

Importantly, however, even without all of the answers to the above issues, the existing data provide opportunities for modulation of regeneration. For example, it has been shown that a Kv1.3 (KCNA3) and Kv3.1 (KCNC1) blockade increases neural progenitor cell proliferation (Liebau et al., 2006); likewise, induction of H^+ flux induces regeneration of a complex appendage (Adams et al., 2007) and blockade of gap-junction-mediated signals results in the formation of a complete, properly patterned head in a planarian tail blastema (Nogi and Levin, 2005). These techniques can already be integrated into efforts to augment regeneration. The recent development of light-gated ion transporters has been particularly exciting; while these have so far been mainly used for neurobiological studies (Arrenberg et al., 2009; Banghart et al., 2004; Chambers et al., 2006; Gradinaru et al., 2008), optogenetics offers the potential of high-resolution spatiotemporal control of bioelectrical changes in cells during regeneration and development.

Three specific directions are being pursued in our group to provide additional opportunities for the field. One is the generation of mutant model species (e.g., *Xenopus*) expressing fluorescent proteins that report pH (Miesenbock et al., 1998) or voltage (Tsutsui et al., 2008), which will greatly augment the ability to study bioelectric properties of cells and tissues in a multitude of regenerative and disease states, or under molecular or pharmacological modulation. Another is the generation of mutants ubiquitously expressing light-gated ion transporters (Wang et al., 2007; Zhang et al., 2007; Zhao et al., 2008), which will allow unprecedented spatiotemporal control over bioelectric states in any tissue/organ of interest. Finally, in collaboration with bioengineers, we are working on the construction of regenerative sleeves (Figure 3.9)—bioreactors to be applied to wounds (e.g., stump amputations) in which the physiological state of wound cells

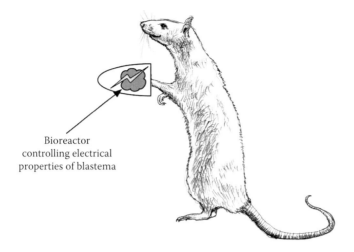

Bioreactor
controlling electrical
properties of blastema

FIGURE 3.9 A schematic of the regenerative sleeve: application to limb regeneration. Once sufficient quantitative data are available about the specific bioelectric states that promote regeneration, it will be necessary to develop sophisticated bioreactors, such as that pictured on the limb amputation wound in the rat model. These bioreactors will use microfluidics and light delivery to control, using pharmacological, genetic, and optical means, the physiological properties of the wound. This is one vision of how information on bioelectrical controls of cell behavior can be transitioned into applications in regenerative biomedicine (Nuccitelli et al., 2008, 2009; Shapiro et al., 2005). (From Levin, *Semin. Cell. Dev. Biol.*, 20, 552, 2009.)

can be precisely controlled by pharmacological, optical, electrical, and genetic means to trigger regeneration and control patterning.

The widely conserved, multiscale, instructive capacity of bioelectric events, coupled with their ability to induce complex downstream patterning cascades, makes ion flow an extremely powerful control modality. Recent discoveries have shed light on the genetic response elements that are activated by ionic signals. The development of specific strategies for modulation of physiological state (whether through gene therapy with controllable transporters or by targeting endogenously expressed channels), in combination with efforts focused on biochemical factors, is sure to open exciting new vistas in regenerative medicine.

ACKNOWLEDGMENTS

M.L. thanks members of the Levin lab, Richard Nuccitelli, Ken Robinson, Richard Borgens, and Lionel Jaffe, for numerous useful discussions. Dany S. Adams provided very helpful comments on the manuscript. Figure 3.9 was drawn by Jay Dubb; phenotypes in Figure 3.5 were obtained by Sherry Aw and Junji Morokuma. Data in Figure 3.1A,B were obtained by Dany S. Adams. This work was supported by the National Institutes of Health (HD055850, GM078484), DARPA (W911NF-07-1-0572), and NHTSA (DTNH22-06-G-00001).

REFERENCES

Adams, D. S. 2008. A new tool for tissue engineers: Ions as regulators of morphogenesis during development and regeneration. *Tissue Eng Part A* 14:1461–68.

Adams, D. S., Levin, M. 2006a. Inverse drug screens: A rapid and inexpensive method for implicating molecular targets. *Genesis* 44:530–40.

Adams, D. S., Levin, M. 2006b. Strategies and techniques for investigation of biophysical signals in patterning. In *Analysis of growth factor signaling in embryos*, ed. M. Whitman, A. K. Sater, 177–262. Boca Raton, FL: Taylor and Francis Books.

Adams, D. S., Masi, A., Levin, M. 2007. H+ pump-dependent changes in membrane voltage are an early mechanism necessary and sufficient to induce *Xenopus* tail regeneration. *Development* 134:1323–35.

Adams, D. S., Robinson, K. R., Fukumoto, T., Yuan, S., Albertson, R. C., Yelick, P., Kuo, L., McSweeney, M., Levin, M. 2006. Early, H+-V-ATPase-dependent proton flux is necessary for consistent left-right patterning of non-mammalian vertebrates. *Development* 133:1657–71.

Aeckerle, S., Wurster, B., Malchow, D. 1985. Oscillations and cyclic AMP-induced changes of the K+ concentration in *Dictyostelium discoideum*. *EMBO J* 4:39–43.

Andersen, T., Newman, R., Otter, T. 2009. Shape homeostasis in virtual embryos. *Artif Life* 15:161–83.

Anderson, J. D. 1951. Galvanotaxis of slime mold. *J Gen Physiol* 35:1–16.

Arcangeli, A. 2005. Expression and role of hERG channels in cancer cells. *Novartis Found Symp* 266:225–32; discussion, 232–34.

Arcangeli, A., Becchetti, A. 2006. Complex functional interaction between integrin receptors and ion channels. *Trends Cell Biol* 16:631–39.

Arcangeli, A., Carla, M., Bene, M., Becchetti, A., Wanke, E., Olivotto, M. 1993. Polar/apolar compounds induce leukemia cell differentiation by modulating cell-surface potential. *Proc Natl Acad Sci USA* 90:5858–62.

Arcangeli, A., Crociani, O., Lastraioli, E., Masi, A., Pillozzi, S., Becchetti, A. 2009. Targeting ion channels in cancer: A novel frontier in antineoplastic therapy. *Curr Med Chem* 16:66–93.

Arrenberg, A. B., Del Bene, F., Baier, H. 2009. Optical control of zebrafish behavior with halorhodopsin. *Proc Natl Acad Sci USA* 106:17968–73.

Asashima, M., Shimada, K., Pfeiffer, C. J. 1991. Magnetic shielding induces early developmental abnormalities in the newt, *Cynops pyrrhogaster*. *Bioelectromagnetics* 12:215–24.

Astigiano, S., Damonte, P., Fossati, S., Boni, L., Barbieri, O. 2005. Fate of embryonal carcinoma cells injected into postimplantation mouse embryos. *Differentiation* 73:484–90.

Aw, S., Adams, D. S., Qiu, D., Levin, M. 2008. H,K-ATPase protein localization and Kir4.1 function reveal concordance of three axes during early determination of left-right asymmetry. *Mech Dev* 125:353–72.

Aw, S., Levin, M. 2009. Is left-right asymmetry a form of planar cell polarity? *Development* 136:355–66.

Balana, B., Nicoletti, C., Zahanich, I., Graf, E. M., Christ, T., Boxberger, S., Ravens, U. 2006. 5-Azacytidine induces changes in electrophysiological properties of human mesenchymal stem cells. *Cell Res* 16:949–60.

Banghart, M., Borges, K., Isacoff, E., Trauner, D., Kramer, R. H. 2004. Light-activated ion channels for remote control of neuronal firing. *Nat Neurosci* 7:1381–86.

Barclay, T. H., Schapira, D. V. 1983. Malignant tumors of the small intestine. *Cancer* 51:878–81.

Barkai, N., Ben-Zvi, D. 2009. 'Big frog, small frog'—Maintaining proportions in embryonic development. *FEBS J* 276:1196–207.

Barth, L. G., Barth, L. J. 1974a. Ionic regulation of embryonic induction and cell differentiation in *Rana pipiens*. *Dev Biol* 39:1–22.

Barth, L. J., Barth, L. G. 1974b. Effect of the potassium ion on induction of notochord from gastrula ectoderm of *Rana pipiens*. *Biol Bull* 146:313–25.

Basanta, D., Miodownik, M., Baum, B. 2008. The evolution of robust development and homeostasis in artificial organisms. *PLoS Comput Biol* 4:e1000030.

Beck, C. W., Izpisua Belmonte, J. C., Christen, B. 2009. Beyond early development: *Xenopus* as an emerging model for the study of regenerative mechanisms. *Dev Dyn* 238:1226–48.

Becker, R. O. 1967. The electrical control of growth processes. *Med Times* 95:657–69.

Becker, R. O. 1972. Stimulation of partial limb regeneration in rats. *Nature* 235:109–11.

Becker, R. O. 1974. The basic biological data transmission and control system influenced by electrical forces. *Ann NY Acad Sci* 238:236–41.

Beloussov, L. V. 2001. Morphogenetic fields: Outlining the alternatives and enlarging the context. *Riv Biol* 94:219–35.

Beloussov, L. V., Opitz, J. M., Gilbert, S. F. 1997. Life of Alexander G. Gurwitsch and his relevant contribution to the theory of morphogenetic fields. *Int J Dev Biol* 41:771–77; comment, 778–79.

Bentrup, F., Sandan, T., Jaffe, L. 1967. Induction of polarity in *Fucus* eggs by potassium ion gradients. *Protoplasma* 64:254.

Bentrup, F. W., Jaffe, L. F. 1968. Analyzing the "group effect": Rheotropic responses of developing *Fucus* eggs. *Protoplasma* 65:25–35.

Biagiotti, T., D'Amico, M., Marzi, I., Di Gennaro, P., Arcangeli, A., Wanke, E., Olivotto, M. 2005. Cell renewing in neuroblastoma: Electrophysiological and immunocytochemical characterization of stem cells and derivatives. *Stem Cells* 24:443–53, e-pub ahead of print.

Bianchi, L., Wible, B., Arcangeli, A., Taglialatela, M., Morra, F., Castaldo, P., Crociani, O., Rosati, B., Faravelli, L., Olivotto, M., Wanke, E. 1998. herg encodes a K+ current highly conserved in tumors of different histogenesis: A selective advantage for cancer cells? *Cancer Res* 58:815–22.

Binggeli, R., Weinstein, R. 1986. Membrane potentials and sodium channels: Hypotheses for growth regulation and cancer formation based on changes in sodium channels and gap junctions. *J Theor Biol* 123:377–401.

Blackiston, D. J., McLaughlin, K. A., Levin, M. 2009. Bioelectric controls of cell proliferation: Ion channels, membrane voltage, and the cell cycle. *Cell Cycle* 8(21):3527–36.

Bohrmann, J., Gutzeit, H. 1987. Evidence against electrophoresis as the principal mode of protein transport in vitellogenic ovarian follicles of *Drosophila*. *Development* 101:279–88.

Borgens, R., Metcalf, M., Shi, R. 1994. Endogenous ionic currents and voltages in amphibian embryos. *J Exp Zool* 268:307–22.

Borgens, R., Robinson, K., Vanable, J., McGinnis, M. 1989. *Electric fields in vertebrate repair*. New York: Alan R. Liss.

Borgens, R. B. 1982. What is the role of naturally produced electric current in vertebrate regeneration and healing. *Int Rev Cytol* 76:245–98.

Borgens, R. B. 1983. The role of ionic current in the regeneration and development of the amphibian limb. *Prog Clin Biol Res* 110(Pt A):597–608.

Borgens, R. B. 1984. Are limb development and limb regeneration both initiated by an integumentary wounding? A hypothesis. *Differentiation* 28:87–93.

Borgens, R. B. 1986. The role of natural and applied electric fields in neuronal regeneration and development. *Prog Clin Biol Res* 210:239–50.

Borgens, R. B., Blight, A. R., McGinnis, M. E. 1987. Behavioral recovery induced by applied electric fields after spinal cord hemisection in guinea pig. *Science* 238:366–69.

Borgens, R. B., Blight, A. R., McGinnis, M. E. 1990. Functional recovery after spinal cord hemisection in guinea pigs: The effects of applied electric fields. *J Comp Neurol* 296:634–53.

Borgens, R. B., Blight, A. R., Murphy, D. J. 1986. Axonal regeneration in spinal cord injury: A perspective and new technique. *J Comp Neurol* 250:157–67.

Borgens, R. B., McGinnis, M. E., Vanable, J. W., Jr., Miles, E. S. 1984. Stump currents in regenerating salamanders and newts. *J Exp Zool* 231:249–56.

Borgens, R. B., Roederer, E., Cohen, M. J. 1981. Enhanced spinal cord regeneration in lamprey by applied electric fields. *Science* 213:611–17.

Borgens, R. B., Shi, R. 1995. Uncoupling histogenesis from morphogenesis in the vertebrate embryo by collapse of the transneural tube potential. *Dev Dyn* 203:456–67.

Borgens, R. B., Toombs, J. P., Breur, G., Widmer, W. R., Waters, D., Harbath, A. M., March, P., Adams, L. G. 1999. An imposed oscillating electrical field improves the recovery of function in neurologically complete paraplegic dogs. *J Neurotrauma* 16:639–57.

Borgens, R. B., Vanable, J. W., Jr., Jaffe, L. F. 1977. Bioelectricity and regeneration. I. Initiation of frog limb regeneration by minute currents. *J Exp Zool* 200:403–16.

Boutillier, A. L., Kienlen-Campard, P., Loeffler, J. P. 1999. Depolarization regulates cyclin D1 degradation and neuronal apoptosis: A hypothesis about the role of the ubiquitin/proteasome signalling pathway. *Eur J Neurosci* 11:441–48.

Bowman, E. J., O'Neill, F. J., Bowman, B. J. 1997. Mutations of pma-1, the gene encoding the plasma membrane H+-ATPase of *Neurospora crassa*, suppress inhibition of growth by concanamycin A, a specific inhibitor of vacuolar ATPases. *J Biol Chem* 272:14776–86.

Brackenbury, W. J., Djamgoz, M. B. 2006. Activity-dependent regulation of voltage-gated Na+ channel expression in Mat-LyLu rat prostate cancer cell line. *J Physiol* 573:343–56.

Brockes, J. P. 1998. Regeneration and cancer. *Biochim Biophys Acta* 1377:M1–11.

Brooks, R. A., Woodruff, R. I. 2004. Calmodulin transmitted through gap junctions stimulates endocytic incorporation of yolk precursors in insect oocytes. *Dev Biol* 271:339–49.

Bryant, P. J., Bryant, S. V., French, V. 1977. Biological regeneration and pattern formation. *Sci Am* 237:66–76.

Bryant, S. V., Fyfe, D., Singer, M. 1971. The effects of denervation on the ultrastructure of young limb regenerates in the newt, *Triturus*. *Dev Biol* 24:577–95.

Burr, H. S. 1941a. Changes in the field properties of mice with transplanted tumors. *Yale J Biol Med* 13:783–88.

Burr, H. S. 1941b. Field properties of the developing frog's egg. *Proc Natl Acad Sci USA* 27:276–81.

Burr, H. S. 1944. The meaning of bioelectric potentials. *Yale J Biol Med* 16:353.

Burr, H. S., Northrop, F. 1935. The electrodynamic theory of life. *Q Rev Biol* 10:322–33.

Burr, H. S., Northrop, F. S. C. 1939. Evidence for the existence of an electro dynamic field in living organisms. *Proc Natl Acad Sci USA* 25:284–88.

Bustamante, J. O. 1994. Nuclear electrophysiology. *J Membr Biol* 138:105–12.

Bustamante, J. O., Hanover, J. A., Liepins, A. 1995. The ion channel behavior of the nuclear pore complex. *J Membr Biol* 146:239–51.

Bustamante, J. O., Liepins, A., Hanover, J. A. 1994. Nuclear pore complex ion channels (review). *Mol Membr Biol* 11:141–50.

Butler, E. G., O'Brien, J. P. 1942. Effects of localized x-irradiation on regeneration of the urodele limb. *Anat Rec* 84:407–13.

Cannata, S. M., Bagni, C., Bernardini, S., Christen, B., Filoni, S. 2001. Nerve-independence of limb regeneration in larval *Xenopus laevis* is correlated to the level of fgf-2 mRNA expression in limb tissues. *Dev Biol* 231:436–46.

Certal, A. C., Almeida, R. B., Carvalho, L. M., Wong, E., Moreno, N., Michard, E., Carneiro, J., Rodriguez-Leon, J., Wu, H. M., Cheung, A. Y., Feijo, J. A. 2008. Exclusion of a proton ATPase from the apical membrane is associated with cell polarity and tip growth in *Nicotiana tabacum* pollen tubes. *Plant Cell* 20:614–34.

Chafai, M., Louiset, E., Basille, M., Cazillis, M., Vaudry, D., Rostene, W., Gressens, P., Vaudry, H., Gonzalez, B. J. 2006. PACAP and VIP promote initiation of electrophysiological activity in differentiating embryonic stem cells. *Ann NY Acad Sci* 1070:185–89.

Chambers, J. J., Banghart, M. R., Trauner, D., Kramer, R. H. 2006. Light-induced depolarization of neurons using a modified Shaker K(+) channel and a molecular photoswitch. *J Neurophysiol* 96:2792–96.

Chao, P. H., Lu, H. H., Hung, C. T., Nicoll, S. B., Bulinski, J. C. 2007. Effects of applied DC electric field on ligament fibroblast migration and wound healing. *Connect Tissue Res* 48:188–97.

Chera, S., Ghila, L., Dobretz, K., Wenger, Y., Bauer, C., Buzgariu, W., Martinou, J. C., Galliot, B. 2009. Apoptotic cells provide an unexpected source of Wnt3 signaling to drive hydra head regeneration. *Dev Cell* 17:279–89.

Cherubini, A., Hofmann, G., Pillozzi, S., Guasti, L., Crociani, O., Cilia, E., Di Stefano, P., Degani, S., Balzi, M., Olivotto, M., Wanke, E., Becchetti, A., Defilippi, P., Wymore, R., Arcangeli, A. 2005. Human ether-a-go-go-related gene 1 channels are physically linked to beta1 integrins and modulate adhesion-dependent signaling. *Mol Biol Cell* 16:2972–83.

Chifflet, S., Hernandez, J. A., Grasso, S. 2005. A possible role for membrane depolarization in epithelial wound healing. *Am J Physiol Cell Physiol* 288:C1420–30.

Chifflet, S., Hernandez, J. A., Grasso, S., Cirillo, A. 2003. Nonspecific depolarization of the plasma membrane potential induces cytoskeletal modifications of bovine corneal endothelial cells in culture. *Exp Cell Res* 282:1–13.

Child, C. M. 1941. *Patterns and problems of development*. Chicago: University of Chicago Press.

Chin, L. S., Park, C. C., Zitnay, K. M., Sinha, M., DiPatri, A. J., Jr., Perillan, P., Simard, J. M. 1997. 4-Aminopyridine causes apoptosis and blocks an outward rectifier K+ channel in malignant astrocytoma cell lines. *J Neurosci Res* 48:122–27.

Chittajallu, R., Chen, Y., Wang, H., Yuan, X., Ghiani, C. A., Heckman, T., McBain, C. J., Gallo, V. 2002. Regulation of Kv1 subunit expression in oligodendrocyte progenitor cells and their role in G1/S phase progression of the cell cycle. *Proc Natl Acad Sci USA* 99:2350–55.

Chute, J. P. 2006. Stem cell homing. *Curr Opin Hematol* 13:399–406.

Cone, C. D. 1974. The role of the surface electrical transmembrane potential in normal and malignant mitogenesis. *Ann NY Acad Sci* 238:420–35.

Cone, C. D., Cone, C. M. 1976. Induction of mitosis in mature neurons in central nervous system by sustained depolarization. *Science*.192:155–58.

Cone, C. D., Tongier, M. 1971. Control of somatic cell mitosis by simulated changes in the transmembrane potential level. *Oncology* 25:168–82.

Cone, C. D., Tongier, M. 1973. Contact inhibition of division: Involvement of the electrical transmembrane potential. *J Cell Physiol* 82:373–86.

Conti, M. 2004. Targeting K+ channels for cancer therapy. *J Exp Ther Oncol* 4:161–66.

Cooper, M. S. 1984. Gap junctions increase the sensitivity of tissue cells to exogenous electric fields. *J Theor Biol* 111:123–30.

Davies, A., Douglas, L., Hendrich, J., Wratten, J., Tran Van Minh, A., Foucault, I., Koch, D., Pratt, W. S., Saibil, H. R., Dolphin, A. C. 2006. The calcium channel alpha2delta-2 subunit partitions with CaV2.1 into lipid rafts in cerebellum: Implications for localization and function. *J Neurosci* 26:8748–57.

Denker, S. P., Barber, D. L. 2002a. Cell migration requires both ion translocation and cytoskeletal anchoring by the Na-H exchanger NHE1. *J Cell Biol* 159:1087–96.

Denker, S. P., Barber, D. L. 2002b. Ion transport proteins anchor and regulate the cytoskeleton. *Curr Opin Cell Biol* 14:214–20.

De Robertis, E. M., Morita, E. A., Cho, K. W. 1991. Gradient fields and homeobox genes. *Development* 112:669–78.

Dimmitt, J., Marsh, G. 1952. Electrical control of morphogenesis in regenerating *Dugesiatigrina*. 2. Potential gradient vs current density as control factors. *J Cell Comp Physiol* 40:11–23.

Diss, J. K., Stewart, D., Pani, F., Foster, C. S., Walker, M. M., Patel, A., Djamgoz, M. B. 2005. A potential novel marker for human prostate cancer: Voltage-gated sodium channel expression *in vivo*. *Prostate Cancer Prostatic Dis* 8:266–73.

Djamgoz, M. B. A., Mycielska, M., Madeja, Z., Fraser, S. P., Korohoda, W. 2001. Directional movement of rat prostate cancer cells in direct-current electric field: Involvement of voltage-gated Na+ channel activity. *J Cell Sci* 114:2697–705.

Dolberg, D. S., Bissell, M. J. 1984. Inability of Rous sarcoma virus to cause sarcomas in the avian embryo. *Nature* 309:552–56.

Donaldson, D. J., Mason, J. M. 1975. Cancer-related aspects of regeneration research: A review. *Growth* 39:475–96.

Driesch, H. 1929. *The science and philosophy of the organism*. London: A. & C. Black.

Dryselius, S., Lund, P. E., Gylfe, E., Hellman, B. 1994. Variations in ATP-sensitive K+ channel activity provide evidence for inherent metabolic oscillations in pancreatic beta-cells. *Biochem Biophys Res Commun* 205:880–85.

Erskine, L., McCaig, C. D. 1995a. The effects of lyotropic anions on electric field-induced guidance of cultured frog nerves. *J Physiol* 486:229–36.

Erskine, L., McCaig, C. D. 1995b. Growth cone neurotransmitter receptor activation modulates electric field-guided nerve growth. *Dev Biol* 171:330–39.

Erskine, L., Stewart, R., McCaig, C. D. 1995. Electric field-directed growth and branching of cultured frog nerves: Effects of aminoglycosides and polycations. *J Neurobiol* 26:523–36.

Esser, A. T., Smith, K. C., Weaver, J. C., Levin, M. 2006. Mathematical model of morphogen electrophoresis through gap junctions. *Dev Dyn* 235:2144–59.

Fang, K. S., Ionides, E., Oster, G., Nuccitelli, R., Isseroff, R. R. 1999. Epidermal growth factor receptor relocalization and kinase activity are necessary for directional migration of keratinocytes in DC electric fields. *J Cell Sci* 112:1967–78.

Farias, L. M., Ocana, D. B., Diaz, L., Larrea, F., Avila-Chavez, E., Cadena, A., Hinojosa, L. M., Lara, G., Villanueva, L. A., Vargas, C., Hernandez-Gallegos, E., Camacho-Arroyo, I., Duenas-Gonzalez, A., Perez-Cardenas, E., Pardo, L. A., Morales, A., Taja-Chayeb, L., Escamilla, J., Sanchez-Pena, C., Camacho, J. 2004. Ether a go-go potassium channels as human cervical cancer markers. *Cancer Res* 64:6996–7001.

Farinella-Ferruzza, N. 1953. Risultati di trapianti di bottone codale di urodeli su anuri e vice versa. *Rivista di Biologia*. 45:523–27.

Farinella-Ferruzza, N. 1956. The transformation of a tail into a limb after xenoplastic transformation. *Experientia* 15:304–5.

Feijo, J. A., Sainhas, J., Hackett, G. R., Kunkel, J. G., Hepler, P. K. 1999. Growing pollen tubes possess a constitutive alkaline band in the clear zone and a growth-dependent acidic tip. *J Cell Biol* 144:483–96.

Feijo, J. A., Sainhas, J., Holdaway-Clarke, T., Cordeiro, M. S., Kunkel, J. G., Hepler, P. K. 2001. Cellular oscillations and the regulation of growth: The pollen tube paradigm. *Bioessays* 23:86–94.

Filek, M., Holda, M., Machackova, I., Krekule, J. 2005. The effect of electric field on callus induction with rape hypocotyls. *Z Naturforsch C* 60:876–82.

Fischbarg, J., Diecke, F. P. 2005. A mathematical model of electrolyte and fluid transport across corneal endothelium. *J Membr Biol* 203:41–56.

Fitzharris, G., Baltz, J. M. 2006. Granulosa cells regulate intracellular pH of the murine growing oocyte via gap junctions: Development of independent homeostasis during oocyte growth. *Development* 133:591–9.

Flanagan, L. A., Lu, J., Wang, L., Marchenko, S. A., Jeon, N. L., Lee, A. P., Monuki, E. S. 2008. Unique dielectric properties distinguish stem cells and their differentiated progeny. *Stem Cells* 26:656–65.

Forrester, J. V., Lois, N., Zhao, M., McCaig, C. 2007. The spark of life: The role of electric fields in regulating cell behaviour using the eye as a model system. *Ophthalmic Res* 39:4–16.

Frame, S., Balmain, A. 2000. Integration of positive and negative growth signals during ras pathway activation *in vivo*. *Curr Opin Genet Dev* 10:106–13.

Fraser, S. P., Diss, J. K., Chioni, A. M., Mycielska, M. E., Pan, H., Yamaci, R. F., Pani, F., Siwy, Z., Krasowska, M., Grzywna, Z., Brackenbury, W. J., Theodorou, D., Koyuturk, M., Kaya, H., Battaloglu, E., De Bella, M. T., Slade, M. J., Tolhurst, R., Palmieri, C., Jiang, J., Latchman, D. S., Coombes, R. C., Djamgoz, M. B. 2005. Voltage-gated sodium channel expression and potentiation of human breast cancer metastasis. *Clin Cancer Res* 11:5381–89.

Fraser, S. P., Grimes, J. A., Djamgoz, M. B. 2000. Effects of voltage-gated ion channel modulators on rat prostatic cancer cell proliferation: Comparison of strongly and weakly metastatic cell lines. *Prostate* 44:61–76.

French, V., Bryant, P. J., Bryant, S. V. 1976. Pattern regulation in epimorphic fields. *Science* 193:969–81.

Frohlich, K. O. 1950. [Mitogenetic radiation]. *Dtsch Gesundheitsw* 5:1131–36.

Fukumoto, T., Blakely, R., Levin, M. 2005a. Serotonin transporter function is an early step in left-right patterning in chick and frog embryos. *Dev Neurosci* 27:349–63.

Fukumoto, T., Kema, I. P., Levin, M. 2005b. Serotonin signaling is a very early step in patterning of the left-right axis in chick and frog embryos. *Curr Biol* 15:794–803.

Funk, R. H., Monsees, T., Ozkucur, N. 2009. Electromagnetic effects—From cell biology to medicine. *Prog Histochem Cytochem* 43:177–264.

Furusawa, C., Kaneko, K. 1998a. Emergence of multicellular organisms with dynamic differentiation and spatial pattern. *Artif Life* 4:79–93.

Furusawa, C., Kaneko, K. 1998b. Emergence of rules in cell society: Differentiation, hierarchy, and stability. *Bull Math Biol* 60:659–87.

Furusawa, C., Kaneko, K. 2000a. Complex organization in multicellularity as a necessity in evolution. *Artif Life* 6:265–81.

Furusawa, C., Kaneko, K. 2000b. Origin of complexity in multicellular organisms. *Phys Rev Lett* 84:6130–33.

Furusawa, C., Kaneko, K. 2001. Theory of robustness of irreversible differentiation in a stem cell system: Chaos hypothesis. *J Theor Biol* 209:395–416.

Furusawa, C., Kaneko, K. 2002. Origin of multicellular organisms as an inevitable conse-quence of dynamical systems. *Anat Rec* 268:327–42.

Furusawa, C., Kaneko, K. 2003. Robust development as a consequence of generated posi-tional information. *J Theor Biol* 224:413–35.

Gallo, V., Zhou, J. M., McBain, C. J., Wright, P., Knutson, P. L., Armstrong, R. C. 1996. Oligodendrocyte progenitor cell proliferation and lineage progression are regulated by glutamate receptor-mediated K+ channel block. *J Neurosci* 16:2659–70.

Ghiani, C. A., Yuan, X., Eisen, A. M., Knutson, P. L., DePinho, R. A., McBain, C. J., Gallo, V. 1999. Voltage-activated K+ channels and membrane depolarization regulate accu-mulation of the cyclin-dependent kinase inhibitors p27(Kip1) and p21(CIP1) in glial progenitor cells. *J Neurosci* 19:5380–92.

Gilbert, M. S., Saad, A. H., Rupnow, B. A., Knox, S. J. 1996. Association of BCL-2 with membrane hyperpolarization and radioresistance. *J Cell Physiol* 168:114–22.

Gillespie, J. I. 1983. The distribution of small ions during the early development of *Xenopus laevis* and *Ambystoma mexicanum* embryos. *J Physiol* 344:359–77.

Giulian, D., Diacumakos, E. G. 1977. The electrophysiological mapping of compartments within a mammalian cell. *J Cell Biol* 72:86–103.

Gow, N. A., Morris, B. M. 1995. The electric fungus. *Bot J Scotland* 47:263–77.

Gradinaru, V., Thompson, K. R., Deisseroth, K. 2008. eNpHR: A Natronomonas halorho-dopsin enhanced for optogenetic applications. *Brain Cell Biol* 36:129–39.

Grahn, J. C., Reilly, D. A., Nuccitelli, R. L., Isseroff, R. R. 2003. Melanocytes do not migrate directionally in physiological DC electric fields. *Wound Repair Regen* 11:64–70.

Grapengiesser, E., Berts, A., Saha, S., Lund, P. E., Gylfe, E., Hellman, B. 1993. Dual effects of Na/K pump inhibition on cytoplasmic Ca2+ oscillations in pancreatic beta-cells. *Arch Biochem Biophys* 300:372–77.

Grey, R. D., Bastiani, M. J., Webb, D. J., Schertel, E. R. 1982. An electrical block is required to prevent polyspermy in eggs fertilized by natural mating of *Xenopus laevis*. *Dev Biol* 89:475–84.

Gruler, H., Nuccitelli, R. 1991. Neural crest cell galvanotaxis—New data and a novel-approach to the analysis of both galvanotaxis and chemotaxis. *Cell Motility Cytoskel* 19:121–33.

Guyenot, E. 1927. Le probleme morphogenetique dans la regeneration des urodeles: deter-mination et potentialites des regenerats. *Rev Suisse Zool* 34:127–55.

Guyenot, E., Schotte, O. E. 1927. Greffe de regenerat et differenciation induite. *C R Soc Phys His Nat Geneve* 44:21–23.

Habela, C. W., Olsen, M. L., Sontheimer, H. 2008. ClC3 is a critical regulator of the cell cycle in normal and malignant glial cells. *J Neurosci* 28:9205–17.

Han, I. S., Seo, T. B., Kim, K. H., Yoon, J. H., Yoon, S. J., Namgung, U. 2007. Cdc2-mediated Schwann cell migration during peripheral nerve regeneration. *J Cell Sci* 120:246–55.

Harrington, D. B., Becker, R. O. 1973. Electrical stimulation of RNA and protein synthesis in the frog erythrocyte. *Exp Cell Res* 76:95–98.

Hegle, A. P., Marble, D. D., Wilson, G. F. 2006. A voltage-driven switch for ion-independent signaling by ether-a-go-go K+ channels. *Proc Natl Acad Sci USA* 103:2886–91.

Higashimori, H., Sontheimer, H. 2007. Role of Kir4.1 channels in growth control of glia. *Glia* 55:1668–79.

Holcombe, M., Paton, R. 1998. *Information processing in cells and tissues*. New York: Plenum Press.

Hotary, K. B., Robinson, K. R. 1990. Endogenous electrical currents and the resultant volt-age gradients in the chick embryo. *Dev Biol* 140:149–60.

Hotary, K. B., Robinson, K. R. 1991. The neural tube of the *Xenopus* embryo maintains a potential difference across itself. *Brain Res Dev Brain Res* 59:65–73.

Hotary, K. B., Robinson, K. R. 1992. Evidence of a role for endogenous electrical fields in chick embryo development. *Development* 114:985–96.

Hotary, K. B., Robinson, K. R. 1994. Endogenous electrical currents and voltage gradients in *Xenopus* embryos and the consequences of their disruption. *Dev Biol* 166:789–800.

Houillon, C. 1989. Experimental teratomas from xenogenic embryonic implants in the urodele amphibian, *Pleurodeles waltlii* Michah. *C R Acad Sci III* 308:229–36.

Hulser, D. F., Lauterwasser, U. 1982. Membrane potential oscillations in homokaryons. An endogenous signal for detecting intercellular communication. *Exp Cell Res* 139:63–70.

Huttenlocher, A., Horwitz, A. R. 2007. Wound healing with electric potential. *N Engl J Med* 356:303–4.

Hwang, J. S., Kobayashi, C., Agata, K., Ikeo, K., Gojobori, T. 2004. Detection of apoptosis during planarian regeneration by the expression of apoptosis-related genes and TUNEL assay. *Gene* 333:15–25.

Hyman, L., Bellamy, A. 1922. Studies on the correlation between metabolic gradients, electrical gradients, and galvanotaxis I. *Biol Bull.* XLIII:313–47.

Illmensee, K., Mintz, B. 1976. Totipotency and normal differentiation of single teratocarcinoma cells cloned by injection into blastocysts. *Proc Natl Acad Sci USA* 73:549–53.

Jaffe, L. 1981. The role of ionic currents in establishing developmental pattern. *Philos Trans R Soc B* 295:553–66.

Jaffe, L. 1982. Developmental currents, voltages, and gradients. In *Developmental order: Its origin and regulation*, ed. S. Subtelny, 183–215. New York: Alan R. Liss.

Jaffe, L. A., Schlichter, L. C. 1985. Fertilization-induced ionic conductances in eggs of the frog, *Rana pipiens*. *J Physiol* 358:299–319.

Jaffe, L. F. 1966. Electrical currents through the developing *Fucus* egg. *Proc Natl Acad Sci USA* 56:1102–9.

Jaffe, L. F. 1968. Localization in the developing *Fucus* egg and the general role of localizing currents. *Adv Morphogr* 7:295–328.

Jaffe, L. F., Nuccitelli, R. 1974. An ultrasensitive vibrating probe for measuring steady extracellular currents. *J Cell Biol* 63:614–28.

Jaffe, L. F., Nuccitelli, R. 1977. Electrical controls of development. *Ann Rev Biophys Bioeng* 6:445–76.

Jenkins, L. S., Duerstock, B. S., Borgens, R. B. 1996. Reduction of the current of injury leaving the amputation inhibits limb regeneration in the red spotted newt. *Dev Biol* 178:251–62.

Jirsch, J., Deeley, R. G., Cole, S. P., Stewart, A. J., Fedida, D. 1993. Inwardly rectifying K+ channels and volume-regulated anion channels in multidrug-resistant small cell lung cancer cells. *Cancer Res* 53:4156–60.

Johnson, N. C., Dillard, M. E., Baluk, P., McDonald, D. M., Harvey, N. L., Frase, S. L., Oliver, G. 2008. Lymphatic endothelial cell identity is reversible and its maintenance requires Prox1 activity. *Genes Dev* 22:3282–91.

Jones, S., Ribera, A. 1994. Overexpression of a potassium channel gene perturbs neural differentiation. *J Neurosci* 14:2789–99.

Jouhou, H., Yamamoto, K., Homma, A., Hara, M., Kaneko, A., Yamada, M. 2007. Depolarization of isolated horizontal cells of fish acidifies their immediate surrounding by activating V-ATPase. *J Physiol* 585:401–12.

Kalmijn, A. J. 1971. The electric sense of sharks and rays. *J Exp Biol* 55:371–83.

Kalmijn, A. J. 1982. Electric and magnetic field detection in elasmobranch fishes. *Science* 218:916–18.

Kaneko, K., Yomo, T. 1997. Isologous diversification: A theory of cell differentiation. *Bull Math Biol* 59:139–96.

Kaneko, K., Yomo, T. 1999. Isologous diversification for robust development of cell society. *J Theor Biol* 199:243–56.

Kauffman, S. A. 1993. *The origins of order: Self-organization and selection in evolution.* New York: Oxford University Press.

Kim, C. J., Cho, Y. G., Jeong, S. W., Kim, Y. S., Kim, S. Y., Nam, S. W., Lee, S. H., Yoo, N. J., Lee, J. Y., Park, W. S. 2004. Altered expression of KCNK9 in colorectal cancers. *APMIS* 112:588–94.

Klezovitch, O., Fernandez, T. E., Tapscott, S. J., Vasioukhin, V. 2004. Loss of cell polarity causes severe brain dysplasia in Lgl1 knockout mice. *Genes Dev* 18:559–71.

Klezovitch, O., Risk, M., Coleman, I., Lucas, J. M., Null, M., True, L. D., Nelson, P. S., Vasioukhin, V. 2008. A causal role for ERG in neoplastic transformation of prostate epithelium. *Proc Natl Acad Sci USA* 105:2105–10.

Kline, D., Robinson, K. R., Nuccitelli, R. 1983. Ion currents and membrane domains in the cleaving *Xenopus* egg. *J Cell Biol* 97:1753–61.

Konig, S., Beguet, A., Bader, C. R., Bernheim, L. 2006. The calcineurin pathway links hyperpolarization (Kir2.1)-induced Ca2+ signals to human myoblast differentiation and fusion. *Development* 133:3107–14.

Kosari, F., Sheng, S., Kleyman, T. R. 2006. Biophysical approach to determine the subunit stoichiometry of the epithelial sodium channel using the *Xenopus laevis* oocyte expression system. *Methods Mol Biol* 337:53–63.

Krotz, F., Riexinger, T., Buerkle, M. A., Nithipatikom, K., Gloe, T., Sohn, H. Y., Campbell, W. B., Pohl, U. 2004. Membrane-potential-dependent inhibition of platelet adhesion to endothelial cells by epoxyeicosatrienoic acids. *Arterioscler Thromb Vasc Biol* 24:595–600.

Kucheryavykh, Y. V., Kucheryavykh, L. Y., Nichols, C. G., Maldonado, H. M., Baksi, K., Reichenbach, A., Skatchkov, S. N., Eaton, M. J. 2007. Downregulation of Kir4.1 inward rectifying potassium channel subunits by RNAi impairs potassium transfer and glutamate uptake by cultured cortical astrocytes. *Glia* 55:274–81.

Kumar, A., Godwin, J. W., Gates, P. B., Garza-Garcia, A. A., Brockes, J. P. 2007. Molecular basis for the nerve dependence of limb regeneration in an adult vertebrate. *Science* 318:772–77.

Kunzelmann, K. 2005. Ion channels and cancer. *J Membr Biol* 205:159–73.

Kurtz, I., Schrank, A. R. 1955. Bioelectrical properties of intact and regenerating earthworms *Eisenia foetida. Physiol Zool* 28:322–30.

Lang, F., Foller, M., Lang, K., Lang, P., Ritter, M., Vereninov, A., Szabo, I., Huber, S. M., Gulbins, E. 2007. Cell volume regulatory ion channels in cell proliferation and cell death. *Methods Enzymol* 428:209–25.

Lang, F., Foller, M., Lang, K. S., Lang, P. A., Ritter, M., Gulbins, E., Vereninov, A., Huber, S. M. 2005. Ion channels in cell proliferation and apoptotic cell death. *J Membr Biol* 205:147–57.

Lang, F., Friedrich, F., Kahn, E., Woll, E., Hammerer, M., Waldegger, S., Maly, K., Grunicke, H. 1991. Bradykinin-induced oscillations of cell membrane potential in cells expressing the Ha-ras oncogene. *J Biol Chem* 266:4938–42.

Lastraioli, E., Guasti, L., Crociani, O., Polvani, S., Hofmann, G., Witchel, H., Bencini, L., Calistri, M., Messerini, L., Scatizzi, M., Moretti, R., Wanke, E., Olivotto, M., Mugnai, G., Arcangeli, A. 2004. herg1 gene and HERG1 protein are overexpressed in colorectal cancers and regulate cell invasion of tumor cells. *Cancer Res* 64:606–11.

Lauritzen, I., De Weille, J. R., Lazdunski, M. 1997. The potassium channel opener (-)-cromakalim prevents glutamate-induced cell death in hippocampal neurons. *J Neurochem* 69:1570–79.

Lauritzen, I., Zanzouri, M., Honore, E., Duprat, F., Ehrengruber, M. U., Lazdunski, M., Patel, A. J. 2003. K+-dependent cerebellar granule neuron apoptosis. Role of task leak K+ channels. *J Biol Chem* 278:32068–76.

Lee, M., Vasioukhin, V. 2008. Cell polarity and cancer—Cell and tissue polarity as a non-canonical tumor suppressor. *J Cell Sci* 121:1141–50.

Levin, M. 2006. Is the early left-right axis like a plant, a kidney, or a neuron? The integration of physiological signals in embryonic asymmetry. *Birth Defects Res C Embryo Today* 78:191–223.

Levin, M. 2007a. Gap junctional communication in morphogenesis. *Prog Biophys Mol Biol* 94:186–206.

Levin, M. 2007b. Large-scale biophysics: Ion flows and regeneration. *Trends Cell Biol* 17:262–71.

Levin, M. 2009. Bioelectric mechanisms in regeneration: Unique aspects and future perspectives. *Semin Cell Dev Biol* 20:543–56.

Levin, M., Buznikov, G. A., Lauder, J. M. 2006. Of minds and embryos: Left-right asymmetry and the serotonergic controls of pre-neural morphogenesis. *Dev Neurosci* 28:171–85.

Levin, M., Thorlin, T., Robinson, K. R., Nogi, T., Mercola, M. 2002. Asymmetries in H+/K+-ATPase and cell membrane potentials comprise a very early step in left-right patterning. *Cell* 111:77–89.

Lewis, J. 2008. From signals to patterns: Space, time, and mathematics in developmental biology. *Science* 322:399–403.

Li, L., Liu, F., Salmonsen, R. A., Turner, T. K., Litofsky, N. S., Di Cristofano, A., Pandolfi, P. P., Jones, S. N., Recht, L. D., Ross, A. H. 2002. PTEN in neural precursor cells: Regulation of migration, apoptosis, and proliferation. *Mol Cell Neurosci* 20:21–9.

Liebau, S., Propper, C., Bockers, T., Lehmann-Horn, F., Storch, A., Grissmer, S., Wittekindt, O. H. 2006. Selective blockage of Kv1.3 and Kv3.1 channels increases neural progenitor cell proliferation. *J Neurochem* 99:426–37.

Lin, H., Xiao, J., Luo, X., Wang, H., Gao, H., Yang, B., Wang, Z. 2007. Overexpression HERG K(+) channel gene mediates cell-growth signals on activation of oncoproteins SP1 and NF-kappaB and inactivation of tumor suppressor Nkx3.1. *J Cell Physiol* 212:137–47.

Lin-Liu, S., Adey, W. R., Poo, M. M. 1984. Migration of cell surface concanavalin A receptors in pulsed electric fields. *Biophys J* 45:1211–17.

Lund, E. 1947. *Bioelectric fields and growth.* Austin: University of Texas Press.

MacFarlane, S. N., Sontheimer, H. 2000. Changes in ion channel expression accompany cell cycle progression of spinal cord astrocytes. *Glia* 30:39–48.

Maden, M. 1978. Neurotrophic control of the cell cycle during amphibian limb regeneration. *J Embryol Exp Morphol* 48:169–75.

Malhi, H., Irani, A. N., Rajvanshi, P., Suadicani, S. O., Spray, D. C., McDonald, T. V., Gupta, S. 2000. KATP channels regulate mitogenically induced proliferation in primary rat hepatocytes and human liver cell lines. Implications for liver growth control and potential therapeutic targeting. *J Biol Chem* 275:26050–57.

Maric, D., Maric, I., Wen, X., Fritschy, J. M., Sieghart, W., Barker, J. L., Serafini, R. 1999. GABAA receptor subunit composition and functional properties of Cl– channels with differential sensitivity to zolpidem in embryonic rat hippocampal cells. *J Neurosci* 19:4921–37.

Marsh, G., Beams, H. 1957. Electrical control of morphogenesis in regenerating *Dugesia tigrina*. *J Cell Comp Physiol* 39:191–211.

Marsh, G., Beams, H. W. 1945. The orientation of pollen tubes of *Vinca* in the electric current. *J Cell Comp Physiol* 25:195–204.

Marsh, G., Beams, H. W. 1946. Orientation of chick nerve fibers by direct electric currents. *Anat Rec* 94:370.

Marsh, G., Beams, H. W. 1947a. Electrical control of growth polarity in regenerating *Dugesia-tigrina*. *Federation Proc* 6:163–64.

Marsh, G., Beams, H. W. 1947b. Orientation of growth direction of the onion root in a transverse electric field. *Anat Rec* 99:623.

Marsh, G., Beams, H. W. 1949. Electrical control of axial polarity in a regenerating annelid. *Anat Rec* 105:513–14.

Marsh, G., Beams, H. W. 1950. Electrical control of growth axis in a regenerating annelid. *Anat Rec* 108:512.

Marsh, G., Beams, H. W. 1952. Electrical control of morphogenesis in regenerating *Dugesia tigrina*. 1. Relation of axial polarity to field strength. *J Cell Comp Physiol* 39:191.

Martens, J. R., O'Connell, K., Tamkun, M. 2004. Targeting of ion channels to membrane microdomains: Localization of KV channels to lipid rafts. *Trends Pharmacol Sci* 25:16–21.

Martinez-Frias, M. L., Frias, J. L., Opitz, J. M. 1998. Errors of morphogenesis and developmental field theory. *Am J Med Genet* 76:291–96.

Masuda, C. A., Montero-Lomeli, M. 2000. An NH2-terminal deleted plasma membrane H+-ATPase is a dominant negative mutant and is sequestered in endoplasmic reticulum derived structures. *Biochem Cell Biol* 78:51–58.

Mathews, A. P. 1903. Electrical polarity in the hydroids. *Am J Physiol* 8:294–99.

Mazzanti, M., Bustamante, J. O., Oberleithner, H. 2001. Electrical dimension of the nuclear envelope. *Physiol Rev* 81:1–19.

McCaig, C. D., Dover, P. J. 1993. Raised cyclic-AMP and a small applied electric field influence differentiation, shape, and orientation of single myoblasts. *Dev Biol* 158:172–82.

McCaig, C. D., Rajnicek, A. M., Song, B., Zhao, M. 2002. Has electrical growth cone guidance found its potential? *Trends Neurosci* 25:354–59.

McCaig, C. D., Rajnicek, A. M., Song, B., Zhao, M. 2005. Controlling cell behavior electrically: Current views and future potential. *Physiol Rev* 85:943–78.

McGillivray, A. M., Gow, N. A. 1986. Applied electrical fields polarize the growth of mycelial fungi. *J Gen Microbiol* 132:2515–25.

Megason, S. G., Fraser, S. E. 2007. Imaging in systems biology. *Cell* 130:784–95.

Mescher, A. L., White, G. W., Brokaw, J. J. 2000. Apoptosis in regenerating and denervated, nonregenerating urodele forelimbs. *Wound Repair Regen* 8:110–16.

Messerli, M., Robinson, K. R. 1997. Tip localized Ca2+ pulses are coincident with peak pulsatile growth rates in pollen tubes of *Lilium longiflorum*. *J Cell Sci* 110:1269–78.

Messerli, M. A., Collis, L. P., Smith, P. J. 2009. Ion trapping with fast-response ion-selective microelectrodes enhances detection of extracellular ion channel gradients. *Biophys J* 96:1597–605.

Messerli, M. A., Creton, R., Jaffe, L. F., Robinson, K. R. 2000. Periodic increases in elongation rate precede increases in cytosolic Ca2+ during pollen tube growth. *Dev Biol* 222:84–98.

Messerli, M. A., Danuser, G., Robinson, K. R. 1999. Pulsatile influxes of H+, K+ and Ca2+ lag growth pulses of *Lilium longiflorum* pollen tubes. *J Cell Sci* 112:1497–509.

Messerli, M. A., Robinson, K. R. 1998. Cytoplasmic acidification and current influx follow growth pulses of *Lilium longiflorum* pollen tubes. *Plant J* 16:87–91.

Messerli, M. A., Robinson, K. R. 2003. Ionic and osmotic disruptions of the lily pollen tube oscillator: Testing proposed models. *Planta* 217:147–57.

Michard, E., Alves, F., Feijo, J. A. 2008. The role of ion fluxes in polarized cell growth and morphogenesis: The pollen tube as an experimental paradigm. *Int J Dev Biol* 53(8–10):1609–22.

Miesenbock, G., De Angelis, D. A., Rothman, J. E. 1998. Visualizing secretion and synaptic transmission with pH-sensitive green fluorescent proteins. *Nature* 394:192–95.

Miki, T., Iwanaga, T., Nagashima, K., Ihara, Y., Seino, S. 2001. Roles of ATP-sensitive K+ channels in cell survival and differentiation in the endocrine pancreas. *Diabetes* 50(Suppl 1):S48–51.

Miki, T., Tashiro, F., Iwanaga, T., Nagashima, K., Yoshitomi, H., Aihara, H., Nitta, Y., Gonoi, T., Inagaki, N., Miyazaki, J., Seino, S. 1997. Abnormalities of pancreatic islets by targeted expression of a dominant-negative KATP channel. *Proc Natl Acad Sci USA* 94:11969–73.

Miller, G. 2006. Optogenetics. Shining new light on neural circuits. *Science* 314:1674–76.

Mintz, B., Illmensee, K. 1975. Normal genetically mosaic mice produced from malignant teratocarcinoma cells. *Proc Natl Acad Sci USA* 72:3585–89.

Moriarty, L. J., Borgens, R. B. 2001. An oscillating extracellular voltage gradient reduces the density and influences the orientation of astrocytes in injured mammalian spinal cord. *J Neurocytol* 30:45–57.

Morley, G. E., Taffet, S. M., Delmar, M. 1996. Intramolecular interactions mediate pH regulation of connexin43 channels. *Biophys J* 70:1294–302.

Morokuma, J., Blackiston, D., Adams, D. S., Seebohm, G., Trimmer, B., Levin, M. 2008a. Modulation of potassium channel function confers a hyperproliferative invasive phenotype on embryonic stem cells. *Proc Natl Acad Sci USA* 105:16608–13.

Morokuma, J., Blackiston, D., Levin, M. 2008b. KCNQ1 and KCNE1 K+ channel components are involved in early left-right patterning in *Xenopus laevis* embryos. *Cell Physiol Biochem* 21:357–72.

Morris, B. M., Reid, B., Gow, N. A. 1992. Electrotaxis of zoospores of *Phytophthora palmivora* at physiologically-relevant fields strengths. *Plant Cell Environ* 15:645–53.

Moschou, E. A., Chaniotakis, N. A. 2000. Ion-partitioning membrane-based electrochemical sensors. *Anal Chem* 72:1835–42.

Mu, D., Chen, L., Zhang, X., See, L. H., Koch, C. M., Yen, C., Tong, J. J., Spiegel, L., Nguyen, K. C., Servoss, A., Peng, Y., Pei, L., Marks, J. R., Lowe, S., Hoey, T., Jan, L. Y., McCombie, W. R., Wigler, M. H., Powers, S. 2003. Genomic amplification and oncogenic properties of the KCNK9 potassium channel gene. *Cancer Cell* 3:297–302.

Murata, Y., Iwasaki, H., Sasaki, M., Inaba, K., Okamura, Y. 2005. Phosphoinositide phosphatase activity coupled to an intrinsic voltage sensor. *Nature* 435:1239–43.

Mycielska, M. E., Djamgoz, M. B. 2004. Cellular mechanisms of direct-current electric field effects: Galvanotaxis and metastatic disease. *J Cell Sci* 117:1631–39.

Nadal-Ginard, B., Kajstura, J., Anversa, P., Leri, A. 2003. A matter of life and death: Cardiac myocyte apoptosis and regeneration. *J Clin Invest* 111:1457–59.

Needham, J. 1936. New advances in the chemistry and biology of organized growth. *Proc R Soc Med* 29:1577–626.

Nicholson, B. J. 2003. Gap junctions—From cell to molecule. *J Cell Sci* 116:4479–81.

Nishimura, K. Y., Isseroff, R. R., Nuccitelli, R. 1996. Human keratinocytes migrate to the negative pole in direct current electric fields comparable to those measured in mammalian wounds. *J Cell Sci* 109(Pt 1):199–207.

Nogi, T., Levin, M. 2005. Characterization of innexin gene expression and functional roles of gap-junctional communication in planarian regeneration. *Dev Biol* 287:314–35.

Novák, B., Bentrup, F. W. 1972. An electrophysiological study of regeneration in *Acetabularia mediterranea*. *Planta* 108:227–44.

Novak, B., Sirnoval, C. 1975. Inhibition of regeneration of A*cetabularia mediterranea* enucleated posterior stalk segments by electrical isolation. *Plant Sci Lett* 5:183–88.

Nuccitelli, R. 1980. Vibrating probe—High spatial-resolution extracellular current measurement. *Federation Proc* 39:2129.

Nuccitelli, R. 1992. Endogenous ionic currents and DC electric-fields in multicellular animal-tissues. *Bioelectromagnetics* Suppl. 1:147–57.

Nuccitelli, R. 2003. A role for endogenous electric fields in wound healing. *Curr Top Dev Biol* 58:1–26.

Nuccitelli, R., Chen, X., Pakhomov, A. G., Baldwin, W. H., Sheikh, S., Pomicter, J. L., Ren, W., Osgood, C., Swanson, R. J., Kolb, J. F., Beebe, S. J., Schoenbach, K. H. 2009. A new pulsed electric field therapy for melanoma disrupts the tumor's blood supply and causes complete remission without recurrence. *Int J Cancer* 125:438–45.

Nuccitelli, R., Nuccitelli, P., Ramlatchan, S., Sanger, R., Smith, P. J. 2008. Imaging the electric field associated with mouse and human skin wounds. *Wound Repair Regen* 16:432–41.

Nuckels, R. J., Ng, A., Darland, T., Gross, J. M. 2009. The vacuolar-ATPase complex regulates retinoblast proliferation and survival, photoreceptor morphogenesis, and pigmentation in the zebrafish eye. *Invest Ophthalmol Vis Sci* 50:893–905.

Olsen, M. L., Sontheimer, H. 2004. Mislocalization of Kir channels in malignant glia. *Glia* 46:63–73.

Onganer, P. U., Djamgoz, M. B. 2005. Small-cell lung cancer (human): Potentiation of endocytic membrane activity by voltage-gated Na(+) channel expression *in vitro*. *J Membr Biol* 204:67–75.

Onganer, P. U., Seckl, M. J., Djamgoz, M. B. 2005. Neuronal characteristics of small-cell lung cancer. *Br J Cancer* 93:1197–201.

Opitz, J. M. 1986. Developmental field theory and observations—Accidental progress? *Am J Med Genet Suppl* 2:1–8.

Ouadid-Ahidouch, H., Ahidouch, A. 2008. K+ channel expression in human breast cancer cells: Involvement in cell cycle regulation and carcinogenesis. *J Membr Biol* 221:1–6.

Ousingsawat, J., Spitzner, M., Schreiber, R., Kunzelmann, K. 2008. Upregulation of colonic ion channels in APC (Min/+) mice. *Pflugers Arch* 456:847–55.

Oviedo, N. J., Beane, W. S. 2009. Regeneration: The origin of cancer or a possible cure? *Semin Cell Dev Biol* 20:557–64.

Oviedo, N. J., Levin, M. 2007. smedinx-11 is a planarian stem cell gap junction gene required for regeneration and homeostasis. *Development* 134:3121–31.

Oviedo, N. J., Morokuma, J., Walentek, P., Kema, I. P., Gu, M. B., Ahn, J. M., Hwang, J. S., Gojobori, T., Levin, M. 2010. Long-range neural and gap junction protein-mediated cues control polarity during planarian regeneration. *Dev Biol* 339:188–99.

Oviedo, N. J., Newmark, P. A., Sanchez Alvarado, A. 2003. Allometric scaling and proportion regulation in the freshwater planarian *Schmidtea mediterranea*. *Dev Dyn* 226:326–33.

Ozkucur, N., Epperlein, H. H., Funk, R. H. 2010. Ion imaging during axolotl tail regeneration *in vivo*. *Dev Dyn* 239:2048–57.

Palacios-Prado, N., Bukauskas, F. F. 2009. Heterotypic gap junction channels as voltage-sensitive valves for intercellular signaling. *Proc Natl Acad Sci USA* 106(35):14855–60.

Pandiella, A., Magni, M., Lovisolo, D., Meldolesi, J. 1989. The effect of epidermal growth factor on membrane potential. Rapid hyperpolarization followed by persistent fluctuations. *J Biol Chem* 264:12914–21.

Pardo, L. A., del Camino, D., Sanchez, A., Alves, F., Bruggemann, A., Beckh, S., Stuhmer, W. 1999. Oncogenic potential of EAG K(+) channels. *EMBO J* 18:5540–47.

Park, K. S., Jung, K. H., Kim, S. H., Kim, K. S., Choi, M. R., Kim, Y., Chai, Y. G. 2007. Functional expression of ion channels in mesenchymal stem cells derived from umbilical cord vein. *Stem Cells* 25:2044–52.

Paul, R., Chatterjee, R., Tuszynski, J. A., Fritz, O. G. 1983. Theory of long-range coherence in biological systems. I. The anomalous behaviour of human erythrocytes. *J Theor Biol* 104:169–85.

Pawlowski, A., Weddell, G. 1967. Induction of tumors in denervated skin. *Nature* 213:1234–36.

Pei, L., Wiser, O., Slavin, A., Mu, D., Powers, S., Jan, L. Y., Hoey, T. 2003. Oncogenic potential of TASK3 (Kcnk9) depends on K+ channel function. *Proc Natl Acad Sci USA* 100(13):7803–7.

Perona, R., Serrano, R. 1988. Increased pH and tumorigenicity of fibroblasts expressing a yeast proton pump. *Nature* 334:438–40.

Pflugfelder, V. O. 1938. *Verhandl Deutsch Zool Ges-zool Anz* 11(Suppl.).

Pflugfelder, V. O. 1954. *Strahlentherapie* 93.

Pfulgfelder, V. O. 1939. *Z Wiss Zool* 152:129.

Pfulgfelder, V. O. 1950. *Z Krebsforsch* 56:107.

Pineda, R. H., Svoboda, K. R., Wright, M. A., Taylor, A. D., Novak, A. E., Gamse, J. T., Eisen, J. S., Ribera, A. B. 2006. Knockdown of Nav1.6a Na+ channels affects zebrafish motoneuron development. *Development* 133:3827–36.

Pizzarello, D. J., Wolsky, A. 1966. Carcinogenesis and regeneration in newts. *Experientia* 22:387–88.

Pokorny, J., Jandova, A., Kobilkova, J., Heyberger, K., Hraba, T. 1983. Frohlich electromagnetic radiation from human leukocytes: Implications for leukocyte adherence inhibition test. *J Theor Biol* 102:295–305.

Polezhaev, L. V. 1972. *Loss and restoration of regenerative capacity in tissues and organs of animals*. Cambridge, MA: Harvard University Press.

Poo, M., Robinson, K. R. 1977. Electrophoresis of concanavalin A receptors along embryonic muscle cell membrane. *Nature*. 265:602–5.

Popp, F. A. 2009. Cancer growth and its inhibition in terms of coherence. *Electromagn Biol Med* 28:53–60.

Popp, F. A., Li, K. H., Gu, Q. 1992. *Recent advances in biophoton research and its applications*. River Edge, N.J.: Singapore.

Pullar, C. E., Baier, B. S., Kariya, Y., Russell, A. J., Horst, B. A., Marinkovich, M. P., Isseroff, R. R. 2006. beta4 integrin and epidermal growth factor coordinately regulate electric field-mediated directional migration via Rac1. *Mol Biol Cell* 17:4925–35.

Pullar, C. E., Isseroff, R. R. 2005. Cyclic AMP mediates keratinocyte directional migration in an electric field. *J Cell Sci* 118:2023–34.

Putney, L. K., Barber, D. L. 2003. Na-H exchange-dependent increase in intracellular pH times G2/M entry and transition. *J Biol Chem* 278:44645–49.

Raja, Sivamani, K., Garcia, M. S., Isseroff, R. R. 2007. Wound re-epithelialization: Modulating keratinocyte migration in wound healing. *Front Biosci* 12:2849–68.

Rajnicek, A. M., Foubister, L. E., McCaig, C. D. 2006. Growth cone steering by a physiological electric field requires dynamic microtubules, microfilaments and Rac-mediated filopodial asymmetry. *J Cell Sci* 119:1736–45.

Rajnicek, A. M., Foubister, L. E., McCaig, C. D. 2007. Prioritising guidance cues: Directional migration induced by substratum contours and electrical gradients is controlled by a rho/cdc42 switch. *Dev Biol* 312:448–60.

Rajnicek, A. M., McCaig, C. D., Gow, N. A. 1994. Electric fields induced curved growth of *Enterobacter cloacae, E. coli,* and B. *subtilis* cells. *J Bacteriol* 176:703–13.

Rathore, K. S., Cork, R. J., Robinson, K. R. 1991. A cytoplasmic gradient of Ca2+ is correlated with the growth of lily pollen tubes. *Dev Biol* 148:612–19.

Ravens, U. 2006. Electrophysiological properties of stem cells. *Herz* 31:123–26.

Reid, B., Nuccitelli, R., Zhao, M. 2007. Non-invasive measurement of bioelectric currents with a vibrating probe. *Nat Protoc* 2:661–69.

Reid, B., Song, B., McCaig, C. D., Zhao, M. 2005. Wound healing in rat cornea: The role of electric currents. *Faseb J* 19:379–86.

Reid, B., Song, B., Zhao, M. 2009. Electric currents in *Xenopus* tadpole tail regeneration. *Dev Biol* 335:198–207.

Reshkin, S. J., Bellizzi, A., Albarani, V., Guerra, L., Tommasino, M., Paradiso, A., Casavola, V. 2000. Phosphoinositide 3-kinase is involved in the tumor-specific activation of human breast cancer cell Na(+)/H(+) exchange, motility, and invasion induced by serum deprivation. *J Biol Chem* 275:5361–69.

Ribera, A. 1998. Ion channel activity drives ion channel expression. *J Physiol* 511:645.

Robinson, K. R. 1989. Endogenous and applied electrical currents: Their measurement and application. In *Electric fields in vertebrate repair*, 1–25. New York: Alan R. Liss.

Robinson, K. R., Cormie, P. 2008. Electric field effects on human spinal injury: Is there a basis in the *in vitro* studies? *Dev Neurobiol* 68:274–80.

Robinson, K. R., Messerli, M. A. 1996. Electric embryos: The embryonic epithelium as a generator of developmental information. In *Nerve growth and guidance*, ed. C. D. McCaig, 131–50. London: Portland Press.

Robinson, K. R., Messerli, M. A. 2002. *Pulsating ion fluxes and growth at the pollen tube tip.* Sci STKE 2002, PE51.

Rose, S. M. 1972. Correlation between bioelectrical and morphogenetic polarity during regeneration in tubularia. *Dev Biol* 28:274.

Rose, S. M. 1974. Bioelectric control of regeneration in tubularia. *Am Zool* 14:797–803.

Rose, S. M., Wallingford, H. M. 1948. Transformation of renal tumors of frogs to normal tissues in regenerating limbs of salamanders. *Science* 107:457.

Rouzaire-Dubois, B., Gerard, V., Dubois, J. M. 1993. Involvement of K+ channels in the quercetin-induced inhibition of neuroblastoma cell growth. *Pflugers Arch* 423:202–5.

Ruben, L. N., Balls, M., Stevens, J. 1966. Cancer and super-regeneration in *Triturus viridescens* limbs. *Experientia* 22:260–61.

Rubin, H. 1985. Cancer as a dynamic developmental disorder. *Cancer Res* 45:2935–42.

Rubin, H. 1990. The significance of biological heterogeneity. *Cancer Metastasis Rev* 9:1–20.

Russo, R. E., Reali, C., Radmilovich, M., Fernandez, A., Trujillo-Cenoz, O. 2008. Connexin 43 delimits functional domains of neurogenic precursors in the spinal cord. *J Neurosci* 28:3298–309.

Salo, E., Baguna, J. 1985. Cell movement in intact and regenerating planarians. Quantitation using chromosomal, nuclear and cytoplasmic markers. *J Embryol Exp Morphol* 89:57–70.

Sasaki, M., Gonzalez-Zulueta, M., Huang, H., Herring, W. J., Ahn, S., Ginty, D. D., Dawson, V. L., Dawson, T. M. 2000. Dynamic regulation of neuronal NO synthase transcription by calcium influx through a CREB family transcription factor-dependent mechanism. *Proc Natl Acad Sci USA* 97:8617–22.

Scharrer, B. 1945. Experimental tumors after nerve section in an insect. *Proc Soc Exp Biol Med* 60:184–89.

Scharrer, B. 1953. Insect tumors induced by nerve severance: Incidence and mortality. *Cancer Res* 13:73–6.

Scharrer, B., Lochhead, M. S. 1950. Tumors in the invertebrates: A review. *Cancer Res* 10:403–19.

Schiffmann, Y. 1989. The second messenger system as the morphogenetic field. *Biochem Biophys Res Commun* 165:1267–71.

Schiffmann, Y. 1991. An hypothesis: Phosphorylation fields as the source of positional information and cell differentiation—(cAMP, ATP) as the universal morphogenetic Turing couple. *Prog Biophys Mol Biol* 56:79–105.

Schnabel, R., Bischoff, M., Hintze, A., Schulz, A. K., Hejnol, A., Meinhardt, H., Hutter, H. 2006. Global cell sorting in the *C. elegans* embryo defines a new mechanism for pattern formation. *Dev Biol* 294:418–31.

Schwab, A. 2001. Function and spatial distribution of ion channels and transporters in cell migration. *Am J Physiol Renal Physiol* 280:F739–47.

Schwab, A., Gabriel, K., Finsterwalder, F., Folprecht, G., Greger, R., Kramer, A., Oberleithner, H. 1995. Polarized ion transport during migration of transformed Madin-Darby canine kidney cells. *Pflugers Arch* 430:802–7.

Seilern-Aspang, F., Kratochwill, L. 1962. Induction and differentiation of an epithelial tumour in the newt (*Triturus cristatus*). *J Embryol Exp Morphol* 10:337–56.

Seilern-Aspang, F., Kratochwill, L. 1965. Relation between regeneration and tumor growth. *Regeneration in animals and related problems*, 452–73. Amsterdam: North-Holland Publishing Company.

Shapiro, S., Borgens, R., Pascuzzi, R., Roos, K., Groff, M., Purvines, S., Rodgers, R. B., Hagy, S., Nelson, P. 2005. Oscillating field stimulation for complete spinal cord injury in humans: A phase 1 trial. *J Neurosurg Spine* 2:3–10.

Sharma, K. K., Niazi, I. A. 1990. Restoration of limb regeneration ability in frog tadpoles by electrical stimulation. *Indian J Exp Biol* 28:733–38.

Shi, R., Borgens, R. B. 1995. Three-dimensional gradients of voltage during development of the nervous system as invisible coordinates for the establishment of embryonic pattern. *Dev Dyn* 202:101–14.

Shimeld, S. M., Levin, M. 2006. Evidence for the regulation of left-right asymmetry in *Ciona intestinalis* by ion flux. *Dev Dyn* 235:1543–53.

Shuba, Y. M., Prevarskaya, N., Lemonnier, L., Van Coppenolle, F., Kostyuk, P. G., Mauroy, B., Skryma, R. 2000. Volume-regulated chloride conductance in the LNCaP human prostate cancer cell line. *Am J Physiol Cell Physiol* 279:C1144–54.

Simons, M., Gault, W. J., Gotthardt, D., Rohatgi, R., Klein, T. J., Shao, Y., Lee, H. J., Wu, A. L., Fang, Y., Satlin, L. M., Dow, J. T., Chen, J., Zheng, J., Boutros, M., Mlodzik, M. 2009. Electrochemical cues regulate assembly of the frizzled/dishevelled complex at the plasma membrane during planar epithelial polarization. *Nat Cell Biol* 11:286–94.

Sisken, B. F. 1992. Electrical-stimulation of nerves and their regeneration. *Bioelectrochem Bioenergetics* 29:121–26.

Sisken, B. F., Fowler, I. 1981. Induction of limb regeneration in the chick-embryo. *Anat Rec* 199:A238–39.

Sisken, B. F., Fowler, I., Romm, S. 1984. Response of amputated rat limbs to fetal nerve tissue implants and direct current. *J Orthop Res* 2:177–89.

Sisken, B. F., Walker, J., Orgel, M. 1993. Prospects on clinical-applications of electrical-stimulation for nerve regeneration. *J Cell Biochem* 51:404–9.

Slack, J. M. 1980. A serial threshold theory of regeneration. *J Theor Biol* 82:105–40.

Slack, J. M. 1982. Regeneration and the second anatomy of animals. In *Developmental order: Its origin and regulation*, ed. Subtelny, S. 423–36. New York: Alan R. Liss.

Slack, J. M. 1985. Homoeotic transformations in man: Implications for the mechanism of embryonic development and for the organization of epithelia. *J Theor Biol* 114:463–90.

Smith, P. J. S., Sanger, R. S., Messerli, M. A. 2007. Principles, development and applications of self-referencing electrochemical microelectrodes to the determination of fluxes at cell membranes. In *Methods and new frontiers in neuroscience*, ed. A. C. Michael, 373–405. Boca Raton, FL: CRC Press.

Smith, S. D. 1967. Induction of partial limb regeneration in *Rana pipiens* by galvanic stimulation. *Anat Rec* 158:89.

Smith, S. D. 1970. Effects of electrical fields upon regeneration in the metazoa. *Am Zool* 10:133–40.

Smith, S. D. 1981. The role of electrode position in the electrical induction of limb regeneration in subadult rats. *Bioelectrochem Bioenergetics* 8:661–70.

Sollars, S. I., Smith, P. C., Hill, D. L. 2002. Time course of morphological alterations of fungiform papillae and taste buds following chorda tympani transection in neonatal rats. *J Neurobiol* 51:223–36.

Song, B., Gu, Y., Pu, J., Reid, B., Zhao, Z., Zhao, M. 2007. Application of direct current electric fields to cells and tissues *in vitro* and modulation of wound electric field *in vivo*. *Nat Protoc* 2:1479–89.

Soto, A. M., Sonnenschein, C. 2004. The somatic mutation theory of cancer: Growing problems with the paradigm? *Bioessays* 26:1097–107.

Spemann, H. 1938. *Embryonic development and induction*. New Haven, CT: Yale University Press.

Steinberg, B. E., Touret, N., Vargas-Caballero, M., Grinstein, S. 2007. *In situ* measurement of the electrical potential across the phagosomal membrane using FRET and its contribution to the proton-motive force. *Proc Natl Acad Sci USA* 104:9523–28.

Stern, C. 1982. Experimental reversal of polarity in chick embryo epiblast sheets *in vitro*. *Exp Cell Res* 140:468–71.

Stewart, R., Allan, D. W., McCaig, C. D. 1996. Lectins implicate specific carbohydrate domains in electric field stimulated nerve growth and guidance. *J Neurobiol* 30:425–37.

Stillwell, E. F., Cone, C. M., Cone, C. D. 1973. Stimulation of DNA synthesis in CNS neurones by sustained depolarisation. *Nat New Biol* 246:110–11.

Stokes, C. L., Rinzel, J. 1993. Diffusion of extracellular K+ can synchronize bursting oscillations in a model islet of Langerhans. *Biophys J* 65:597–607.

Stuhmer, W., Alves, F., Hartung, F., Zientkowska, M., Pardo, L. A. 2006. Potassium channels as tumour markers. *FEBS Lett.* 580:2850–52.

Stump, R. F., Robinson, K. R. 1983. *Xenopus* neural crest cell migration in an applied electrical field. *J Cell Biol* 97:1226–33.

Sundelacruz, S., Levin, M., Kaplan, D. L. 2008. Membrane potential controls adipogenic and osteogenic differentiation of mesenchymal stem cells. *PLoS One* 3:e3737.

Sundelacruz, S., Levin, M., Kaplan, D. L. 2009. Role of membrane potential in the regulation of cell proliferation and differentiation. *Stem Cell Rev Rep* 5:231–46.

Svoboda, K. R., Linares, A. E., Ribera, A. B. 2001. Activity regulates programmed cell death of zebrafish Rohon-Beard neurons. *Development* 128:3511–20.

Takagi, H., Kaneko, K., Yomo, T. 2000. Evolution of genetic codes through isologous diversification of cellular states. *Artif Life* 6:283–305.

Tao, Y., Yan, D., Yang, Q., Zeng, R., Wang, Y. 2006. Low K+ promotes NF-kappaB/DNA binding in neuronal apoptosis induced by K+ loss. *Mol Cell Biol* 26:1038–50.

Tataria, M., Perryman, S. V., Sylvester, K. G. 2006. Stem cells: Tissue regeneration and cancer. *Semin Pediatr Surg* 15:284–92.

Telfer, W., Woodruff, R., Huebner, E. 1981. Electrical polarity and cellular differentiation in meroistic ovaries. *Am Zool.* 21:675–86.

Teng, G. Q., Zhao, X., Lees-Miller, J. P., Quinn, F. R., Li, P., Rancourt, D. E., London, B., Cross, J. C., Duff, H. J. 2008. Homozygous missense N629D hERG (KCNH2) potassium channel mutation causes developmental defects in the right ventricle and its outflow tract and embryonic lethality. *Circ Res* 103:1483–91.

Trinh, N. T., Prive, A., Kheir, L., Bourret, J. C., Hijazi, T., Amraei, M. G., Noel, J., Brochiero, E. 2007. Involvement of KATP and KvLQT1 K+ channels in EGF-stimulated alveolar epithelial cell repair processes. *Am J Physiol Lung Cell Mol Physiol* 293:L870–82.

Trinh, N. T., Prive, A., Maille, E., Noel, J., Brochiero, E. 2008. EGF and K+ channel activity control normal and cystic fibrosis bronchial epithelia repair. *Am J Physiol Lung Cell Mol Physiol* 295:L866–80.

Tseng, A. S., Adams, D. S., Qiu, D., Koustubhan, P., Levin, M. 2007. Apoptosis is required during early stages of tail regeneration in *Xenopus laevis*. *Dev Biol* 301:62–69.

Tseng, A. S., Levin, M. 2008. Tail regeneration in *Xenopus laevis* as a model for understanding tissue repair. *J Dent Res* 87:806–16.

Tsonis, P. A. 1983. Effects of carcinogens on regenerating and non-regenerating limbs in amphibia (review). *Anticancer Res* 3:195–202.

Tsonis, P. A. 1987. Embryogenesis and carcinogenesis: Order and disorder. *Anticancer Res* 7:617–23.

Tsutsui, H., Karasawa, S., Okamura, Y., Miyawaki, A. 2008. Improving membrane voltage measurements using FRET with new fluorescent proteins. *Nat Methods* 5:683–5.

Tyner, K. M., Kopelman, R., Philbert, M. A. 2007. "Nanosized voltmeter" enables cellular-wide electric field mapping. *Biophys J* 93:1163–74.

Uzman, J. A., Patil, S., Uzgare, A. R., Sater, A. K. 1998. The role of intracellular alkalinization in the establishment of anterior neural fate in *Xenopus*. *Dev Biol* 193:10–20.

Valenzuela, S. M., Mazzanti, M., Tonini, R., Qiu, M. R., Warton, K., Musgrove, E. A., Campbell, T. J., Breit, S. N. 2000. The nuclear chloride ion channel NCC27 is involved in regulation of the cell cycle. *J Physiol* 529(Pt 3):541–52.

van Lunteren, E., Sankey, C., Moyer, M., Snajdar, R. M. 2002. Role of K+ channels in L-6 myoblast migration. *J Muscle Res Cell Motil* 23:197–204.

van West, P., Morris, B. M., Reid, B., Appiah, A. A., Osborne, M. C., Campbell, T. A., Shepherd, S. J. 2002. Oomycete plant pathogens use electric fields to target roots. *Mol Plant Microbe Interact* 15:790–98.

Viczian, A. S., Solessio, E. C., Lyou, Y., Zuber, M. E. 2009. Generation of functional eyes from pluripotent cells. *PLoS Biol* 7:e1000174.

Waddington, C. H. 1935. Cancer and the theory of organisers. *Nature* 135:606–8.

Wallace, R. 2007. Neural membrane microdomains as computational systems: Toward molecular modeling in the study of neural disease. *Biosystems* 87:20–30.

Walters, Z. S., Haworth, K. E., Latinkic, B. V. 2009. NKCC1 (SLC12a2) induces a secondary axis in *Xenopus laevis* embryos independently of its co-transporter function. *J Physiol* 587:521–29.

Wang, E., Zhao, M., Forrester, J. V., CD, M. C. 2000. Re-orientation and faster, directed migration of lens epithelial cells in a physiological electric field. *Exp Eye Res* 71:91–98.

Wang, H., Peca, J., Matsuzaki, M., Matsuzaki, K., Noguchi, J., Qiu, L., Wang, D., Zhang, F., Boyden, E., Deisseroth, K., Kasai, H., Hall, W. C., Feng, G., Augustine, G. J. 2007. High-speed mapping of synaptic connectivity using photostimulation in Channelrhodopsin-2 transgenic mice. *Proc Natl Acad Sci USA* 104:8143–48.

Wang, H., Zhang, Y., Cao, L., Han, H., Wang, J., Yang, B., Nattel, S., Wang, Z. 2002. HERG K+ channel, a regulator of tumor cell apoptosis and proliferation. *Cancer Res* 62:4843–48.

Wang, K., Xue, T., Tsang, S. Y., Van Huizen, R., Wong, C. W., Lai, K. W., Ye, Z., Cheng, L., Au, K. W., Zhang, J., Li, G. R., Lau, C. P., Tse, H. F., Li, R. A. 2005. Electrophysiological properties of pluripotent human and mouse embryonic stem cells. *Stem Cells* 23:1526–34.

Wang, L., Zhou, P., Craig, R. W., Lu, L. 1999. Protection from cell death by mcl-1 is mediated by membrane hyperpolarization induced by K(+) channel activation. *J Membr Biol* 172:113–20.

Wang, Z. 2004. Roles of K+ channels in regulating tumour cell proliferation and apoptosis. *Pflugers Arch* 448:274–86.

Weiss, P. A. 1939. *Principles of development; a text in experimental embryology.* New York: H. Holt and Company.

Wenisch, S., Trinkaus, K., Hild, A., Hose, D., Heiss, C., Alt, V., Klisch, C., Meissl, H., Schnettler, R. 2006. Immunochemical, ultrastructural and electrophysiological investigations of bone-derived stem cells in the course of neuronal differentiation. *Bone* 38:911–21.

White, R. M., Zon, L. I. 2008. Melanocytes in development, regeneration, and cancer. *Cell Stem Cell* 3:242–52.

Wolff, C., Fuks, B., Chatelain, P. 2003. Comparative study of membrane potential-sensitive fluorescent probes and their use in ion channel screening assays. *J Biomol Screen* 8:533–43.

Wolsky, A. 1978. Regeneration and cancer. *Growth* 42:425–26.

Wonderlin, W. F., Strobl, J. S. 1996. Potassium channels, proliferation and G1 progression. *J Membr Biol* 154:91–107.

Woodruff, R., Kulp, J., LaGaccia, E. 1988. Electrically mediated protein movement in *Drosophila* follicles. *Roux's Arch Dev Biol* 197:231–38.

Woodruff, R., Telfer, W. 1980. Electrophoresis of proteins in intercellular bridges. *Nature* 286:84–6.

Woodruff, R. I. 2005. Calmodulin transit via gap junctions is reduced in the absence of an electric field. *J Insect Physiol* 51:843–52.

Wyart, C., Del Bene, F., Warp, E., Scott, E. K., Trauner, D., Baier, H., Isacoff, E. Y. 2009. Optogenetic dissection of a behavioural module in the vertebrate spinal cord. *Nature* 461:407–10.

Yan, X., Han, J., Zhang, Z., Wang, J., Cheng, Q., Gao, K., Ni, Y., Wang, Y. 2009. Lung cancer A549 cells migrate directionally in DC electric fields with polarized and activated EGFRs. *Bioelectromagnetics* 30:29–35.

Yang, S. J., Liang, H. L., Ning, G., Wong-Riley, M. T. 2004. Ultrastructural study of depolarization-induced translocation of NRF-2 transcription factor in cultured rat visual cortical neurons. *Eur J Neurosci* 19:1153–62.

Yasuda, T., Bartlett, P. F., Adams, D. J. 2008. K(ir) and K(v) channels regulate electrical properties and proliferation of adult neural precursor cells. *Mol Cell Neurosci* 37:284–97.

Yntema, C. L. 1959. Blastema formation in sparsely innervated and aneurogenic forelimbs of amblystoma larvae. *J Exp Zool* 142:423–39.

Yoshida, H., Kaneko, K. 2009. Unified description of regeneration by coupled dynamical systems theory: Intercalary/segmented regeneration in insect legs. *Dev Dyn* 238:1974–83.

Yu, K., Ruan, D. Y., Ge, S. Y. 2002. Three electrophysiological phenotypes of cultured human umbilical vein endothelial cells. *Gen Physiol Biophys* 21:315–26.

Yun, Z., Zhengtao, D., Jiachang, Y., Fangqiong, T., Qun, W. 2007. Using cadmium telluride quantum dots as a proton flux sensor and applying to detect H9 avian influenza virus. *Anal Biochem* 364:122–27.

Zallen, J. A. 2007. Planar polarity and tissue morphogenesis. *Cell* 129:1051–63.

Zhan, L., Rosenberg, A., Bergami, K. C., Yu, M., Xuan, Z., Jaffe, A. B., Allred, C., Muthuswamy, S. K. 2008. Deregulation of scribble promotes mammary tumorigenesis and reveals a role for cell polarity in carcinoma. *Cell* 135:865–78.

Zhang, F., Wang, L. P., Brauner, M., Liewald, J. F., Kay, K., Watzke, N., Wood, P. G., Bamberg, E., Nagel, G., Gottschalk, A., Deisseroth, K. 2007. Multimodal fast optical interrogation of neural circuitry. *Nature* 446:633–39.

Zhang, W., Kone, B. C. 2002. NF-kappaB inhibits transcription of the H(+)-K(+)-ATPase alpha(2)-subunit gene: Role of histone deacetylases. *Am J Physiol Renal Physiol* 283:F904–11.

Zhang, Y., Levin, M. 2009. Particle tracking model of electrophoretic morphogen movement reveals stochastic dynamics of embryonic gradient. *Dev Dyn* 238:1923–35.

Zhao, M. 2009. Electrical fields in wound healing—An overriding signal that directs cell migration. *Semin Cell Dev Biol* 20:674–82.

Zhao, M., Agius-Fernandez, A., Forrester, J. V., McCaig, C. D. 1996. Orientation and directed migration of cultured corneal epithelial cells in small electric fields are serum dependent. *J Cell Sci* 109:1405–14.

Zhao, M., Dick, A., Forrester, J. V., McCaig, C. D. 1999a. Electric field-directed cell motility involves up-regulated expression and asymmetric redistribution of the epidermal growth factor receptors and is enhanced by fibronectin and laminin. *Mol Biol Cell* 10:1259–76.

Zhao, M., Forrester, J. V., McCaig, C. D. 1999b. A small, physiological electric field orients cell division. *Proc Natl Acad Sci USA* 96:4942–46.

Zhao, M., McCaig, C. D., Agius-Fernandez, A., Forrester, J. V., Araki-Sasaki, K. 1997. Human corneal epithelial cells reorient and migrate cathodally in a small applied electric field. *Curr Eye Res* 16:973–84.

Zhao, M., Song, B., Pu, J., Wada, T., Reid, B., Tai, G., Wang, F., Guo, A., Walczysko, P., Gu, Y., Sasaki, T., Suzuki, A., Forrester, J. V., Bourne, H. R., Devreotes, P. N., McCaig, C. D., Penninger, J. M. 2006. Electrical signals control wound healing through phosphatidylinositol-3-OH kinase-gamma and PTEN. *Nature* 442:457–60.

Zhao, S., Cunha, C., Zhang, F., Liu, Q., Gloss, B., Deisseroth, K., Augustine, G. J., Feng, G. 2008. Improved expression of halorhodopsin for light-induced silencing of neuronal activity. *Brain Cell Biol* 36:141–54.

Zimmermann, M. R., Maischak, H., Mithofer, A., Boland, W., Felle, H. H. 2009. System potentials, a novel electrical long distance apoplastic signal in plants, induced by wounding. *Plant Physiol* 149(3):1593–600.

Zupanc, G. K. 2006. Neurogenesis and neuronal regeneration in the adult fish brain. *J Comp Physiol A Neuroethol Sens Neural Behav Physiol* 192:649–70.

Zupanc, M. M., Wellbrock, U. M., Zupanc, G. K. 2006. Proteome analysis identifies novel protein candidates involved in regeneration of the cerebellum of teleost fish. *Proteomics* 6:677–96.

4 Stem Cell Physiological Responses to Noninvasive Electrical Stimulation

Igor Titushkin, Shan Sun, Vidya Rao,
and Michael Cho
Department of Bioengineering
University of Illinois
Chicago, Illinois

CONTENTS

INTRODUCTION: STEM CELL TYPES

Stem cells are distinct unspecialized cells with unique features found in most multicellular organisms. They are characterized by the ability to renew themselves and to differentiate into a diverse range of specialized cell types (Cedar 2006; Moore and Lemischka 2006). The most known types of mammalian stem cells are embryonic stem cells, adult stem cells, and induced pluripotent stem cells (Jaishankar and Vrana 2009; Kao et al. 2008). Embryonic stem cells (ESCs), isolated from the inner cell mass of blastocysts, are pluripotent, meaning they can become virtually any cell in the body, and can conceivably be grown and differentiated into replacement cells for therapeutic need (Grivennikov 2008; Jensen 2009). However, the seemingly unlimited therapeutic potential associated with human ESCs is tempered with safety (e.g., teratoma formation), legal, and ethical issues (Mertes and Pennings 2009).

In contrast, adult or somatic stem cells exist throughout the body after embryonic development and are found in different types of tissues, such as the brain, bone marrow, blood, blood vessels, skeletal muscles, fat, skin, and liver (Bernardo et al. 2009; Mizuno 2009). They remain in a quiescent or nondividing state for years until activated by disease or tissue injury. Adult stem cells can renew themselves and can differentiate to yield the major tissue or organ-specific cell types with special functions (Kuci et al. 2009). The recent advances in human adult stem cell research demonstrate that these pluripotent cells are capable of differentiating into both excitable and nonexcitable tissue phenotypes (Chamberlain et al. 2007; Tropel et al. 2006). In many tissues these cells serve as an internal repair system, dividing essentially without limit to replenish other cells as long as the organism is still alive.

Induced pluripotent stem cells (iPSCs) are adult specialized cells that have been genetically reprogrammed to an embryonic stem cell-like state by being forced to express genes and factors important for maintaining the defining properties of embryonic stem cells (Colman and Dreesen 2009). Using genetic reprogramming with protein transcription factors, pluripotent stem cells equivalent to embryonic stem cells have been derived, for example, from human adult skin and other tissues (Romano 2008). Although iPSCs' impressive differentiation potential is comparable to embryonic stem cell potency, induced pluripotent stem cells do not rely on embryos, and therefore may avoid many ethical concerns (Hochedlinger and Plath 2009). iPSCs and adult stem cells are currently considered the best means to create personalized cells for regenerative medicine because the resulting cells are highly compatible with the person's immune system (Le Blanc et al. 2003).

Stem cell-based therapeutic applications in regenerative medicine have the potential to dramatically change the treatment of many human diseases. A number of adult stem cell therapies already exist, particularly bone marrow transplants that are used to treat leukemia (Gahrton and Bjorkstrand 2000). In the future, stem cell-derived technologies are expected to be used in the treatment of cancer, Parkinson's disease, spinal cord injuries, multiple sclerosis, and muscle damage, and a wide variety of other diseases (Goldman and Windrem 2006; Nikolic et al. 2009; Sensebe and Bourin 2009). However, realization of the exciting potential for stem cell-based biomedical and therapeutic applications requires a careful characterization of their unique biological and biophysical properties, as well as cell-cell and cell-environment interactions. For example, recent efforts have been focused on the manipulation of stem cells so that they possess the necessary characteristics for successful differentiation, transplantation, and engraftment using inductive soluble factors, designing a suitable mechanical environment, and by applying noninvasive physical forces (e.g., mechanical stress, ultrasound, or electrical stimulation) (Genovese et al. 2009; Kurpinski and Li 2007; Stolberg and McCloskey 2009).

CONTROL OF STEM CELL BEHAVIOR BY ELECTRICAL STIMULATION

In particular, control of the behavior and the intended fate of stem cells might be feasible through manipulation of the cellular properties by noninvasive electric fields. Indeed, external electrical stimulation has been shown to induce a variety of cellular and molecular responses, including, to name just a few, microfilament reorganization (Cho et al. 1996; Li and Kolega 2002), cell surface receptor redistribution (Cho et al. 1994; Zhao et al. 1999), changes in intracellular calcium dynamics (Cho et al. 1999; Khatib et al. 2004; Titushkin 2004), galvanotropic cell migration and orientation (Chao et al. 2007; Wang, Zhao, et al. 2003), enhanced differentiation (Sauer et al. 1999; Sun et al. 2007), proliferation (Chang et al. 2004), protein biosynthesis (MacGinitie et al. 1994), and angiogenesis (Bai et al. 2004; Zhao et al. 2004). These results are of particular interest for the emerging field of tissue engineering that is likely to revolutionize the current medical care approaches and improve quality of life for millions of people by restoring, maintaining, or enhancing tissue and organ function (Fodor 2003; Park et al. 2008). Moreover, stem cell-based tissue engineering is expected to play an increasingly important role in both therapeutic and diagnostic applications (Grayson et al. 2009; Salgado et al. 2006; Warren et al. 2004). Advances in electric field utilization for control of both adult and embryonic stem cell behavior and fate will certainly lead to enhanced cell differentiation techniques for tissue engineering. A successful artificial tissue assembly from its component cells using dielectrophoresis and other electrokinetic techniques has been recently demonstrated

(Markx 2008). Physical factors such as both mechanical and electrical stimulation have been applied separately and combined with soluble factors to facilitate stem cell lineage specification and commitment to particular phenotypes (Altman et al. 2002; Datta et al. 2006; Sun et al. 2007). Finally, in addition to demonstrating possible synergistic stem cell differentiation in response to a combination of soluble factors and externally applied electrical stimuli, elucidation of the potential coupling mechanisms is required. In this review, we will present some interesting electrically induced biochemical and biomechanical responses in stem cells (e.g., altered intracellular calcium dynamics, modulation of cell cytoskeleton and plasma membrane mechanics). This electrically assisted regulation of distinctive stem cell properties has important implications for stem cell differentiation, scaffold integration, and other potential biomedical applications. Besides, the diversity of bioelectric responses induced by the whole electromagnetic spectrum has many intriguing ramifications for mode-dependent electrical control of stem cell behaviors (Haddad et al. 2007; Kloth 2005). Further elucidation of the electro-coupling mechanisms is expected to establish a rational paradigm for electrically mediated control of stem cell physiology.

CALCIUM DYNAMICS AND STEM CELLS DIFFERENTIATION REGULATION BY ELECTRICAL STIMULATION

CALCIUM DYNAMICS

Calcium (Ca^{2+}) is a versatile second messenger that participates in multiple cell signaling cascades. The concentration of cytosolic free calcium ($[Ca^{2+}]_i$) regulates important cellular and molecular processes such as proliferation (Dave and Bordey 2009), differentiation (Resende et al. 2008), cytoskeletal reorganization (Cho et al. 1996), gene expression (Wu et al. 2006), and metabolism (Brough et al. 2005). Transient elevations in cytosolic calcium concentration, often referred to as spikes, are a nearly universal mode of signaling in both excitable and non-excitable cells (Thul et al. 2008). While the role of oscillatory Ca^{2+} signals is not fully understood, it is evident that spatial and temporal patterns of Ca^{2+} dynamics (e.g., spiking amplitude and frequency, spatial distribution) are important characteristics of cellular regulatory pathways. Understanding the molecular information embedded in the $[Ca^{2+}]_i$ oscillations is expected to lead to elucidation of the complex intracellular signaling mechanisms involved in the regulation of cellular and molecular interactions (Putney and Bird 2008). Cells may recognize the $[Ca^{2+}]_i$ oscillations through sophisticated mechanisms to use the information encoded in the Ca^{2+} dynamics. For example, whereas rapid and localized changes of $[Ca^{2+}]_i$ (i.e., calcium spikes) regulate fast subcellular and molecular responses, intercellular or intracellular $[Ca^{2+}]_i$ elevation propagations (i.e., calcium waves) control slower responses (Berridge et al. 2003). In addition, the frequency of the Ca^{2+} oscillations is known to reflect the strength of the extracellular stimulus (Berridge 2007). For example, the calcium-binding proteins such as troponin C in skeletal muscle cells and calmodulin in all eukaryotic cells may serve as transducers of the Ca^{2+} signals by changing their activity as a function of the $[Ca^{2+}]_i$ spiking frequency (Chawla 2002). Such frequency-modulated Ca^{2+} signaling mechanisms determine the qualitative and quantitative nature of the genomic response, and can be translated to frequency-dependent cell responses such as proliferation and differentiation (Schreiber 2005).

Calcium oscillation also plays a key role in human mesenchymal stem cells (hMSCs) homeostasis (Foreman et al. 2006). Extracted from adult mesenchymal tissues such as bone marrow, muscle, synovium, and adipose tissue, hMSCs demonstrate a great therapeutic potential due to their multiple differentiation capabilities (Barry and Murphy 2004; Bernardo et al. 2009; Short et al. 2003). These cells have long been shown to differentiate into different phenotypic lineages in the presence of inductive biochemical factors or extracellular matrix proteins such as collagen or vitronectin (Salasznyk et al. 2004). Physical forces, including electrical stimulation, and substrate stiffness have also been shown to alter the Ca^{2+} dynamics (Kim et al. 2009; Titushkin 2004), and to promote stem cell differentiation (Engler 2006; Sauer 1999). However, the role of the Ca^{2+} activities for hMSC differentiation has not been clearly established. While hMSCs are known to exhibit robust spontaneous intracellular calcium oscillations (Foreman et al. 2006), the $[Ca^{2+}]_i$ spiking in hMSCs no doubt represents a complex dynamical process that reflects Ca^{2+} transportation to and from the exterior of the cell, cytosol, and intracellular stores, exchange between cells, or diffusion and buffering due to the binding of Ca^{2+} to proteins (Berridge 2007).

THE REGULATION OF CA^{2+} OSCILLATIONS

The $[Ca^{2+}]_i$ oscillations are likely mediated by at least several cell type-dependent Ca^{2+} influx/efflux pathways. First, Ca^{2+} entry across the plasma membrane is possible via voltage-operated Ca^{2+} channels (VGCCs), stretch-activated cation channels, and agonist-dependent channels (Mohan and Gandhi 2008). Several pharmacologically distinguishable VGCCs operated by the plasma membrane electric potential have been identified in excitable and nonexcitable cell types (e.g., high-voltage-activated L-type, low-voltage-activated T-type, and nervous-tissue-specific N-type channels (Adams and Snutch 2007; Mohan and Gandhi 2008; Zamponi et al. 2009)). Second, Ca^{2+} release from the intracellular stores (e.g., endoplasmic and sarcoplasmic reticulum, mitochondria) is controlled by inositol 1,4,5-triphosphate IP_3-gated channels and ryanodine receptors (Hamilton and Serysheva 2009; Hanson et al. 2004). It is generally thought that the endoplasmic reticulum represents the main Ca^{2+} store in the cell, and calcium release from the endoplasmic reticulum plays a predominant role in generating sustained $[Ca^{2+}]_i$ oscillations by IP_3-gated channels and the ryanodine receptors (Berridge 2007). IP_3 is known to be generated as a result of phosphatidylinositol 4,5-bisphosphate (PIP_2) hydrolysis by phospholipase C (PLC), a G-protein-associated enzyme. Among many other cell types, the calcium dynamics in mesenchymal stem cells appear to have a stronger dependence on the IP_3-gated channels (Kawano et al. 2002). Third, the excess Ca^{2+} is pumped back from cytosol into internal stores or extruded to extracellular medium by specific Ca^{2+}-ATPase pumps (Hovnanian 2007). Fourth, the important role of autocrine and paracrine interactions in the regulation of Ca^{2+} oscillations was recently established. In this signaling pathway the ATP secreted via hemi-gap-junction channels leads to stimulation of metabotropic P2Y purinergic receptors, which in turn activate PLC enzyme to produce IP_3 (Kawano et al. 2002, 2006). Alternately, ATP can induce Ca^{2+} influx by activation of ionotropic P2X receptors that are ATP-dependent, ligand-gated channels permeable to sodium, potassium, and calcium (Coppi et al. 2007).

The biochemical cascades for tissue-specific gene activation may be associated with the altered Ca^{2+} dynamics (Sinner et al. 2006). Thus, alterations in the intracellular Ca^{2+} activities have recently been shown to induce embryonic stem cells to differentiate to neuronal phenotypes (Ulrich and Majumder 2006). We also found that typical temporal $[Ca^{2+}]_i$ oscillation profiles are distinctively different in hMSCs that are biochemically induced into neurogenic or osteogenic phenotypes (Figure 4.1). For example, addition of osteoinductive soluble factors significantly decreases the spiking activity in hMSCs. In contrast, cells undergoing neurodifferentiation exhibit a pattern that consists of numerous $[Ca^{2+}]_i$ spikes of both small and large amplitude. This type of pattern is unique to the neurogenic cells and not found in any other cell types we have tested thus far, including undifferentiated human or rat mesenchymal stem cells, osteoblasts, fibroblasts, and even primary myocytes. In an independent study, Kawano et al. (2006) found that as the hMSCs differentiated into adipocytes, the spontaneous cytosolic

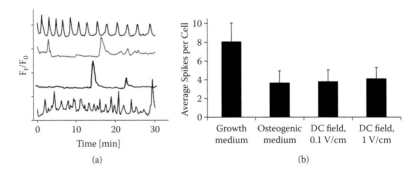

(a)   (b)

FIGURE 4.1 Effects of electrical stimulation and soluble induction factors on calcium oscillations in human mesenchymal stem cells. (A) Typical regular calcium spiking in undifferentiated hMSCs (top trace) is suppressed during osteodifferentiation by soluble factors (second trace), or a dc electric field exposure (1 V/cm; third trace) to a level normally found in fully differentiated osteoblasts. For comparison, hMSCs undergoing neurodifferentiation display yet another distinct pattern of calcium oscillations (bottom trace) with multiple $[Ca^{2+}]_i$ spikes whose amplitudes vary substantially. (B) Average number of $[Ca^{2+}]_i$ spikes per cell observed within thirty minutes in undifferentiated stem cells is decreased significantly ($p < 0.05$) by a dc electrical stimulation and the soluble factors in osteogenic medium.

$[Ca^{2+}]_i$ oscillations disappeared. These findings provide evidence that the altered Ca^{2+} dynamics may be unique to the tissue type, and perhaps be used as an early indicator to predict hMSC differentiation.

CONTROL OF CA^{2+} DYNAMICS BY ELECTRICAL STIMULATION

REGULATION OF DIFFERENTIATION

While molecular and cellular interactions, gene expression profiles, and cell differentiation can be linked to the altered intracellular Ca^{2+} dynamics, intracellular calcium activities themselves can be regulated by a variety of external physical stimuli, including electrical exposure (Cho et al. 2002; Titushkin et al. 2004). This is an interesting observation, because electrical stimulation was recently found to regulate cell differentiation, with the calcium ion apparently playing a significant role in mediating electric field effects (Tonini et al. 2001; Tsai et al. 2009). For example, pulsed electrical stimulation has been shown to strongly influence embryonal stem cells to assume a neuronal fate (Yamada et al. 2007). The electrical induction of calcium ion influx played an important role in this differentiation system. In another study, electrical stimulation of embryoid bodies derived from pluripotent embryonic stem cells was found to affect myogenic differentiation (Sauer et al. 1999). Significant effects of a single electric pulse applied for 90 s on cardiomyocyte differentiation were achieved with the field strengths of 2.5 and 5 V/cm, which increased both the number of cardiomyocytes beating foci and the size of the beating foci (Sauer et al. 1999). In yet another report by Lohmann et al. (2000), a short electromagnetic stimulus repeated at 15 Hz for eight hours per day caused a reduction in osteoblast-like cell proliferation, but increased cellular alkaline phosphatase-specific activity, osteocalcin synthesis, and collagen production indicative of osteogenic commitment (Schwartz et al. 2009). Enhanced osteogenic differentiation as the net effect of this electromagnetic signal appears to be mediated by the changes in calcium-dependent local factor production, such as prostaglandin E$_2$ and TGF-β1 (Ciombor et al. 2002; Zhuang et al. 1997). Therefore, $[Ca^{2+}]_i$ oscillations seem to play a bidirectional role to serve as an indicator of cell type-specific commitment and to mediate stem cell differentiation.

REGULATION OF CA^{2+} OSCILLATIONS

Our laboratory has recently demonstrated that $[Ca^{2+}]_i$ oscillations in hMSCs are dramatically altered in response to the osteoinductive soluble factors and a direct current (dc) electrical stimulation (Sun et al. 2007). In response to osteoinductive factors, $[Ca^{2+}]_i$ oscillation frequency decreases to a level similar to that found in the terminally differentiated human osteoblasts (Figure 4.1). Similarly, a 1 V/cm dc electrical exposure also alters the $[Ca^{2+}]_i$ oscillation pattern in undifferentiated hMSC so as to resemble the typical pattern found in mature osteoblasts. In fact, application of either a 0.1 or 1 V/cm dc stimulus or a 1 V/cm, 1 Hz sinusoidal oscillatory stimulus induces the similar effect by reducing the $[Ca^{2+}]_i$ spikes in undifferentiated hMSCs. When exposed to a noninvasive electrical stimulation in the presence of the osteoinductive factors, the $[Ca^{2+}]_i$ spiking is further suppressed. After a ten-day culture in osteoinductive medium with daily electrical exposure, the expression levels of several specific osteogenic molecular markers were significantly higher than when using the osteoinductive factors alone. In other words, a combination of osteogenic soluble factors with an electrical stimulation synergistically increases differentiation of hMSCs into osteoblasts (Sun et al. 2007). However, application of an electrical stimulation by itself fails to induce hMSC differentiation, suggesting that an externally applied physical cue may amplify but not necessarily activate the mechanisms involved in osteogenic commitment. It remains to be determined, however, how an electrical stimulation can influence the $[Ca^{2+}]_i$ oscillations and facilitate osteodifferentiation in hMSCs. The potential mechanisms that could mediate the altered Ca^{2+} dynamics in hMSCs are presented in Figure 4.2. The inductive soluble factors can diffuse across the cell membrane and produce immediate effects in the $[Ca^{2+}]$ oscillation, as we have demonstrated earlier.

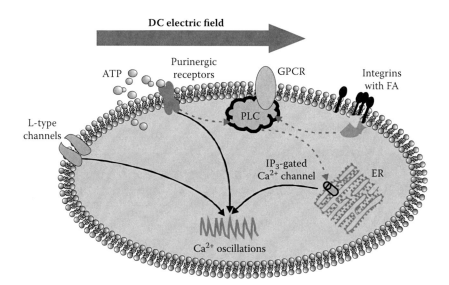

FIGURE 4.2 See color insert. Potential electrocoupling mechanisms of hMSC calcium dynamics modulated by electrical stimulation. The coupling mechanisms of dc and low-frequency oscillatory fields are likely confined to the cell surface, and might include multiple types of calcium channels (voltage-gated calcium channels (VGCCs) and stretch-activated cation channels (SACCs)), membrane receptors (e.g., G-protein-coupled receptors), and integrins. As a result, an electrical stimulus leads to the PLC-mediated signaling through the IP_3 pathway, intracellular ATP secretion, and ATP-sensitive purinergic receptor activation. All these pathways likely contribute to $[Ca^{2+}]_i$ oscillation modulation by an external electric field.

THE ROLE OF ION CHANNELS

Electrical stimulation can couple to the cell via at least several mechanisms and alter the Ca^{2+} dynamics. First, because the Ca^{2+} channels are known to sustain $[Ca^{2+}]_i$ oscillations, hMSCs express these channels. Electrically operated voltage-gated calcium channels (VGCCs) are excellent candidates for mediating Ca^{2+} influx across the cell membrane induced by endogenous electric fields (Cho 2002). Indeed, functional L-type channels can be identified in undifferentiated hMSCs (Heubach et al. 2004). Treatment of cells with verapamil (an L-type VGCC inhibitor) or depletion of extracellular Ca^{2+} decreased the $[Ca^{2+}]_i$ oscillations to a level comparable to that found in terminally differentiated osteoblasts (Sun et al. 2007). These findings led us to formulate a postulate that extracellular Ca^{2+} is required to sustain the $[Ca^{2+}]_i$ spiking via the L-type Ca^{2+} channels. As hMSCs undergo osteodifferentiation, the $[Ca^{2+}]_i$ oscillations are decreased. Therefore, the role of L-type Ca^{2+} channels would be expected to diminish, suggesting that these channels may be downregulated. It is plausible to assume that the membrane potential is altered in response to electrical stimulation, and that the Ca^{2+} fluxes are induced through voltage-gated calcium channels. But it is not clear how cells can sense and respond to an electrical stimulus as small as 0.1 V/cm (Robinson 1985), which cannot activate VGCCs. For example, a cell ~50 μm in diameter exposed to a 0.1 V/cm dc field would experience a ~0.25 mV change in the membrane potential, which is insufficient to cause VGCC activation. Using a membrane-potential-sensitive dye, $DiBAC_4(3)$, we verified that application of an electrical stimulation as large as 2 V/cm in magnitude does not significantly alter the membrane potential. In contrast, a large invasive stimulus (e.g., 10 V/cm) depolarizes the plasma membrane, and the membrane potential is not fully restored upon removal of electrical stimulation. Interestingly, the treatment of hMSCs with Li^{3+} has been found to block the $[Ca^{2+}]_i$ oscillations (Kawano et al. 2003). Since Li^{3+} is known to inhibit the stretch-activated cation channels (SACCs) (Savio-Galimberti and Ponce-Hornos 2006), it appears plausible that SACCs are required to sustain the $[Ca^{2+}]_i$ oscillations in the undifferentiated hMSCs and to mediate electrically altered Ca^{2+} activities. SACCs can respond to electrically mediated cell morphological changes by the influx of cations, including Ca^{2+}, as has been shown for several other cell types (Cho et al. 1999; Khatib et al. 2004). It must be noted that SACCs are not Ca^{2+} specific and allow other cations, including Na^+, which will further depolarize the cell membrane.

THE ROLE OF PHOSPHOLIPASE C

Second, we have shown that when the phospholipase C activity is inhibited, the $[Ca^{2+}]_i$ oscillations are essentially abolished in hMSCs. This leads to another hypothesis that, as hMSCs undergo osteodifferentiation, the PLC activity is reduced. Concomitant reduction of L-type channels and PLC activity may be coordinated to suppress $[Ca^{2+}]_i$ oscillations in hMSCs undergoing osteodifferentiation.

One plausible molecular mechanism mediating the altered $[Ca^{2+}]_i$ spiking pattern in response to noninvasive electrical stimulation also involves PLC (Figure 4.2). A PLC-coupling mechanism was proposed that is initiated by electrically induced activation of G-protein-coupled receptors (Cho 2002; Khatib et al. 2004). The PLC-mediated signaling through the release of internal Ca^{2+} and protein kinase C activation can potentially couple to the mitogen kinase protein (MAP) kinase cascades, which are known to be involved in the cell differentiation (Luttrell 2002; Takeda et al. 2004).

THE ROLE OF INTEGRINS

Third, integrins have been shown to regulate the Ca^{2+} dynamics, which in turn controls integrin-mediated adhesion (Sjaastad and Nelson 1997), and cell migration through phosphorylation of focal adhesion kinase (FAK) (Giannone et al. 2002). This suggests a strong correlation between differential focal adhesions formed in response to electrical stimulation and electrically altered Ca^{2+} dynamics. As we will show in the next section, electrical exposure causes significant reorganization of actin microfilament bundles (e.g., stress fibers) closely associated with abundant focal adhesion contacts in hMSCs. We note that, in response to a 0.1 V/cm electrical stimulation, the maximal change in the $[Ca^{2+}]_i$ is expected to require an eighty-minute exposure (Khatib et al. 2004), which may be long enough to induce cellular responses via mechanisms other than direct changes in $[Ca^{2+}]_i$, such as cell surface receptor redistribution (e.g., electroosmosis). For example, an electrical stimulation has been shown to induce integrin redistribution, clustering, and activation of some focal adhesion proteins, such as focal adhesion kinase, paxillin, vinculin, and src (Cho 2002). Enzymatic activity of these proteins could then promote MAP kinase activation (Miranti and Brugge 2002). Such involvement of adhesion proteins may mimic the integrin-matrix interactions, which are known to regulate the MAP kinase signaling cascades. MAP kinase regulation by PLC or focal adhesions-mediated pathways alters Ca^{2+} dynamics and facilitates a synergistic hMSC differentiation. Additional studies are under way to elucidate the involvement of the MAP kinases in the electrically mediated cell differentiation.

THE ROLE OF ATP

Fourth, hMSCs were found to secrete ATP (Kawano et al. 2006), and these autocrine and paracrine signaling pathways modulate $[Ca^{2+}]_i$ oscillations. Through this mechanism, ATP also promotes activation and nuclear translocation of the transcription factor NFAT, suggesting a pivotal role of extracellular ATP in the hMSC differentiation. Besides, spontaneously released ATP was found to decrease hMSC proliferation rate in culture and to produce two different electrophysiological responses in the cells, including ionic conductance modulation (Coppi et al. 2007). Inward and outward ATP-sensitive currents appear to be Ca^{2+} dependent and mediated by P2X and P2Y purinergic receptor activation. Interestingly, we recently detected and characterized intracellular ATP depletion in response to an electrical stimulation (Titushkin and Cho 2009). This result indicates that an external electric field may alter the Ca^{2+} dynamics in part by regulating the ATP release pathways, which themselves can depend on the calcium influx, thus creating complex ATP-Ca^{2+} feedback electrocoupling mechanisms.

SUMMARY OF ELECTROCOUPLING MECHANISMS

An integrated electrocoupling model presented in Figure 4.2 is not intended to cover an exhaustive listing of possible electrocoupling mechanisms, which can depend on stem cell type and origin, modality of applied electrical stimulus (e.g., dc or oscillatory, frequency, intensity, duration, etc.), and the intended cell specialization (e.g., excitable or nonexcitable cell types). For example, in contrast to osteogenic commitment, neurogenic differentiation is expected to upregulate the N-type calcium channels' expression, which will play a major role in mediating electrically modulated Ca^{2+} activities. Unlike dc or low-frequency oscillatory fields, which are confined to the cell surface, high-frequency electromagnetic signals penetrate inside the cell and may directly couple to the intracellular processes. This would represent yet another degree of complexity, because the molecular mechanisms at the cell surface, in the cytoplasm, and in the nucleus now have to be determined and integrated. Therefore, it is clear that careful selection of the electric field parameters is required for efficient regulation of stem cell differentiation into a particular phenotype. Besides, electrical stimulation is known to affect multiple calcium-dependent and -independent subcellular processes simultaneously, including calcium dynamics, cellular mechanics, membrane receptor upregulation, local growth factors production, and other processes (Bai et al. 2004; Chang et al. 2004; Chao et al. 2007; Cho et al. 1994, 1996, 1999; Khatib et al. 2004; Li and Kolega 2002; MacGinitie et al. 1994; Sauer et al. 1999; Sun et al. 2007; Titushkin et al. 2004; Wang, Jia, et al. 2003; Zhao et al. 1999, 2004). The net effects of an electrical stimulation on stem cell differentiation will be a result of the intricate interplay between multiple possibly cooperative or competitive mechanisms that can eventually impact the stem cell fate. Although comprehensive characterization of the multifaceted effects of electric fields is still pending, the current results strongly suggest that, in addition to serving as an indicator of stem cell differentiation, the electrically altered Ca^{2+} dynamics is likely to be involved in mediating the stem cell commitment, and the calcium activity modulation by noninvasive electrical stimulation may provide a novel technique for synergistic stem cell differentiation into tissue-specific lineages.

CONTROL OF STEM CELL MECHANICS BY NONINVASIVE ELECTRIC FIELD

Biomechanics is known to play an important role in cell metabolism and behaviors (Huang et al. 2004). Cell physiology, tissue-specific functions, and fate critically depend on the cell mechanical properties and extracellular mechanical environment. In different tissues the cells adapt their biomechanics to the extra-cellular matrix and respond to numerous environmental mechanical cues. Cell mechanical characteristics, such as cytoskeleton elasticity, membrane tension, and adhesion strength, may play an important role in cell homeostasis and dif-ferentiation (McBeath et al. 2004; Meyers et al. 2005). For example, matching mechanical properties of differentiating stem cells to those of the tissue-specific descendant cells is crucial for the development of a functional load-bearing tissue. Moreover, manipulation of the cellular mechanics in tissue engineering applica-tions may increase the efficacy of cell differentiation, integration with scaffolds, and eventual tissue substitute maturation.

A wide spectrum of mechanical properties demonstrated by cells of meso-dermal origin (Kuznetsova et al. 2007; Van Vliet et al. 2003) opens an excit-ing perspective to regulate cell homeostasis by controlling its biomechanics using external physical stimuli. Indeed, specific types of mechanical stimula-tion (e.g., shear stress, cyclic stretching, and compressive loading) can be used for modulation of gene expression, matrix components secretion, cell differen-tiation, and ultimately engineering a functional load-bearing tissue (Janmey and McCulloch 2007; Shieh and Athanasiou 2003). Therefore, systematic investiga-tions of cell mechanical modulation by external physical forces (e.g., electric field) could allow targeted regulation of mechanically mediated cell functions (McCaig et al. 2005). Exogenous electric fields are known to induce a variety of cellular and molecular responses, which can be both cell type and stimulus mode specific. As discussed above, for example, electrically induced modulation of cal-cium activities appears to have profound impact on stem cell differentiation (Sun et al. 2007). In this section, we will describe the stem cell mechanical responses to an electrical exposure, including actin reorganization, membrane-cytoskeleton separation, and adhesion modulation. Careful characterization of electrically mediated mechanical responses is expected to lead to facilitated stem cell dif-ferentiation and other tissue engineering applications.

REGULATION OF CELL ELASTICITY BY ELECTRICAL STIMULATION

One of the most important cellular mechanical parameters is cell elasticity. Cell compliance is an essential characteristic, as cells in many load-bearing tissues must undergo multiple deformations without losing their integrity (Janmey and McCulloch 2007; Lim et al. 2006). Substantial structural and functional differ-ences between the cytoskeletons of mesoderm-derived cells (myocytes, osteo-blasts, endothelial cells, kidney cells, etc. (Kuznetsova et al. 2007)) imply that the cytoskeleton may also participate in embryogenesis and cell differentiation. Besides providing cell shape and stability, the cytoskeleton plays an impor-tant role in signaling pathways that regulate intracellular processes and protein expression in response to a changing biomechanical environment. The role of the cytoskeleton as a mechanotransducer has been long recognized and articulated in studies of cellular responses to changing substrate stiffness, cell shape, stretching, and shear stress (Ingber 2006). Several important cytoskeleton signaling pathways include the Rho family GTPases (e.g., RhoA, Cdc42, Rac) (Orr et al. 2006). Complex interactions between these molecular switches, triggered by external physical signals, cause an activation of numerous downstream effectors (e.g., ROCK). The resulting global structural rearrangement of the cytoskeleton itself or altered gene transcription profiles may affect cell adhesion, protein biosynthesis, and cell homeostasis (Wang et al. 2007). For example, McBeath et al. (2004) have recently demonstrated that stem cell lineage commitment is mediated by RhoA activity, specifically via its effect on ROCK-mediated cytoskeletal tension. The ability of RhoA-ROCK signaling to direct osteogenic vs. adipogenic hMSC differentiation requires nonmuscle myosin II activity, emphasizing the role of cytoskeletal tension and the contractile forces generated by the cell in mechanotransduction (Guilak et al. 2009; McBeath et al. 2004).

THE ACTIN CYTOSKELETON

The cell cytoskeleton is a dynamic structure capable of reorganization as required by the cell type, morphogenesis stage, and environmental conditions (Janmey et al. 2009). We have recently reported, for example, considerable mechanical and structural differences between actin cytoskeletons of human mesenchymal stem cells (hMSCs) and their default descendants—osteoblasts (Titushkin and Cho 2007). Whereas in hMSCs actins are organized as thick bundles (stress fibers) traversing the cell cytoplasm, in osteoblasts they are arranged as a thin, dense microfilament meshwork filling the cell interior (Figure 4.3). This difference is reflected in the cell elasticity: Young's modulus of hMSCs (3.2 ± 1.4 kPa) is almost twofold higher than that of osteoblasts (1.7 ± 1.0 kPa), as determined with the atomic force microscope (AFM) microindentation technique. hMSC stiffness is regulated by actin organization, with microtubules providing only a minor contribution to the cytoskeleton elasticity (Titushkin and Cho 2007). This result is consistent with other studies suggesting that microfilaments, but not microtubules or intermediate filaments, are mainly responsible for the elastic properties of cultured chondrocytes, myocytes, or fibroblasts (Collinsworth et al. 2002; Takai et al. 2005; Trickey et al. 2004). During biochemically induced hMSC osteogenic commitment, the actin cytoskeleton structure and mechanics are altered correspondingly. hMSC cytoskeleton elasticity decreases to a value similar to that found in mature osteoblasts. For example, after a ten-day culture in osteogenic medium, the elastic modulus of hMSCs is only ~20% higher than that of normal osteoblasts. Cellular mechanics modulation is associated with a remarkable actin cytoskeleton remodeling observed in hMSCs undergoing osteogenic differentiation (Rodriguez et al. 2004; Titushkin and Cho 2007). Following a ten-day induction with the osteogenic soluble factors, the thick actin stress fibers in hMSCs are gradually replaced with a network of thinner microfilaments typical for mature osteoblasts.

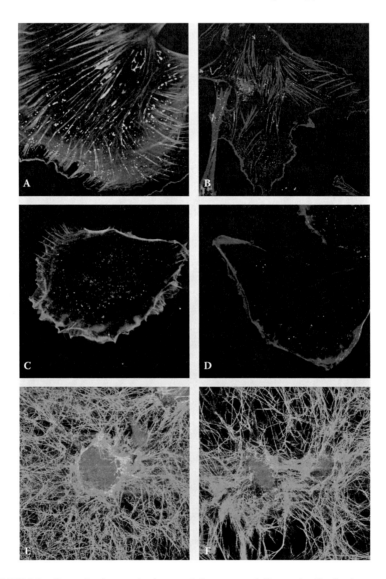

FIGURE 4.3   See color insert. Actin cytoskeleton remodeling and cell adhesion modulation by dc electric field exposure. (A–D) Immunofluorescent images of a thin filamentous actin meshwork (red) and vinculin (green) showed that hMSCs contain thick actin stress fibers and multiple large adhesion contacts (A). In contrast, mature osteoblasts showed fewer and smaller focal adhesions (C). Actins and focal adhesions were partially disassembled in both hMSCs (B) and osteoblasts (D) after exposure to a 2 V/cm dc field for 60 min in serum-free saline. (E,F) Stem cells in three-dimensional collagen gel (1 mg/ml) were visualized using second harmonic generation signals from collagen (green) and cell tracker dye fluorescence (red). Collagen fiber bundles are involved in tight stem cell adhesion in the three-dimensional scaffold (E), and only partially relax in response to a 10 V/cm electrical stimulus without significant cell reorientation (F).

The Effect of Electrical Stimulation on the Actin Cytoskeleton

Interestingly, stem cell elasticity also decreases during exposure to an external 2 V/cm dc electric field, likely due to substantial actin cytoskeleton reorganization (Figure 4.3; Titushkin and Cho 2009). As we have already mentioned earlier, direct current and low-frequency oscillating electric fields are unable to penetrate into the cell interior due to high resistivity of the cell membrane. Since lipid bylayer conductivity is about 10^6 to 10^8 times smaller than that of the cytoplasm, the plasma membrane can be treated as a nonconductive insulator to these fields (Poo 1981). As the direct dc field coupling to actins is excluded, molecular signaling pathways involved in the regulation of cell mechanics are likely initiated at the cell surface. Partial actin disassembly may be attributed to an electrically induced increase in the average intracellular calcium concentration. Changes in $[Ca^{2+}]_i$ can be mediated by a variety of well-characterized mechanisms, as discussed in detail in the previous section. Both Ca^{2+} influx from extracellular medium and Ca^{2+} release from intracellular stores can contribute to the modulation of Ca^{2+} activities in response to electrical exposure. The details of calcium-mediated actin dynamics depend on the original actin cytoskeleton structure. For example, stiff actin stress fibers in hMSCs seem to be relatively more susceptible to an external electric field than softer but apparently more stable actin meshwork in osteoblasts. As a result, the elasticity decreases more rapidly in stem cells than in osteoblasts during the same electric field exposure. This could occur because the differential expression profiles of actin-binding proteins in hMSCs and bone cells are responsible for distinct F-actin structural patterns. Interestingly, Yourek et al. (2007) recently reported that after osteogenic and chondrogenic differentiation, the hMSC cytoskeleton becomes more stable and resistant to actin-disrupting drugs (e.g., cytochalasin). In addition, the magnitude and dynamics of $[Ca^{2+}]_i$ elevation in response to electrical stimulation can be very different in these cell types. For example, robust $[Ca^{2+}]_i$ oscillations are evident in unstimulated hMSCs but not in osteoblasts (Sun et al. 2007).

The Combined Effect of Growth Factors and Electrical Stimulation

The effect of electrical stimulation on actin cytoskeleton becomes more complicated in the presence of growth factors, hormones, and cytokines and other signaling molecules. Unlike simple electrically induced actin disassembly in serum-free conditions, localized redistribution of polymeric actin takes place when the cell is exposed to a dc field in a medium containing 15% fetal bovine serum. We observed polymeric actin accumulation on the cathode side of the cell with concomitant actin depolymerization on the anode side (Titushkin and Cho 2009). This actin redistribution might indicate directional serum-dependent cell electromigration toward the cathode, which has been previously reported (Pullar and Isseroff 2005; Zhao et al. 1999). Many cell membrane receptors are redistributed in response to a dc electric field (Cho et al. 1994), and many growth factors (e.g., EGF, FGF, TGF-β1) may bind to appropriate receptors to trigger signaling pathways and to produce local changes in actin dynamics. The resulting cell galvanotaxis requires serum-derived growth factors, and involves asymmetric actin polymerization/depolymerization (Farboud et al. 2000; Li and Kolega 2002; Zhao et al. 1999). Generally, stem cells demonstrate less electromigration activity than osteoblasts or fibroblasts under the similar experiment conditions. This may be explained by the stronger hMSC adhesion to the substrate compared to osteoblasts, as evidenced by the higher number and increased size of focal adhesions in hMSCs (Figure 4.3). Since focal adhesions are closely associated with the stress fibers, the dynamics of reorganization of focal contacts is expected to follow the temporal patterns of actin remodeling during hMSC exposure to the electric field.

MEMBRANE-CYTOSKELETON COUPLING

Another important mechanical component of the cell is its plasma membrane. Separating and protecting the interior of the cell from the external environment, the membrane participates in the inward-outward trafficking, cell adhesion, motility, and cell-cell interactions, and is likely involved in cell differentiation (Jena 2007; Raucher and Sheetz 2000). Surface tension is the major membrane mechanical characteristic known to regulate many intracellular events, and multiple mechanisms are involved in its maintenance, including lipid material recycling, control of membrane rigidity (e.g., by cholesterol content), and membrane interaction with the underlying cytoskeleton (Morris and Homann 2001; Raucher and Sheetz 1999; Sun et al. 2007). The membrane is physically attached to the actin cytoskeleton at focal adhesion sites as well as by specific linker proteins such as spectrin, myosin-I, and ezrin, radixin, moesin (ERM) family proteins (Sheetz et al. 2006). ERM linker proteins are abundantly present in the cell cytoplasm of most eukaryotic cells and are the most likely candidates for membrane-cytoskeleton coupling (Fievet et al. 2007; Louvet-Vallee 2000; Mangeat et al. 1999). Upon phosphorylation these small (~80 kDa) molecules can bind to both F-actins and integral transmembrane proteins. Inhibition of the linker proteins by energy depletion was reported to result in membrane separation from the cytoskeleton and subsequent cell membrane blebbing (Chen and Wagner 2001). As we recently demonstrated, mesenchymal stem cells differ considerably in their membrane-cytoskeleton interactions from mature osteoblasts (Titushkin and Cho 2006). This is evidenced by the much longer optically extracted membrane tethers in hMSCs, compared to osteoblasts. Thick actin stress fibers in stem cells provide relatively few available binding sites for the ERM linkers. In contrast, a closely packed actin network in osteoblasts ensures a high surface density of binding sites for the ERM proteins, as has been immunofluorescently visualized (Titushkin and Cho 2009). The resulting membrane-cytoskeleton interaction is much stronger in osteoblasts than in hMSCs. In osteoblasts, either ATP depletion (leading to ERM dephosphorylation and inhibition) or actin microfilament depolymerization causes a significant increase in the membrane tether length, suggesting membrane separation from the cytoskeleton. In contrast, in hMSCs this treatment does not cause any further significant increase in the membrane tether length, primarily due to originally very weak membrane-cytoskeleton interaction in these cells (Titushkin and Cho 2006). Membrane-cytoskeleton coupling increases during hMSC osteogenic commitment in accordance with osteogenically mediated actin remodeling (Titushkin and Cho 2007). Functionally, a strong membrane-cytoskeleton adhesion should be beneficial to keep the structural integrity of osteoblasts subjected to continuous stress cycles. On the other hand, in hMSCs, a lower membrane tension may better facilitate endo- and exocytosis and contribute to a higher sensitivity of these cells to various soluble biochemical environmental stimuli.

THE ROLE OF ELECTRICAL STIMULATION IN MEMBRANE-CYTOSKELETON COUPLING

The electric field significantly modulates the mechanical characteristics of the plasma membrane. Electrical stimulation appears to mediate ATP depletion in hMSCs and osteoblasts, which leads to dephosphorylation and inhibition of ERM linker proteins (Chen and Wagner 2001; Titushkin and Cho 2009). Western blot experiments confirm a reduced level of ERM protein phosphorylation in both cell types following an electrical stimulation and biochemical ATP depletion. This electrically mediated decrease in the intracellular ATP level, inhibition of ERM proteins, and membrane separation from the cytoskeleton are especially obvious in mature osteoblasts, due to the originally stronger membrane-cytoskeleton interaction, compared to in hMSCs. Clearly, disruption of the actin cytoskeleton itself also contributes to the membrane dissociation from the cytoskeleton (Titushkin and Cho 2009).

The exact mechanism of ATP depletion in response to an electrical stimulation is not clear. First, a decrease in the intracellular ATP may be due to transiently intensive ATP consumption by the cellular biomolecular machinery (e.g., transmembrane ion pumps) in response to electric-field-mediated changes in cell metabolism. Second, electrically induced ATP release from cells has been reported (Sauer et al. 2002; Seegers et al. 2002). ATP can be released through exocytosis mechanisms (e.g., secretory vesicles), a specific ATP-transporting system such as anion channels, hemi-gap-junction channels, or even transient electroporetic membrane damage (Katsuragi et al. 1993; Wang et al. 1996). In fact, multiple feedback loops in the electric-field-induced cell biomechanical changes can be speculated. For example, both ATP-dependent P2X ligand-gated channels (Seegers et al. 2002) and morphologically sensitive stretch-activated cation channels (Cho et al. 1999) can contribute to Ca^{2+} influx into the cell during electrical stimulation. In turn, the influx of Ca^{2+} may interfere with glycolysis in the cytoplasm and aerobic respiration in mitochondria (Wojtczak 1996). The details of the specific electrocoupling mechanisms responsible for biomechanics modulation in electric-field-exposed stem cells remain to be elucidated. Overall, the effect of an electric field on cellular mechanical properties is a result of an intricate interplay of events involving two major molecular mediators: Ca^{2+} and ATP. Therefore, the changes in the membrane and the cytoskeleton mechanics are concurrent during cell exposure to an electric field.

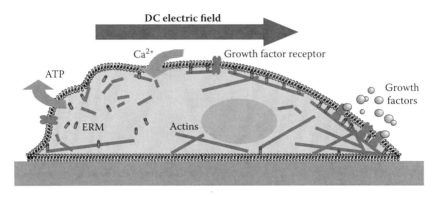

FIGURE 4.4 Schematic view of electrical stimulation effects on hMSC biomechanics. An external electric field induces an increase in the average cytosolic calcium concentration, which leads to partial actin cytoskeleton disassembly and a decrease in the cell elasticity. If present, growth factors could bind to electrically redistributed plasma membrane receptors and trigger local actin polymerization to mediate serum-dependent cell electromigration. In addition, an electrical stimulation causes intracellular ATP depletion, which results in inhibition of the ERM linker protein binding and its dissociation from the membrane and the actin cytoskeleton. The resultant membrane separation from the cytoskeleton and the decreased membrane tension are attributed both to electrically induced downregulation of active ERM proteins and actin depolymerization.

The cell biomechanical responses to electrical stimulation can be regulated by changing the field strength and mode. For example, the effect of low-frequency (e.g., 1 to 10 Hz) oscillatory electric fields on stem cell mechanical properties is qualitatively similar to the effects of a 1 to 3 V/cm dc field. Generally, a lower dc field strength requires more time to achieve similar modulation of cellular mechanical properties than a higher field strength. The proposed model of dc field effect on cellular mechanics is schematically presented in Figure 4.4. It includes actin cytoskeleton disassembly mediated by intracellular Ca^{2+} elevation and membrane-cytoskeleton separation caused by ERM linkers' inhibition through intracellular ATP depletion. Membrane protein redistribution, asymmetric actin polymerization, and growth factor–dependent galvanotaxis are further outcomes of dc field exposure. The exact details of these electrocoupling mechanisms may vary in different cell types.

SUMMARY

One important conclusion from these findings is that the effect of electrical stimulation is cell type dependent and reversible. This observation may be used to explain the synergistic osteogenic hMSC differentiation by daily application of a low-intensity electrical stimulation (Sun et al. 2007). For example, since the stress fibers appear less stable than the thin microfilaments during a dc field exposure (see above), they may be disassembled first under electrical exposure. This could bring the cell elastic and structural properties closer to those of fully differentiated osteoblasts. Electrically induced membrane dissociation from the cytoskeleton and the consequent decrease in membrane tension can then enhance endocytosis and transmembrane trafficking of soluble osteogenic factors. Cell recovery in the osteogenic medium after each short-term electrical exposure will result in further rearrangement of actins and ERM proteins into the osteogenic-type pattern to facilitate hMSC osteogenic commitment. In contrast, neuronal-like cells have a very weak actin cytoskeleton and relatively loose plasma membrane as indicated by tether extraction experiments (Dai et al. 1998). Therefore, a higher-strength electrical stimulation might be required to facilitate hMSC neurogenic differentiation, which would maximally disrupt actin cytoskeleton and inhibit ERM linkers. Thus, the electrical parameters of an electrical stimulus may be precisely controlled for the selective manipulation of the mechanical properties for particular and preselected cell phenotypes.

In addition to potentially facilitating stem cell commitment to a particular lineage, based on their biomechanics, electrical stimulation may also be useful for cell integration into an engineered extracellular environment with defined mechanical properties, including control of cell distribution patterns based on the directional electromigration, dielectyrophoresis, electro-orientation, and electrorotation (Markx 2008). However, an in-depth understanding of the cell-extracellular matrix mechanical and biophysical interactions in three-dimensional systems, as well as elucidation of electrocoupling mechanisms, is critical for effective use of these electrokinetic techniques in tissue engineering application.

CONTROL OF STEM CELL ADHESION AND ORIENTATION IN THREE-DIMENSIONAL SCAFFOLD BY ELECTRICAL STIMULATION

Cell orientation and adhesion are known to affect cell behaviors and functions in both natural and engineered tissues (Thery and Bornens 2006; Toyoshima and Nishida 2007). In engineered tissue models the cells are typically seeded in a three-dimensional scaffold, which provides not only the anchorage for the cell, but also a suitable cellular microenvironment that regulates cell differentiation, metabolic activity, and cell-cell signaling (Robert 2001; Streuli 2009). Oriented and spatially patterned cells, for instance, can exhibit favorable adhesion to extracellular matrix (Eisenbarth et al. 2002), and also determine the alignment of collagenous matrix in healing ligaments and tendons (Wang et al. 2003). One of the ultimate goals in tissue engineering is therefore to mimic the natural cell shape and orientation in the engineered tissue constructs using, for example, external physical stimuli. Although induced cell adhesion, migration, and orientation in response to an electrical stimulus have been well documented using two-dimensional cultured cells (Cho et al. 1999; Farboud et al. 2000; Li and Kolega 2002; Pullar and Isseroff 2005), electrically mediated cellular behaviors in a three-dimensional environment remain much less explored, and may differ significantly from two-dimensional cultured cells (Pedersen and Swartz 2005; Serena et al. 2008; Sun et al. 2004). Because cells embedded in a three-dimensional matrix are more representative of native biological tissues, these cell cultures may provide a better model for studying electrically induced cellular responses.

THE EFFECT OF ELECTRICAL STIMULATION ON MESENCHYMAL STEM CELLS WITHIN A COLLAGEN SCAFFOLD

We have recently shown in our laboratory that rat mesenchymal stem cells (rMSCs) seeded in a collagen scaffold respond differently to 7 to 10 V/cm dc electrical stimuli than fibroblasts (Sun et al. 2006). For example, while fibroblasts are induced to reorient themselves perpendicularly to the axis of electrical stimulus, rMSCs exhibit only a limited reorientation and motility (Figure 4.3). At least two physical mechanisms can be envisioned that mediate cell orientation in the three-dimensional collagen scaffold. First, alignment of the collagen fibers in response to an electrical stimulus could induce cell reorientation. However, since no discernable collagen fibril reorganization in the bulk of electrically exposed cell-free collagen scaffold was observed, this mechanism is unlikely responsible for electrically altered cell shape and alignment. Second, active cells themselves may rearrange the local microenvironment by cell-collagen interactions. It is very likely that localized collagen fiber rearrangement by the cells rather than global scaffold alignment is responsible for the cell reorientation.

Specific biomolecular mechanisms that are involved in the regulation of rMSC adhesion and orientation in response to noninvasive electrical stimulus remain to be identified. Electrically regulated Ca^{2+}-dependent subcellular processes, including cytoskeletal reorganization, are likely to cause changes in the cell morphology and reorientation (Cho et al. 1996).

THE ROLE OF INTEGRINS

In addition to growth factors and the GTP-exchanger factor that are required in cell orientation (Nern and Arkowitz 1998; Zhao et al. 1999), integrins appear to be important in this process (Docheva et al. 2007). Based on the images that show unusually strong adhesion between rMSC and collagen fibers (Figure 4.3), and clustering of integrins, it is plausible that integrin-mediated adhesions found in rMSCs differ from those typically observed in terminally differentiated cells. We have mentioned earlier that hMSCs adhere more strongly than osteoblasts to a two-dimensional substrate, as confirmed by plenty of large focal adhesion complexes in the stem cells (Titushkin and Cho 2007). Integrins play a pivotal role in hMSC adhesion and differentiation in three-dimensional matrices as well, although quite different expression patterns for several types of integrins ($\alpha_2\beta_1$, $\alpha_5\beta_1$, $\alpha_V\beta_3$) have been reported in three-dimensional cultures compared to their two-dimensional counterparts (Grayson et al. 2004). In a three-dimensional reconstituted collagen scaffold, integrin interaction with the collagen fibers surrounding the cells is likely responsible for differential cell adhesion. While an entangled network of collagen fibers bound to the periphery of fibroblasts is induced to preferentially align with electrically reoriented cells, the physically strong adhesion originally observed between rMSCs and collagen fiber bundles appears to be only partially relaxed in response to an electrical stimulus without significant collagen realignment (Sun et al. 2006). Blocking of integrins with specific antibodies causes the stem cells to assume a round morphology and to lose physical connections with the collagen fibers. Such cells show a lack of adhesion sites or bundles of collagen fibers, and may have been only spatially confined within the scaffold without molecular attachment. In addition, both fibroblasts and rMSCs lose their ability to respond to an electrical stimulus when integrins are blocked, indicating that integrin-mediated adhesion is likely to mediate three-dimensional cell morphology and orientation. Moreover, integrins on the cell surface may serve as mechanical transducers of low-frequency electric fields (Hart 2006). In this proposed model, the force exerted on a charged integrin molecule by a traverse electric field of physiological strength (~1 V/cm) is similar in magnitude to that produced by a ~1 N/m^2 fluid shear (~1 fN). The mechanical torque of either force is directly transmitted by integrins to the actin cytoskeleton. Interestingly, this model suggests that some electric field effects may be ultimately mechanical in nature (Hart 2006, 2008).

Weak membrane coupling to the cytoskeleton in stem cells supports the postulate of differential adhesion mechanisms. Because a close interaction between the membrane and cytoskeletal organization is required to mediate changes in the cellular morphology and cell reorientation (Raucher et al. 2000), the loosely connected membrane and cytoskeleton observed in stem cells may prevent electrically induced responses such as those observed in fibroblasts. Due to strong integrin-mediated adhesion and weak membrane-cytoskeleton mechanical coupling in stem cells, the usual electrocoupling mechanisms that are sufficient to explain the galvanotatic responses of terminally differentiated fibroblasts may not be directly applicable to stem cells. This result may be attributed at least partially to differential cell adhesion mechanisms. In summary, application of the optimized electrical stimulus could offer a promising engineering technique to regulate cell-type-dependent cellular shape and orientation that are known to be involved in stem cell proliferation and differentiation.

CONCLUSIONS

Stem cells have recently emerged as a valuable therapeutical tool in the rapidly developing fields of regenerative medicine and functional tissue engineering (Ringe et al. 2002; Salgado et al. 2006). Ultimately, the clinical utility of stem cell-based therapies will depend on their performance and cost. Cost-effective physicochemical manipulation of stem cells (e.g., electrical stimulation) will undoubtedly advance the current effort for stem cell-based therapeutics. To this end, recent efforts have focused on using noninvasive electric fields to harness the unique biological and biophysical properties of stem cells for various tissue engineering applications. Although multiple biophysical stem cell responses to electrical stimulation have been reported and carefully characterized, many important issues remain to be solved to develop effective electrically mediated stem cell manipulation techniques. First, stem cells from various origins (e.g., embryonic, adult, and induced stem cells) may respond differently to an external electrical stimulation according to different endogenous electrical activities in their host tissues and organs. Careful consideration of biomolecular changes associated with stem cell differentiation into a particular phenotype is required for characterization of electrically induced responses. For example, expression of the N-type voltage-gated channels found abundantly in neuronal tissue but not in uncommitted mesenchymal or embryonic stem cells may significantly modulate cellular responses during electrically assisted neurogenic differentiation. Second, stem cells grown in three-dimensional scaffolds display tissue development patterns and cellular responses to exogenous electrical stimuli that are distinct from their two-dimensional counterparts (Serena et al. 2008). These differences should be taken into consideration while designing three-dimensional scaffolds for electrically assisted development of functional tissue-engineered constructs. Third, cellular and molecular responses rely strongly on the careful selection of the stimulation mode (e.g., dc, ac, pulsed electric fields, noncontact high-frequency irradiation) and electric field parameters such as intensity, frequency, and exposure duration (Cho 2002). Unfortunately, many biochemical and biomolecular electrocoupling mechanisms remain not fully identified, in part due to the strong dependence on electrical stimulation modality. For example, the dc or low-frequency oscillatory electric fields are likely confined to the cell surface, and initiate biochemical signal transduction cascades. A wide choice of available electromagnetic stimuli suggests plausible effective approaches for precisely controlling the electric field characteristics for selective manipulation of stem cell biophysical properties into a particular and preselected cell phenotype. Optimal use of the electrical stimulus may lead to development of strategies for tissue engineering by manipulating cell differentiation, mobility, and cell incorporation into engineered scaffolds, and eventual maturation of tissue substitute. Furthermore, elucidation of mode- and stem cell origin–dependent electrocoupling mechanisms will enable the researchers to better control the intended stem cell fate and behaviors for general stem cell biology and *in vitro* drug development applications.

REFERENCES

Adams, P. J., and T. P. Snutch. 2007. Calcium channelopathies: Voltage-gated calcium channels. *Subcell Biochem* 45:215–51.
Altman, G. H., R. L. Horan, I. Martin, J. Farhadi, P. R. Stark, V. Volloch, J. C. Richmond, G. Vunjak-Novakovic, and D. L. Kaplan. 2002. Cell differentiation by mechanical stress. *FASEB J* 16(2):270–72.
Bai, H., C. D. McCaig, J. V. Forrester, and M. Zhao. 2004. DC electric fields induce distinct preangiogenic responses in microvascular and macrovascular cells. *Arterioscler Thromb Vasc Biol* 24(7):1234–39.

Barry, F. P., and J. M. Murphy. 2004. Mesenchymal stem cells: Clinical applications and biological characterization. *Int J Biochem Cell Biol* 36(4):568–84.

Bernardo, M. E., F. Locatelli, and W. E. Fibbe. 2009. Mesenchymal stromal cells. *Ann NY Acad Sci* 1176:101–17.

Berridge, M. J. 2007. Inositol trisphosphate and calcium oscillations. *Biochem Soc Symp* 74:1–7.

Berridge, M. J., M. D. Bootman, and H. L. Roderick. 2003. Calcium signalling: Dynamics, homeostasis and remodelling. *Nat Rev Mol Cell Biol* 4(7):517–29.

Brough, D., M. J. Schell, and R. F. Irvine. 2005. Agonist-induced regulation of mitochondrial and endoplasmic reticulum motility. *Biochem J* 392(Pt 2):291–97.

Cedar, S. H. 2006. The function of stem cells and their future roles in healthcare. *Br J Nurs* 15(2):104–7.

Chamberlain, G., J. Fox, B. Ashton, and J. Middleton. 2007. Concise review: Mesenchymal stem cells: Their phenotype, differentiation capacity, immunological features, and potential for homing. *Stem Cells* 25(11):2739–49.

Chang, W. H., L. T. Chen, J. S. Sun, and F. H. Lin. 2004. Effect of pulse-burst electromagnetic field stimulation on osteoblast cell activities. *Bioelectromagnetics* 25(6):457–65.

Chao, P. H., H. H. Lu, C. T. Hung, S. B. Nicoll, and J. C. Bulinski. 2007. Effects of applied DC electric field on ligament fibroblast migration and wound healing. *Connect Tissue Res* 48(4):188–97.

Chawla, S. 2002. Regulation of gene expression by Ca2+ signals in neuronal cells. *Eur J Pharmacol* 447(2–3):131–40.

Chen, J., and M. C. Wagner. 2001. Altered membrane-cytoskeleton linkage and membrane blebbing in energy-depleted renal proximal tubular cells. *Am J Physiol Renal Physiol* 280(4):F619–27.

Cho, M. R. 2002. A review of electrocoupling mechanisms mediating facilitated wound healing. *IEEE Trans Plasma Sci* 30:1504–15.

Cho, M. R., J. P. Marler, H. S. Thatte, and D. E. Golan. 2002. Control of calcium entry in human fibroblasts by frequency-dependent electrical stimulation. *Front Biosci* 7:a1–8.

Cho, M. R., H. S. Thatte, R. C. Lee, and D. E. Golan. 1994. Induced redistribution of cell surface receptors by alternating current electric fields. *FASEB J* 8(10):771–76.

Cho, M. R., H. S. Thatte, R. C. Lee, and D. E. Golan. 1996. Reorganization of microfilament structure induced by ac electric fields. *FASEB J* 10(13):1552–58.

Cho, M. R., H. S. Thatte, M. T. Silvia, and D. E. Golan. 1999. Transmembrane calcium influx induced by ac electric fields. *FASEB J* 13(6):677–83.

Ciombor, D. M., G. Lester, R. K. Aaron, P. Neame, and B. Caterson. 2002. Low frequency EMF regulates chondrocyte differentiation and expression of matrix proteins. *J Orthop Res* 20(1):40–50.

Collinsworth, A. M., S. Zhang, W. E. Kraus, and G. A. Truskey. 2002. Apparent elastic modulus and hysteresis of skeletal muscle cells throughout differentiation. *Am J Physiol Cell Physiol* 283(4):C1219–27.

Colman, A., and O. Dreesen. 2009. Induced pluripotent stem cells and the stability of the differentiated state. *EMBO Rep* 10(7):714–21.

Coppi, E., A. M. Pugliese, S. Urbani, A. Melani, E. Cerbai, B. Mazzanti, A. Bosi, R. Saccardi, and F. Pedata. 2007. ATP modulates cell proliferation and elicits two different electrophysiological responses in human mesenchymal stem cells. *Stem Cells* 25(7):1840–49.

Dai, J., M. P. Sheetz, X. Wan, and C. E. Morris. 1998. Membrane tension in swelling and shrinking molluscan neurons. *J Neurosci* 18(17):6681–92.

Datta, N., Q. P. Pham, U. Sharma, V. I. Sikavitsas, J. A. Jansen, and A. G. Mikos. 2006. *In vitro* generated extracellular matrix and fluid shear stress synergistically enhance 3D osteoblastic differentiation. *Proc Natl Acad Sci USA* 103(8):2488–93.

Dave, K. A., and A. Bordey. 2009. GABA increases Ca2+ in cerebellar granule cell precursors via depolarization: Implications for proliferation. *IUBMB Life* 61(5):496–503.

Docheva, D., C. Popov, W. Mutschler, and M. Schieker. 2007. Human mesenchymal stem cells in contact with their environment: Surface characteristics and the integrin system. *J Cell Mol Med* 11(1):21–38.

Eisenbarth, E., P. Linez, V. Biehl, D. Velten, J. Breme, and H. F. Hildebrand. 2002. Cell orientation and cytoskeleton organisation on ground titanium surfaces. *Biomol Eng* 19(2–6):233–37.

Engler, A. J., S. Sen, H. L. Sweeney, and D. E. Discher. 2006. Matrix elasticity directs stem cell lineage specification. *Cell* 26:677–89.

Foreman, M. A., J. Smith Farboud, B., R. Nuccitelli, I. R. Schwab, and R. R. Isseroff. 2000. DC electric fields induce rapid directional migration in cultured human corneal epithelial cells. *Exp Eye Res* 70(5):667–73.

Fievet, B., D. Louvard, and M. Arpin. 2007. ERM proteins in epithelial cell organization and functions. *Biochim Biophys Acta* 1773(5):653–60.

Fodor, W. L. 2003. Tissue engineering and cell based therapies, from the bench to the clinic: The potential to replace, repair and regenerate. *Reprod Biol Endocrinol* 1:102.

Foreman, M. A., J. Smith, and S. J. Publicover. 2006. Characterisation of serum-induced intracellular Ca2+ oscillations in primary bone marrow stromal cells. *J Cell Physiol* 206(3):664–71.

Gahrton, G., and B. Bjorkstrand. 2000. Progress in haematopoietic stem cell transplantation for multiple myeloma. *J Intern Med* 248(3):185–201.

Genovese, J. A., C. Spadaccio, H. G. Rivello, Y. Toyoda, and A. N. Patel. 2009. Electrostimulated bone marrow human mesenchymal stem cells produce follistatin. *Cytotherapy* 11(4):448–56.

Giannone, G., P. Ronde, M. Gaire, J. Haiech, and K. Takeda. 2002. Calcium oscillations trigger focal adhesion disassembly in human U87 astrocytoma cells. *J Biol Chem* 277(29):26364–71.

Goldman, S. A., and M. S. Windrem. 2006. Cell replacement therapy in neurological disease. *Philos Trans R Soc Lond B Biol Sci* 361(1473):1463–75.

Grayson, W. L., T. Ma, and B. Bunnell. 2004. Human mesenchymal stem cells tissue development in 3D PET matrices. *Biotechnol Prog* 20(3):905–12.

Grayson, W. L., T. P. Martens, G. M. Eng, M. Radisic, and G. Vunjak-Novakovic. 2009. Biomimetic approach to tissue engineering. *Semin Cell Dev Biol* 20(6):665–73.

Grivennikov, I. A. 2008. Embryonic stem cells and the problem of directed differentiation. *Biochemistry* (Mosc) 73(13):1438–52.

Guilak, F., D. M. Cohen, B. T. Estes, J. M. Gimble, W. Liedtke, and C. S. Chen. 2009. Control of stem cell fate by physical interactions with the extracellular matrix. *Cell Stem Cell* 5(1):17–26.

Haddad, J. B., A. G. Obolensky, and P. Shinnick. 2007. The biologic effects and the therapeutic mechanism of action of electric and electromagnetic field stimulation on bone and cartilage: New findings and a review of earlier work. *J Altern Complement Med* 13(5):485–90.

Hamilton, S. L., and I. I. Serysheva. 2009. Ryanodine receptor structure: Progress and challenges. *J Biol Chem* 284(7):4047–51.

Hanson, C. J., M. D. Bootman, and H. L. Roderick. 2004. Cell signalling: IP3 receptors channel calcium into cell death. *Curr Biol* 14(21):R933–35.

Hart, F. X. 2006. Integrins may serve as mechanical transducers for low-frequency electric fields. *Bioelectromagnetics* 27(6):505–8.

Hart, F. X. 2008. The mechanical transduction of physiological strength electric fields. *Bioelectromagnetics* 29(6):447–55.

Heubach, J. F., E. M. Graf, J. Leutheuser, M. Bock, B. Balana, I. Zahanich, T. Christ, S. Boxberger, E. Wettwer, and U. Ravens. 2004. Electrophysiological properties of human mesenchymal stem cells. *J Physiol* 554(Pt 3):659–72.

Hochedlinger, K., and K. Plath. 2009. Epigenetic reprogramming and induced pluripotency. *Development* 136(4):509–23.

Hovnanian, A. 2007. SERCA pumps and human diseases. *Subcell Biochem* 45:337–63.

Huang, H., R. D. Kamm, and R. T. Lee. 2004. Cell mechanics and mechanotransduction: Pathways, probes, and physiology. *Am J Physiol Cell Physiol* 287(1):C1–11.

Ingber, D. E. 2006. Cellular mechanotransduction: Putting all the pieces together again. *FASEB J* 20(7):811–27.

Jaishankar, A., and K. Vrana. 2009. Emerging molecular approaches in stem cell biology. *Biotechniques* 46(5):367–71.

Janmey, P. A., and C. A. McCulloch. 2007. Cell mechanics: Integrating cell responses to mechanical stimuli. *Annu Rev Biomed Eng* 9:1–34.

Janmey, P. A., J. P. Winer, M. E. Murray, and Q. Wen. 2009. The hard life of soft cells. *Cell Motil Cytoskel* 66(8):597–605.

Jena, B. P. 2007. Secretion machinery at the cell plasma membrane. *Curr Opin Struct Biol* 17(4):437–43.

Jensen, J., J. Hyllner, and P. Bjorquist. 2009. Human embryonic stem cell technologies and drug discovery. *J Cell Physiol* 219:513–19.

Kao, C. F., C. Y. Chuang, C. H. Chen, and H. C. Kuo. 2008. Human pluripotent stem cells: Current status and future perspectives. *Chin J Physiol* 51(4):214–25.

Katsuragi, T., T. Tokunaga, M. Ohba, C. Sato, and T. Furukawa. 1993. Implication of ATP released from atrial, but not papillary, muscle segments of guinea pig by isoproterenol and forskolin. *Life Sci* 53(11):961–67.

Kawano, S., K. Otsu, A. Kuruma, S. Shoji, E. Yanagida, Y. Muto, F. Yoshikawa, Y. Hirayama, K. Mikoshiba, and T. Furuichi. 2006. ATP autocrine/paracrine signaling induces calcium oscillations and NFAT activation in human mesenchymal stem cells. *Cell Calcium* 39(4):313–24.

Kawano, S., K. Otsu, S. Shoji, K. Yamagata, and M. Hiraoka. 2003. Ca(2+) oscillations regulated by Na(+)-Ca(2+) exchanger and plasma membrane Ca(2+) pump induce fluctuations of membrane currents and potentials in human mesenchymal stem cells. *Cell Calcium* 34(2):145–56.

Kawano, S., S. Shoji, S. Ichinose, K. Yamagata, M. Tagami, and M. Hiraoka. 2002. Characterization of Ca(2+) signaling pathways in human mesenchymal stem cells. *Cell Calcium* 32(4):165–74.

Khatib, L., D. E. Golan, and M. Cho. 2004. Physiologic electrical stimulation provokes intracellular calcium increase mediated by phospholipase C activation in human osteoblasts. *FASEB J* 18(15):1903–5.

Kim, T. J., J. Seong, M. Ouyang, J. Sun, S. Lu, J. P. Hong, N. Wang, and Y. Wang. 2009. Substrate rigidity regulates Ca2+ oscillation via RhoA pathway in stem cells. *J Cell Physiol* 218(2):285–93.

Kloth, L. C. 2005. Electric stimulation for wound healing: A review of evidence from *in vitro* studies, animal experiments and clinical trials. *Int J Lower Extremity Wounds* 4(1):23–44.

Kuci, S., Z. Kuci, H. Latifi-Pupovci, D. Niethammer, R. Handgretinger, M. Schumm, G. Bruchelt, P. Bader, and T. Klingebiel. 2009. Adult stem cells as an alternative source of multipotential (pluripotential) cells in regenerative medicine. *Curr Stem Cell Res Ther* 4(2):107–17.

Kurpinski, K., and S. Li. 2007. Mechanical stimulation of stem cells using cyclic uniaxial strain. *J Vis Exp* 6:242.

Kuznetsova, T. G., M. N. Starodubtseva, N. I. Yegorenkov, S. A. Chizhik, and R. I. Zhdanov. 2007. Atomic force microscopy probing of cell elasticity. *Micron* 38(8):824–33.

Le Blanc, K., C. Tammik, K. Rosendahl, E. Zetterberg, and O. Ringden. 2003. HLA expression and immunologic properties of differentiated and undifferentiated mesenchymal stem cells. *Exp Hematol* 31(10):890–96.

Li, X., and J. Kolega. 2002. Effects of direct current electric fields on cell migration and actin filament distribution in bovine vascular endothelial cells. *J Vasc Res* 39(5):391–404.

Lim, C. T., E. H. Zhou, and S. T. Quek. 2006. Mechanical models for living cells—A review. *J Biomech* 39(2):195–216.

Lohmann, C. H., Z. Schwartz, Y. Liu, H. Guerkov, D. D. Dean, B. Simon, and B. D. Boyan. 2000. Pulsed electromagnetic field stimulation of MG63 osteoblast-like cells affects differentiation and local factor production. *J Orthop Res* 18(4):637–46.

Louvet-Vallee, S. 2000. ERM proteins: From cellular architecture to cell signaling. *Biol Cell* 92(5):305–16.

Luttrell, L. M. 2002. Activation and targeting of mitogen-activated protein kinases by G-protein-coupled receptors. *Can J Physiol Pharmacol* 80(5):375–82.

MacGinitie, L. A., Y. A. Gluzband, and A. J. Grodzinsky. 1994. Electric field stimulation can increase protein synthesis in articular cartilage explants. *J Orthop Res* 12(2):151–60.

Mangeat, P., C. Roy, and M. Martin. 1999. ERM proteins in cell adhesion and membrane dynamics. *Trends Cell Biol* 9(5):187–92.

Markx, G. H. 2008. The use of electric fields in tissue engineering: A review. *Organogenesis* 4(1):11–17.

McBeath, R., D. M. Pirone, C. M. Nelson, K. Bhadriraju, and C. S. Chen. 2004. Cell shape, cytoskeletal tension, and RhoA regulate stem cell lineage commitment. *Dev Cell* 6(4):483–95.

McCaig, C. D., A. M. Rajnicek, B. Song, and M. Zhao. 2005. Controlling cell behavior electrically: Current views and future potential. *Physiol Rev* 85(3):943–78.

Mertes, H., and G. Pennings. 2009. Cross-border research on human embryonic stem cells: Legal and ethical considerations. *Stem Cell Rev* 5(1):10–7.

Meyers, V. E., M. Zayzafoon, J. T. Douglas, and J. M. McDonald. 2005. RhoA and cytoskeletal disruption mediate reduced osteoblastogenesis and enhanced adipogenesis of human mesenchymal stem cells in modeled microgravity. *J Bone Miner Res* 20(10):1858–66.

Miranti, C. K., and J. S. Brugge. 2002. Sensing the environment: A historical perspective on integrin signal transduction. *Nat Cell Biol* 4(4):E83–90.

Mizuno, H. 2009. Adipose-derived stem cells for tissue repair and regeneration: Ten years of research and a literature review. *J Nippon Med School* 76(2):56–66.

Mohan, C. G., and T. Gandhi. 2008. Therapeutic potential of voltage gated calcium channels. *Mini Rev Med Chem* 8(12):1285–90.

Moore, K. A., and I. R. Lemischka. 2006. Stem cells and their niches. *Science* 311(5769):1880–85.

Morris, C. E., and U. Homann. 2001. Cell surface area regulation and membrane tension. *J Membr Biol* 179(2):79–102.

Nern, A., and R. A. Arkowitz. 1998. A GTP-exchange factor required for cell orientation. *Nature* 391(6663):195–98.

Nikolic, B., S. Faintuch, S. N. Goldberg, M. D. Kuo, and J. F. Cardella. 2009. Stem cell therapy: A primer for interventionalists and imagers. *J Vasc Interv Radiol* 20(8):999–1012.

Orr, A. W., B. P. Helmke, B. R. Blackman, and M. A. Schwartz. 2006. Mechanisms of mechanotransduction. *Dev Cell* 10(1):11–20.

Park, D. H., C. V. Borlongan, D. J. Eve, and P. R. Sanberg. 2008. The emerging field of cell and tissue engineering. *Med Sci Monit* 14(11):RA206–20.

Pedersen, J. A., and M. A. Swartz. 2005. Mechanobiology in the third dimension. *Ann Biomed Eng* 33(11):1469–90.

Poo, M. 1981. *In situ* electrophoresis of membrane components. *Annu Rev Biophys Bioeng* 10:245–76.

Pullar, C. E., and R. R. Isseroff. 2005. Cyclic AMP mediates keratinocyte directional migration in an electric field. *J Cell Sci* 118(Pt 9):2023–34.

Putney, J. W., and G. S. Bird. 2008. Cytoplasmic calcium oscillations and store-operated calcium influx. *J Physiol* 586(13):3055–59.

Raucher, D., and M. P. Sheetz. 1999. Characteristics of a membrane reservoir buffering membrane tension. *Biophys J* 77(4):1992–2002.

Raucher, D., and M. P. Sheetz. 2000. Cell spreading and lamellipodial extension rate is regulated by membrane tension. *J Cell Biol* 148(1):127–36.

Raucher, D., T. Stauffer, W. Chen, K. Shen, S. Guo, J. D. York, M. P. Sheetz, and T. Meyer. 2000. Phosphatidylinositol 4,5-bisphosphate functions as a second messenger that regulates cytoskeleton-plasma membrane adhesion. *Cell* 100(2):221–28.

Resende, R. R., A. S. Alves, L. R. Britto, and H. Ulrich. 2008. Role of acetylcholine receptors in proliferation and differentiation of P19 embryonal carcinoma cells. *Exp Cell Res* 314(7):1429–43.

Ringe, J., C. Kaps, G. R. Burmester, and M. Sittinger. 2002. Stem cells for regenerative medicine: Advances in the engineering of tissues and organs. *Naturwissenschaften* 89(8):338–51.

Robert, L. 2001. Matrix biology: Past, present and future. *Pathol Biol* (Paris) 49(4):279–83.

Robinson, K. R. 1985. The responses of cells to electrical fields: A review. *J Cell Biol* 101(6):2023–27.

Rodriguez, J. P., M. Gonzalez, S. Rios, and V. Cambiazo. 2004. Cytoskeletal organization of human mesenchymal stem cells (MSC) changes during their osteogenic differentiation. *J Cell Biochem* 93(4):721–31.

Romano, G. 2008. Artificial reprogramming of human somatic cells to generate pluripotent stem cells: A possible alternative to the controversial use of human embryonic stem cells. *Drug News Perspect* 21(8):440–45.

Salasznyk, R. M., W. A. Williams, A. Boskey, A. Batorsky, and G. E. Plopper. 2004. Adhesion to vitronectin and collagen I promotes osteogenic differentiation of human mesenchymal stem cells. *J Biomed Biotechnol* 2004(1):24–34.

Salgado, A. J., J. T. Oliveira, A. J. Pedro, and R. L. Reis. 2006. Adult stem cells in bone and cartilage tissue engineering. *Curr Stem Cell Res Ther* 1(3):345–64.

Sauer, H., G. Rahimi, J. Hescheler, and M. Wartenberg. 1999. Effects of electrical fields on cardiomyocyte differentiation of embryonic stem cells. *J Cell Biochem* 75(4):710–23.

Sauer, H., R. Stanelle, J. Hescheler, and M. Wartenberg. 2002. The DC electrical-field-induced Ca(2+) response and growth stimulation of multicellular tumor spheroids are mediated by ATP release and purinergic receptor stimulation. *J Cell Sci* 115 (Pt 16):3265–73.

Savio-Galimberti, E., and J. E. Ponce-Hornos. 2006. Effects of caffeine, verapamil, lithium, and KB-R7943 on mechanics and energetics of rat myocardial bigeminies. *Am J Physiol Heart Circ Physiol* 290(2):H613–23.

Schreiber, R. 2005. Ca2+ signaling, intracellular pH and cell volume in cell proliferation. *J Membr Biol* 205(3):129–37.

Schwartz, Z., M. Fisher, C. H. Lohmann, B. J. Simon, and B. D. Boyan. 2009. Osteoprotegerin (OPG) production by cells in the osteoblast lineage is regulated by pulsed electromagnetic fields in cultures grown on calcium phosphate substrates. *Ann Biomed Eng* 37(3):437–44.

Seegers, J. C., M. L. Lottering, A. M. Joubert, F. Joubert, A. Koorts, C. A. Engelbrecht, and D. H. van Papendorp. 2002. A pulsed DC electric field affects P2-purinergic receptor functions by altering the ATP levels in *in vitro* and *in vivo* systems. *Med Hypotheses* 58(2):171–76.

Sensebe, L., and P. Bourin. 2009. Mesenchymal stem cells for therapeutic purposes. *Transplantation* 87(9 Suppl):S49–53.

Serena, E., M. Flaibani, S. Carnio, L. Boldrin, L. Vitiello, P. De Coppi, and N. Elvassore. 2008. Electrophysiologic stimulation improves myogenic potential of muscle precursor cells grown in a 3D collagen scaffold. *Neurol Res* 30(2):207–14.

Sheetz, M. P., J. E. Sable, and H. G. Dobereiner. 2006. Continuous membrane-cytoskeleton adhesion requires continuous accommodation to lipid and cytoskeleton dynamics. *Annu Rev Biophys Biomol Struct* 35:417–34.

Shieh, A. C., and K. A. Athanasiou. 2003. Principles of cell mechanics for cartilage tissue engineering. *Ann Biomed Eng* 31(1):1–11.

Short, B., N. Brouard, T. Occhiodoro-Scott, A. Ramakrishnan, and P. J. Simmons. 2003. Mesenchymal stem cells. *Arch Med Res* 34(6):565–71.

Sinner, B., O. Friedrich, W. Zink, R. H. Fink, and B. M. Graf. 2006. GABAmimetic intravenous anaesthetics inhibit spontaneous Ca2+-oscillations in cultured hippocampal neurons. *Acta Anaesthesiol Scand* 50(6):742–48.

Sjaastad, M. D., and W. J. Nelson. 1997. Integrin-mediated calcium signaling and regulation of cell adhesion by intracellular calcium. *Bioessays* 19(1):47–55.

Stolberg, S., and K. E. McCloskey. 2009. Can shear stress direct stem cell fate? *Biotechnol Prog* 25(1):10–19.

Streuli, C. H. 2009. Integrins and cell-fate determination. *J Cell Sci* 122(Pt 2):171–77.

Sun, M., N. Northup, F. Marga, T. Huber, F. J. Byfield, I. Levitan, and G. Forgacs. 2007. The effect of cellular cholesterol on membrane-cytoskeleton adhesion. *J Cell Sci* 120(Pt 13):2223–31.

Sun, S., Y. Liu, S. Lipsky, and M. Cho. 2007. Physical manipulation of calcium oscillations facilitates osteodifferentiation of human mesenchymal stem cells. *FASEB J* 21(7):1472–80.

Sun, S., I. Titushkin, and M. Cho. 2006. Regulation of mesenchymal stem cell adhesion and orientation in 3D collagen scaffold by electrical stimulus. *Bioelectrochemistry* 69(2):133–41.

Sun, S., J. Wise, and M. Cho. 2004. Human fibroblast migration in three-dimensional collagen gel in response to noninvasive electrical stimulus. I. Characterization of induced three-dimensional cell movement. *Tissue Eng* 10(9–10):1548–57.

Takai, E., K. D. Costa, A. Shaheen, C. T. Hung, and X. E. Guo. 2005. Osteoblast elastic modulus measured by atomic force microscopy is substrate dependent. *Ann Biomed Eng* 33(7):963–71.

Takeda, K., A. Matsuzawa, H. Nishitoh, K. Tobiume, S. Kishida, J. Ninomiya-Tsuji, K. Matsumoto, and H. Ichijo. 2004. Involvement of ASK1 in Ca2+-induced p38 MAP kinase activation. *EMBO Rep* 5(2):161–66.

Thery, M., and M. Bornens. 2006. Cell shape and cell division. *Curr Opin Cell Biol* 18(6):648–57.

Thul, R., T. C. Bellamy, H. L. Roderick, M. D. Bootman, and S. Coombes. 2008. Calcium oscillations. *Adv Exp Med Biol* 641:1–27.

Titushkin, I., and M. Cho. 2006. Distinct membrane mechanical properties of human mesenchymal stem cells determined using laser optical tweezers. *Biophys J* 90(7):2582–91.

Titushkin, I., and M. Cho. 2007. Modulation of cellular mechanics during osteogenic differentiation of human mesenchymal stem cells. *Biophys J* 93(10):3693–702.

Titushkin, I., and M. Cho. 2009. Regulation of cell cytoskeleton and membrane mechanics by electric field: Role of linker proteins. *Biophys J* 96(2):717–28.

Titushkin, I. A., V. S. Rao, and M. R. Cho. 2004. Mode- and cell-dependent calcium responses induced by electrical stimulus. *IEEE Trans Plasma Sci* 32:1614–19.

Tonini, R., M. D. Baroni, E. Masala, M. Micheletti, A. Ferroni, and M. Mazzanti. 2001. Calcium protects differentiating neuroblastoma cells during 50 Hz electromagnetic radiation. *Biophys J* 81(5):2580–89.

Toyoshima, F., and E. Nishida. 2007. Spindle orientation in animal cell mitosis: Roles of integrin in the control of spindle axis. *J Cell Physiol* 213(2):407–11.

Trickey, W. R., T. P. Vail, and F. Guilak. 2004. The role of the cytoskeleton in the viscoelastic properties of human articular chondrocytes. *J Orthop Res* 22(1):131–39.

Tropel, P., N. Platet, J. C. Platel, D. Noel, M. Albrieux, A. L. Benabid, and F. Berger. 2006. Functional neuronal differentiation of bone marrow-derived mesenchymal stem cells. *Stem Cells* 24(12):2868–76.

Tsai, M. T., W. J. Li, R. S. Tuan, and W. H. Chang. 2009. Modulation of osteogenesis in human mesenchymal stem cells by specific pulsed electromagnetic field stimulation. *J Orthop Res* 27(9):1169–74.

Ulrich, H., and P. Majumder. 2006. Neurotransmitter receptor expression and activity during neuronal differentiation of embryonal carcinoma and stem cells: From basic research towards clinical applications. *Cell Prolif* 39(4):281–300.

Van Vliet, K. J., G. Bao, and S. Suresh. 2003. The biomechanics toolbox: Experimental approaches for living cells and biomolecules. *Acta Materialia* 51:5881–905.

Wang, E., M. Zhao, J. V. Forrester, and C. D. McCaig. 2003. Bi-directional migration of lens epithelial cells in a physiological electrical field. *Exp Eye Res* 76(1):29–37.

Wang, J. H., F. Jia, T. W. Gilbert, and S. L. Woo. 2003. Cell orientation determines the alignment of cell-produced collagenous matrix. *J Biomech* 36(1):97–102.

Wang, J. H., B. P. Thampatty, J. S. Lin, and H. J. Im. 2007. Mechanoregulation of gene expression in fibroblasts. *Gene* 391(1–2):1–15.

Wang, Y., R. Roman, S. D. Lidofsky, and J. G. Fitz. 1996. Autocrine signaling through ATP release represents a novel mechanism for cell volume regulation. *Proc Natl Acad Sci USA* 93(21):12020–25.

Warren, S. M., R. K. Nacamuli, H. M. Song, and M. T. Longaker. 2004. Tissue-engineered bone using mesenchymal stem cells and a biodegradable scaffold. *J Craniofac Surg* 15(1):34–37.

Wojtczak, L. 1996. The crabtree effect: A new look at the old problem. *Acta Biochim Pol* 43(2):361–68.

Wu, X., T. Zhang, J. Bossuyt, X. Li, T. A. McKinsey, J. R. Dedman, E. N. Olson, J. Chen, J. H. Brown, and D. M. Bers. 2006. Local InsP3-dependent perinuclear Ca2+ signaling in cardiac myocyte excitation-transcription coupling. *J Clin Invest* 116(3):675–82.

Yamada, M., K. Tanemura, S. Okada, A. Iwanami, M. Nakamura, H. Mizuno, M. Ozawa, R. Ohyama-Goto, N. Kitamura, M. Kawano, K. Tan-Takeuchi, C. Ohtsuka, A. Miyawaki, A. Takashima, M. Ogawa, Y. Toyama, H. Okano, and T. Kondo. 2007. Electrical stimulation modulates fate determination of differentiating embryonic stem cells. *Stem Cells* 25(3):562–70.

Yourek, G., M. A. Hussain, and J. J. Mao. 2007. Cytoskeletal changes of mesenchymal stem cells during differentiation. *ASAIO J* 53(2):219–28.

Zamponi, G. W., R. J. Lewis, S. M. Todorovic, S. P. Arneric, and T. P. Snutch. 2009. Role of voltage-gated calcium channels in ascending pain pathways. *Brain Res Rev* 60(1):84–89.

Zhao, M., H. Bai, E. Wang, J. V. Forrester, and C. D. McCaig. 2004. Electrical stimulation directly induces pre-angiogenic responses in vascular endothelial cells by signaling through VEGF receptors. *J Cell Sci* 117(Pt 3):397–405.

Zhao, M., A. Dick, J. V. Forrester, and C. D. McCaig. 1999. Electric field-directed cell motility involves up-regulated expression and asymmetric redistribution of the epidermal growth factor receptors and is enhanced by fibronectin and laminin. *Mol Biol Cell* 10(4):1259–76.

Zhuang, H., W. Wang, R. M. Seldes, A. D. Tahernia, H. Fan, and C. T. Brighton. 1997. Electrical stimulation induces the level of TGF-beta1 mRNA in osteoblastic cells by a mechanism involving calcium/calmodulin pathway. *Biochem Biophys Res Commun* 237(2):225–29.

5 Electrical Signals Control Corneal Epithelial Cell Physiology and Wound Repair

Colin D. McCaig
School of Medical Sciences
University of Aberdeen
Aberdeen, Scotland

CONTENTS

INTRODUCTION

The evidence that electrical fields exist within extracellular tissue spaces is strong and wide ranging across animal and plant kingdoms (Jaffe, 1985). Demonstrating unequivocal roles for these electrical signals in regulating cellular physiology has been challenging, and many biologists are not convinced that they have physiological or pathological relevance. Three issues probably explain this:

1. Technological advances. Using string galvanometer measurements and extracellular recording electrodes, much was discovered in the first half of the 1900s of the dynamic, extracellular electrical events that regulate the heartbeat, that arise from the conduction of a compound action potential in peripheral nerves, and that are generated during epileptic seizures, or during cortical spreading depression in the brain. With the development of the intracellular microelectrode, attention shifted to electrical events across a single cell membrane rather than in the extracellular spaces, and the patch clamp technique brought further reductionist concentration on single-channel conductance events within a cell membrane. This had a profound effect in refocussing interest on a different scale (single ion movements rather than the collective effects of many depolarizing axons), on a different location (across a lipid bilayer rather than through the extracellular spaces), and on a different time frame (milliseconds rather than minutes to hours). Consequently, much of the work of our illustrious forebears, in carefully describing the extracellular electrical signatures in a variety of tissues, is unknown.

2. The revolution in molecular biology in the 1960s and 1970s has dominated biology in general and developmental biology in particular. This has attracted much research effort and temporarily left fields like biophysics and physiology relatively neglected. The emergence of systems biology (physiology?) and the need to place the vast molecular genetic knowledge that has been accrued into a postgenomic, whole tissue, whole animal context may restore some of this imbalance.

3. There is a lot of bad science with the good. Examples range from early times through to the present day. Galvani was way ahead of his time in bringing good science to the public (~1790s), since he gave a number of public displays of the bioelectricity intrinsic to the frog nerve and muscle preparations for which he is famous. He would attach frog legs to a lightning conductor during electrical storms and show that the legs twitched. Galvani also was the first to demonstrate directly and convincingly the existence of injury potentials. He touched the cut sciatic nerve stump with its thousands of proximally severed nerves to the thigh muscles of the frog and showed that the leg twitched. He correctly deduced that current must be flowing from the cut nerve stumps to stimulate muscle contraction. Galvani's pioneering studies and public exhibitions arguably establish him as the inventor of bioelectricity. They also excited the Victorian imagination, and around twenty years later, Mary Shelley wrote *Frankenstein*, in which a monster is brought to life from inanimate flesh by electrical stimulation. Galvani's nephew Aldini tainted the family's scientific reputation, since he was at the forefront of a group of practitioners of galvanism that used large voltaic piles (batteries) in public demonstrations of the ability of electricity to make corpses twitch into "life" (Figure 5.1a). More recently, there are those who suggest that we all possess our own external electric field, and that this is the basis of

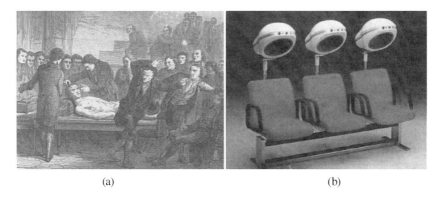

(a)   (b)

FIGURE 5.1   See color insert. (a) A demonstration of galvanism, the raising of a corpse using electricity. The print is called "The Galvanisation of Matthew Clydesdale" (see http://scienceonstreets.phys.strath.ac.uk/galv05play.html). (b) Electrical "therapies" for baldness.

extra-sensory perception. Additionally, you could choose to spend $15K, which would buy you an electrical "helmet" that is claimed to restore hair to bald people (Figure 5.1b). Scepticism, therefore, is understandable.

ELECTRICAL GRADIENTS IN DEVELOPING AND REGENERATING EPITHELIA

In considering the effects of electrical fields on epithelial cells *in vitro* and *in vivo*, it is helpful to keep the following in mind. The extracellular spaces contain gradients of specific guidance molecules, such as growth factors and cytokines. We assume this to be true since it follows intuitively from Fick's principle of diffusion applied to molecules secreted into the extracellular spaces. However, direct demonstrations of this are scant, and it is worth pondering why. Gradients of ions also exist and ionic current flows through the extracellular spaces. This is what the early galvanometers and external recording electrodes were measuring. These arise from localized activation of selective ion pumps or leaks and from the combined effects of activating many channel conductances. We shall return to this point below (Figure 5.2b.) in discussing the source of the electrical signal at epithelial wounds. Here it is crucial to recognize that these extracellular chemical and electrical gradients coexist in extracellular space and time. That is not in doubt. Nor is the fact that they must inevitably interact with each other. Consider a charged protein molecule in the extracellular spaces, a chemotropic molecule, for example, coming under the influence of a steady localized voltage (ionic) gradient. The charged molecule will behave as if it were in an electrophoresis column. It cannot diffuse freely because its net charge will interact electrostatically with the extracellular electrical force. There is one demonstration of this principle of chemical and electrical gradients interacting *in vivo*, although it was only published as an abstract (Messerli and Robinson, 1997). The frog embryo drives

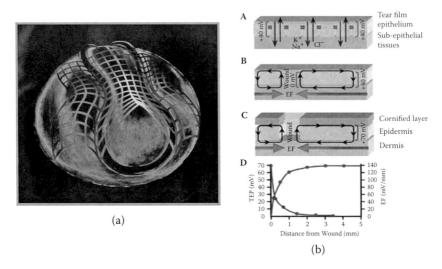

(a)

(b)

FIGURE 5.2   See color insert. (a) The skin potential across amphibian embryos varies spatially and temporally. A pseudocolored map of the skin p.d. is superimposed on the dorsal surface of an amphibian embryo during neurulation. The rostral end of the embryo faces outward. Higher TEP values are shown rostrally (yellow) than caudally, and higher TEP values are evident more medially (yellow) than laterally (pink/purple). (Taken from Shi and Borgens, *Dev. Dyn.*, 202, 101–14, 1995.) (b) Model of transporting corneal epithelium. (A) A single epithelial layer is shown (there are around six in cornea) with each cell separated by tight junctional electrical seals (brown squares). Inward transport of Na^+ and K^- and active efflux of Cl^- create a transepithelial potential (TEP) difference of around +40 mV, inside positive. (B,C) Wounding the epithelial layer in cornea (B) and skin (C) short-circuits the TEP, which falls to 0 mV at the wound, but remains high distal from the wound edge. This induces a flow of ionic current (black arrowheads), and a steady voltage gradient is established with the cathode at the wound (EF red and green arrows). (D) Direct measurements of mammalian skin TEP (red line) as a function of distance from the wound edge. The EF (green line) resulting from the TEP gradient is 140 mV/mm very near the wound edge. The color gradients of the EF arrows in C and D indicate that the EF is strongest very near the wound, and that the potential gradient (hence EF) is steeper in the mammalian skin wound than in the corneal wound, where the TEP is smaller. Therefore, the gradient of TEP per unit distance would be expected to be smaller in the cornea. (Redrawn from Vanable, in *Electric Fields in Vertebrate Repair*, ed. Borgens et al., Alan R. Liss, New York, 1989, pp. 171–224, Figure 3. With permission.)

current out from the area where the hind limb bud will appear, and this current is predictive since it appears several hours before the limb bud (Borgens et al., 1983; Robinson, 1983). The consequence of this is that focally localized current flows out of the limb bud, but there are no such currents across the flank skin close by. Injecting charged fluorescent albumin into the extracellular spaces below flank skin resulted in radial diffusion. Injecting charged albumin into the developing limb bud, by contrast, resulted in a comet-like distribution of protein that was determined by the coexisting electrical signals within the extracellular space (Messerli and Robinson, 1997).

SKIN AND CORNEA SEPARATE CHARGE, WHICH CREATES A TRANSEPITHELIAL POTENTIAL (TEP) DIFFERENCE

Du Bois Reymond (1848) was first to report that a potential difference existed across adult frog skin, and this could be greater than +100 mV, internally positive. Ussing, after whom the chamber is named, largely determined that this resulted from active transport of Na$^+$ inwards (Koefoed-Johnsen and Ussing, 1958). Embryonic epithelia share this property from the early blastocoele stage, since the blastocoele cavity is internally positive with respect to the outside (McCaig and Robinson, 1981). Critically, this transepithelial potential difference is not uniform in all locations, but is regulated spatially and temporally, and this may confer morphogenetic information both in development and in regeneration. For example, the skin potential across the dorsal regions of axolotl tadpoles around neurulation when the neural tube starts to fold over is higher rostrally than caudally, and higher medially than laterally (Shi and Borgens, 1994, 1995; Figure 5.2a). This remains true throughout neurulation, and then once neural tube closure is complete and the earliest axons have grown out, these spatial differences disappear. Crucially, disrupting this spatially variable map pharmacologically, physically (by shunting current through inappropriate regions), or electrically, by placing whole embryos in steady dc EFs, grossly disrupts normal embryonic development, and the nervous system is particularly badly affected (Borgens and Shi, 1995; Hotary and Robinson, 1992; Metcalf and Borgens, 1994). So the electrical signals generated by the embryonic epithelium encode important developmental information (Borgens and Shi, 1995). The same is true of adult epithelium after wounding.

All epithelial and endothelial layers that separate ions to establish a TEP that is maintained by tight junctions will generate a wound-induced EF as follows (see Chapter 1). The stratified mammalian cornea supports a transcorneal potential difference (TCPD) of +30 to +40 mV, internally positive (Figure 5.2b), with the upper layer of epithelial cells all connected by tight junctions that confer barrier properties and electrical resistance to the free flow of ions. Mammalian corneal epithelium transports Na$^+$ and K$^+$ inwards from tears to extracellular fluid, with Cl$^-$ transported out from the extracellular fluid into the tear fluid (Candia, 1973; Klyce, 1975). This charge separation establishes the TCPD. Wounding the epithelium creates an open hole and compromises the high electrical resistance established by the tight junctions. The TCPD falls instantly to zero at the wound, and the epithelium becomes short-circuited, locally. However, because normal ion transport continues in unwounded epithelial cells, the TCPD remains at normal values around 500 µm to 1 mm back from the wound edge. It is this gradient of electrical potential, 0 mV at the short-circuited lesion, +40 mV in unwounded tissue, that instantly establishes a steady, laterally oriented, wound-induced EF where the wound is the cathode. The vector of the wound-induced EF runs under the basal surfaces of the epithelial cells and returns within the tear film across the apical surface of the epithelium (Figure 5.2b). It therefore lies orthogonal to the TCPD generated across the intact epithelium, which has an apical to basal orientation (McCaig et al., 2005).

Steady wound-induced EFs have been measured in two studies. Cuts were made in bovine cornea and in guinea pig and human skin, and the potential difference across the epithelium at different distances back from the wound edge was measured directly using glass microelectrodes. In skin, the peak voltage gradient at the wound edge was 140 mV/mm (Barker et al., 1982), and in cornea 42 mV/mm, although the latter is an underestimate (Chiang et al., 1992). In each tissue the voltage gradient dropped off exponentially from the wound edge. Direct measurements in skin showed voltage gradients of 140 mV/mm in the first 250 µm, 40 mV/mm between 300 and 500 µm, and around 10 mV/mm at 500 to 1,000 µm from the wound edge (Barker et al., 1982).

These observations are important because: (1) The flow of ions at the wound establishes an instantaneous voltage gradient with a vector directed toward the wound edge. This gradient is sustained until wound healing reseals the epithelium electrically and reestablishes a uniformly high electrical resistance, at which point the wound-induced EF drops to zero. (2) Increasing or decreasing the TEP would inevitably increase and decrease the wound-induced voltage gradient and its effects on cell regulation. (3) Cell behaviors controlled by the endogenous EF will be regulated spatially with distance back from the wound, because the voltage gradient drops off exponentially from the wound edge. (4) All cell activities within 500 µm of a wound edge in skin and cornea, and probably in any ion-transporting epithelium or endothelium, gut and blood-brain barrier, for example, inevitably take place within a standing voltage gradient. These include epithelial cell division, epithelial cell migration, nerve sprouting, infiltration of leukocyte and dendritic cells, and endothelial cell remodelling with associated angiogenesis.

Finally, local loops of electrical information must arise, for example, following apoptosis of a single epithelial cell or the sloughing off through damage of a single cell. In each case, single cell loss will generate a local electrical signal that persists until the tight junction electrical seals with neighboring cells re-form. It is especially interesting, therefore, that in stratified epithelia, the cells immediately below the apical layer possess preformed tight junctional proteins that are rapidly assembled to seal a new replacement apical layer cell to its new neighbors within minutes (Hudspeth, 1975). Clearly, epithelia have well-developed mechanisms to reconstruct their barrier function and re-form electrical seals rapidly. In addition, recent evidence shows that tight junctions, which are key to the electrical sealing properties of epithelia, are highly dynamic and change their conductance markedly on a rapid timescale of seconds (Shen et al., 2008). This will create rapidly changing and spatially variable electrical signals between different locations within an intact, uninjured epithelium.

Before describing how cells respond to these wound-induced electrical signals, consider an evolutionary thought experiment: How did the earliest single cell survive a wound in its membrane?

Cells seal large holes in their membranes rapidly. In sea urchin eggs, the influx of Ca^{2+} ions through a hole in the membrane drives a breakup of intracellular membranous organelles into multiple vesicles, which then are driven toward the open wound (Terasaki et al., 1997). Here they create a plug that coalesces to form a new membrane that prevents cellular death from injury. It is not clear why the masses of vesicles move toward the hole, but one possibility is through electrophoresis. Membrane-bound vesicles are negatively charged, and the collapsed potential difference across the cell membrane will make the hole electrically positive with respect to the cytoplasmic side of the remaining lipid bilayer membrane. So both single cells and sheets of cells may use electrical strategies in mounting an immediate wound healing response. In each case, the potential difference becomes short-circuited, either across the single cell membrane or across the whole epithelial sheet, and this initiates an electrical gradient that can act as a guidance cue. This concept applies equally to epithelia and endothelia and may have been preserved from single cells to multicellular organisms with epithelia. Interestingly, one of the earliest organisms with an epithelium is hydra, which sustains a high internally positive TEP of around +60 mV (Josephson and Macklin, 1967).

EPITHELIAL CELLS RESPOND STRONGLY TO AN EF

How do epithelial cells respond to these endogenous electrical signals? A number of tissue culture studies mimic the wound-induced EF in a dish to study cellular responsiveness; a noninclusive summary follows. Bovine and human corneal epithelial cells show robust cathodal migration (Zhao et al. 1996). Cross-perfusion of medium to disrupt any induced chemical gradients confirms that this is a primary response to the electrical vector and not a secondary response to any gradients of chemicals that may have built up in the medium through electrophoresis. Growth factors are essential, since cathodal migration was absent in serum-free medium, but restored by the addition of specific growth factors to serum-free medium, for example, EGF, bFGF, and TGF β 1 (Zhao et al., 1996). Function blocking antibodies to the EGF receptor also prevented cathodal-directed migration, as did blocking MAP kinase activation (Zhao et al., 1999a). One concept, therefore, is that EGF receptors accumulate at the leading cathodal edge by electroosmotic redistribution within the plane of the membrane, and this asymmetrically activates EGF receptor signaling cascades, with leading-edge (cathodal) MAP kinase activation driving cathodal assembly of the cytoskeleton and cathodal-directed migration (Zhao et al., 1999a, 2002; Figure 5.3).

In addition, two genes have been identified in wounded rat cornea and skin that encode PI3K and PTEN and which regulate electrically driven wound healing (Zhao et al., 2006). In keratinocytes from PI3K–/– mice (p110γ gene deletion), EF-induced signaling was decreased, directed movement of the healing epithelium was abolished, and electrotaxis was impaired. Conversely, tissue-specific deletion of the gene encoding PTEN, a negative regulator of PI3K, enhanced both EF-induced signaling and directed keratinocyte migration. Another mediator that couples electric stimuli to intracellular responses is the Na$^+$/H$^+$ exchanger 1 (NHE1), which has been implicated in directional cell migration. NHE1 inhibitors decreased the directedness of cell migration in electric fields (Zhao et al., 2006).

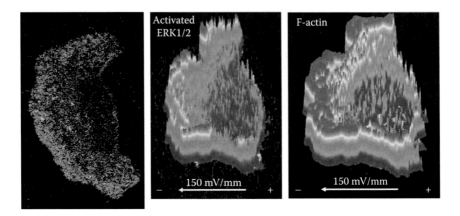

FIGURE 5.3   See color insert. Cathodal accumulation of EGF receptor (left), second messenger molecules (middle), and the F-actin cytoskeleton (right) may underpin cathodally directed cell migration. In all three frames the cathode is at the left and the EF applied was 150mV/mm for three hours. (From Zhao, Pu, Forrester, and McCaig, unpublished.)

Other receptors and signaling mechanisms also are involved, probably in parallel, coactivated signaling pathways. Pullar et al. implicated β4-integrins and β-adrenergic receptors in EF-driven wound healing. They showed that keratinocytes with truncated β4-integrins were blinded to an applied EF in the absence of EGF, and that addition of EGF together with functioning β4-integrins enhanced EF-directed cell migration (Pullar et al., 2006). To explain this, the authors suggested that cooperativity between EF-activated EGF signaling and β4-integrin signaling through the small GTPase Rac might occur, perhaps at focal adhesions (see Chapter 6). Pullar and colleagues have also shown that activation of β-adrenergic receptors inhibits keratinocyte responsiveness to a physiological EF and slows corneal wound healing. By contrast, β-adrenergic receptor antagonists enhanced the migration response to an EF and accelerated corneal wound healing (Pullar et al., 2007).

Importantly, cathodal migration occurs also in corneal epithelial monolayers with scratch wounds (Zhao et al., 2006). Figure 5.4 (see movie 1 in Zhao et al., 2006) highlights an experiment in which multiple signaling cues operate in a hierarchy. In the left column, which shows wound closure in untreated scratches, the directional guidance signals probably include the establishment of local chemical gradient of growth factors and cytokines released by the injured cells at the wound edge. Additionally, cells read and respond to open spaces, because they are released from neighbor-neighbor contacts, and this release of contact inhibition drives directed migration into the wound void. Collectively, these are recognized as default wound healing cues, but in this experiment the wounded monolayer is presented additionally with the electrical signal that would arise *in vivo* at a corneal wound. When this has both a physiological magnitude and a physiological polarity vector (cathode in the wound), wounded monolayers

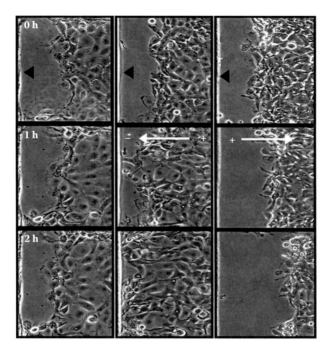

FIGURE 5.4 Scratch monolayer wound healing assays (see movie 1 in Zhao et al., 2006). Left panel: No electrical field. A wound closes gradually over several hours. Middle column: EF 150mV/mm with normal polarity. Wound healing is faster than without EF. Right column: EF 150mV/mm with "wrong" polarity. A wound does not heal, but opens up when the anode is in the wound.

heal faster (middle column). Intriguingly, however, when the magnitude of the signal remains constant, but the polarity is reversed and now is opposite to that which would be present at a corneal wound, the wound does not close, but opens up (right column). In other words, the electrical signal now overrides the default wound healing signals and prevents wound healing. Titrating these signals against one another might be interesting. At what EF strength do the default wound healing conditions prevail and drive wound closure, despite the wrong polarity of the electrical signal?

We know that cyclic nucleotide signaling is important in driving this closure/opening decision in response to the EF cue. Studies on chemotropic turning of neuronal growth cones have implicated the intracellular level of cAMP in the growth cone as a regulator of whether a particular chemotropic molecule acts as a chemo-attractant or as a chemo-repellent cue (Song et al., 1997). Raising or lowering cAMP can turn attractive cues into repellent cues and vice versa. Something similar occurs in EF-driven monolayer wound healing. We have used a protein kinase A agonist, spcAMP, and a protein kinase A antagonist, rpcAMP, to respectively increase and decrease levels of cAMP in the epithelial monolayers. spcAMP-treated monolayers in which we expect cAMP levels to be enhanced now convert the repellent cue (that is, the EF with the wrong polarity) into an attractive electrical signal. Now the wound closes despite an EF of the wrong polarity (Pu, Zhao, and McCaig, unpublished data). The ability to manipulate pharmacologically the attractive/repellent properties of an electrical wound healing cue has completely untapped clinical potential, for example, in the treatment of cancer, spinal cord injury, and wound healing.

CORNEAL WOUND HEALING IN RAT IS REGULATED BY THE WOUND-INDUCED EF

It has been clear for twenty years that electrical signals have a role in amphibian and mammalian wound healing (Rajnicek et al., 1988; Vanable, 1989). More recently our group has shown that directed migration of epithelial cells into the wound, the proliferation of epithelial cells, the axis of division of epithelial cells, the proportion of nerves sprouting at the wound, and the directional growth of nerve sprouts toward the wound edge are all regulated *in vivo* by the wound-induced electrical signal (Song et al., 2002, 2004).

The cellular mechanisms driving Na^+ or K^+ influx and Cl^- efflux across corneal epithelium differ, but are well understood and can be manipulated using a battery of pharmacological agents. Increasing or decreasing the TCPD pharmacologically must increase and reduce, respectively, the wound-induced EF. We have modulated the TCPD using chemicals acting by different cellular mechanisms but sharing the ability to change the TCPD (Song et al., 2004). In rat corneas treated with PGE_2 (0.1 mM), which increases Cl^- efflux, or with aminophylline (10 mM), which inhibits phosphodiesterase breakdown of cAMP and enhances Cl^- efflux, the TCPD increased three- to fourfold. Wounds treated with these drugs healed 2.5 times faster within the first ten hours than control wounds (Song et al., 2002). By contrast, wounds treated with the Na^+/K^+ATPase inhibitor ouabain (10 mM), which reduced the TCPD fivefold to 18% of normal, showed markedly slower wound healing. The rate of wound healing was directly proportional to the size of the TCPD, and therefore to the size of the wound-induced EF (Song et al., 2002).

The natural EF present at experimental wounds in the isolated bovine eye also regulates wound healing (Sta. Iglesia et al., 1998). Reducing the natural EF with the Na^+ channel blocker benzamil (30 mM) or with Na^+-free physiological saline slowed wound healing. The addition of injected current to restore and amplify the endogenous EF at wounds in Na^+-free medium enhanced wound healing (Sta. Iglesia et al., 1998).

THE WOUND-INDUCED EF REGULATES PROLIFERATION OF EPITHELIAL CELLS *IN VIVO*

Cornea respond to a wound by increasing epithelial cell proliferation to provide new cells to fill the wound void. This too is regulated electrically. Cell division is rare in the central area of unwounded corneal epithelium, but prolific within a peripheral ring of stem cells called the limbus, which normally provides cells for epithelial turnover (Cotsarelis et al., 1989). Wounding the corneal epithelium stimulated epithelial cell division, especially near the lesion. Increasing the endogenous wound-induced EF with PGE_2 or aminophylline induced a 40% increase in cell divisions within 600 µm of the wound edge, and suppressing the EF with ouabain caused a 27% suppression of mitoses (Song et al., 2002). Manipulating the EF therefore clearly regulated the cell cycle and altered the frequency of cell division. Because this modulates the population pressure of cells within the epithelium, this is likely to contribute to the rate of wound healing.

THE WOUND-INDUCED EF REGULATES THE AXIS OF EPITHELIAL CELL DIVISION *IN VIVO*

Cultured corneal epithelial cells divide with a cleavage plane that forms perpendicular to the EF vector (Zhao et al., 1999b). In other words, in preparation for cytokinesis the mitotic spindle aligns parallel with the EF vector. The reasons for this are unknown. Importantly, however, the same striking phenomenology occurs *in vivo* (Song et al., 2002). In untreated corneal wounds, which generate their own endogenous EF, the mitotic spindles lie roughly parallel to the EF vector, with cleavage occurring perpendicular to this. Enhancing the wound-generated EF with PGE_2 or aminophylline roughly doubled the proportion of dividing cells whose cleavage planes were perpendicular to the EF vector. In contrast, reducing the wound-generated EF to less than 20% of its endogenous value with ouabain abolished wound-oriented cell divisions.

If the endogenous EF in rat cornea is causal in directing the axis of cell division, then its effects should be highest at the wound edge and decline back from there, because the EF declines exponentially away from the wound. This is the case. Orientation of the mitotic spindle was strongest in the first 200 µm and roughly halved 200 to 400 µm back from the wound. By 600 µm back from the edge, the angle of the mitotic spindle was not different from those seen in the limbus (1,700 µm away), and both were randomly oriented with respect to the wound-generated EF vector (Song et al., 2002). Importantly, 600 µm corresponds to the measured distance that the EF penetrates into the tissue. Oriented division therefore dropped to zero as a function of distance back from the wound edge, and there was no oriented division in the distant limbus where the EF would be zero.

Enhancing the EF with PGE_2 or aminophylline increased the orientation of cell division, with significant orientation now occurring farther back from the wound edge, at 600 µm. Collapsing the EF with ouabain abolished oriented cell division, even within 200 µm of the wound edge. Clearly, the naturally occurring EF controls the orientation of cell divisions *in vivo*.

One clue to potential mechanisms is that phospholipid second messenger signals may transduce the EF into oriented cell division, because the aminoglycoside antibiotic neomycin, which inhibits phospholipase C, but has no effects on the TCPD (or EF), abolished oriented cell divisions *in vivo*. Interestingly, neomycin also prevents EF-induced orientation of embryonic myoblasts and of neuronal growth cones (McCaig and Dover, 1991; Erskine et al., 1995).

THE WOUND-INDUCED EF REGULATES NERVE SPROUTING *IN VIVO*

Nerves sprout in response to wounds in skin (Fitzgerald et al., 1975; Matsuda et al., 1998) and cornea (Rosza et al., 1983; Beuerman and Rosza, 1984). In cornea, with its rich sensory innervation, this is a biphasic process. Early collateral sprouts appear within only a few hours, mostly from intact fibers near the wound. In rabbit cornea these early collateral sprouts show a striking orientation with many parallel nerve bundles growing directly toward the wound edge (Rosza et al., 1983). The early sprouts are transient, and over the following seven days or so, they retract and are replaced by regenerating neurites (Beuerman and Rosza, 1984). The cues guiding growth cones of early sprouts directly toward the wound edge had not been explored before we asked whether the wound-induced EF induced this. Our experiments provided clear evidence of a physiological role for electrical guidance of nerve growth *in vivo* (Rosza et al., 1983). A 4 mm nasal-to-temporal slit wound was made in rat cornea. Early nerve sprouts are evident by sixteen hours, but are not oriented with respect to the wound edge. Between sixteen and twenty-four hours, many more nerve sprouts appear, and most are perpendicular to the wound edge. When the wound-generated EF was enhanced with PGE_2, aminophylline, $AgNO_3$, or ascorbic acid, neurite growth toward the wound was enhanced. More sprouts appeared, sprouts appeared earlier, and sprouts oriented toward the wound edge earlier (within sixteen hours). Collapsing the EF with ouabain or furosemide did not prevent early collateral nerve sprouting, but sprouts were not directed toward the wound edge (Song et al., 2004).

The discovery that the extent of sprouting and the directional growth of the sprouts are controlled by the wound-generated EF is significant because: (1) Effective wound healing in the cornea (and probably in skin and other epithelia and endothelia) requires a normal sensory nerve supply. For example, corneal wound healing is compromised in patients with sensory neuropathies such as occurs in diabetes (Friend and Thoft, 1984). These conditions are characterized by repeated attempts to reepithelialize the wound, but in the absence of strong reinnervation healing is poor and the epithelium sloughs off repeatedly. Important neurotrophic interactions clearly are needed between the epithelial cells and their underlying sensory innervation (Beuerman and Schimmelpfenning, 1980), which enter the wound together with the migrating epithelial cells (Rosza et al., 1983). (2) Because electrical signals arise immediately at a wound and control multiple cell behaviors *in vivo* (cell proliferation, directed cell division, directed epithelial cell migration, and directed nerve sprouting), this suggests that the EF may orchestrate an integrated response of interdependent cell behaviors required coordinately for effective wound healing. Perhaps the instantaneous wound-induced electrical signal "kick-starts" many or all of these cellular processes. These might include additionally the upregulation of growth factor receptors (the EGF receptor is upregulated in corneal epithelial cells by an EF (Zhao et al., 1999, 2002)) and the upregulation of growth factor secretion, which would give rise to chemical gradients (secretion of VEGF by endothelial cells is upregulated by exposure to a physiological EF (Bai et al., 2003)).

CURRENT EPITHELIAL WOUND HEALING TREATMENTS UNWITTINGLY TARGET THE ENDOGENOUS EF

Corneal wounds and skin burns have traditionally been treated with silver nitrate, because of its antimicrobial properties. However, this is a good example of an existing clinical therapy whose mechanism of action may lie also in enhancement of the wound-induced electrical signal. In rabbit cornea, silver nitrate induces an eightfold increase in transepithelial Na^+ transport and a doubling of the TCPD, which is maintained for several hours (Klyce and Marshall,1982), while clean wounds in pig skin healed faster when treated with silver by mechanisms that were not antimicrobial (Geronemus, 1979). A second example is ascorbic acid (vitamin C). It has been known for centuries that scurvy delays wound healing, and topical vitamin C has been used to enhance wound healing with the assumption that the mechanism of action lies in enhanced collagen formation (Bourne 1944). This may be only part of the story, since vitamin C also enhances chloride transport across toad corneal epithelium and produces a 100% increase in the electrical potential difference across the epithelium (TEP; Scott and Cooperstein, 1975). So again, an electrical contribution to the mechanism of action seems likely.

ELECTRICAL WOUND HEALING

Because EF-induced wound healing depends on known ionic transport mechanisms, there is growing interest in approaching wound healing therapies with topically applied agents that target the enhancement or suppression of ion channels or pumps to regulate the transepithelial potential difference. Electrodes embedded in a hydrogel whose ionic or pharmacological content enhances epithelial-driven wound currents would represent a novel approach to wound healing therapies. Suppressing electrical signals that are implicated in promoting the migration of cancer cells that may underpin metastasis also may be possible. A new device (the Dermacorder) that measures electrical fields at wounds in human skin also has been developed and is likely to be of significant clinical use (Nuccitelli et al., 2008), perhaps in combination with some of the above novel approaches.

Electrical therapies to heal bone fractures are commonly used by clinicians. The use of similar treatments for nonhealing skin ulcers has been sporadic, but is increasing markedly in Europe. A device known as Wound-El has been adopted in 160 hospitals with around 3,000 patients treated in Germany, and by 25 hospitals treating 300 to 500 patients in France (since 2009). A better understanding of the physiology of electrical wound healing appears to be paving the way for a greater acceptance of its clinical efficacy.

REFERENCES

Bai, H., Zhao, M., Forrester, J. V., and McCaig, C. D. 2003. Electric stimulation has a direct effect on vascular endothelial cells: Guiding cell elongation, orientation, and migration. *J. Cell Sci.* 117:397–405.

Barker, A. T., Jaffe, L. F., and Vanable, J. W., Jr. 1982. The glabrous epidermis of cavies contains a powerful battery. *Am. J. Physiol.* 242:R358–66.

Beuerman, R. W., and Rosza, A. J. 1984. Collateral sprouts are replaced by regenerating neurites in the wounded corneal epithelium. *Neurosci. Lett.* 44:99–104.

Beuerman, R. W., and Schimmelpfenning, B. 1980. Sensory denervation of the rabbit cornea affects epithelial properties. *Exp. Neurol.* 69:196–201.

Borgens, R. B., Rouleau, M. F., and DeLanney, L. E. 1983. A steady efflux of ionic current predicts hind limb development in the axolotl. *J. Exp. Zool.* 228:491–503.

Borgens, R. B., and Shi, R. 1995. Uncoupling histogenesis from morphogenesis in the vertebrate embryo by collapse of the transneural tube potential. *Dev. Dyn.* 203:456–67.

Bourne, G. H. 1944. Effect of vitamin C deficiency on experimental wounds. Tensile strength and histology. *Lancet* 1:688–92.

Candia, O. A. 1973. Short-circuit current related to active transport of chloride in frog cornea: Effects of furosemide and ethacrynic acid. *Biochim. Biophys. Acta* 298:1011–14.

Chiang, M., Robinson, K. R., and Vanable, J. W., Jr. 1992. Electrical fields in the vicinity of epithelial wounds in the isolated bovine eye. *Exp. Eye Res.* 54:999–1003.

Cotsarelis, G., Cheng, S.-Z., Dong, G., Sun, T.-T., and Lavker, R. M. 1989. Existence of slow-cycling limbal epithelial basal cells that can be preferentially stimulated to proliferate: Implications on epithelial stem cells. *Cell* 57:201–9.

Du Bois Reymond, E. 1848. *Untersuchungen uber tierische elektrizitat.* Berlin, Georg Reimer.

Erskine, L., Stewart, R., and McCaig, C. D. 1995. Electric field directed growth and branching of cultured frog nerves: Effects of aminoglycosides and polycations. *J. Neurobiol.* 26:523–36.

Fitzgerald, M. J. T., Folan, J. C., and O'Brien, T. M. 1975. The innervation of hyperplastic epidermis in the mouse: A light microscopic study. *J. Invest. Dermatol.* 64:169–74.

Friend, J., and Thoft, R. A. 1984. The diabetic cornea. *Int. Ophthalmol. Clin.* 24:111–23.

Geronemus, R. G., Mertz, P. M., and Eaglstein, W. H. 1979. Wound healing. The effects of topical antimicrobial agents. *Arch. Dermatol.* 115:1311–14.

Hotary, K. B., and Robinson, K. R. 1994. Endogenous electrical currents and voltage gradients in *Xenopus* embryos and the consequences of their disruption. *Dev. Biol.* 166:789–800.

Hudspeth, A. J. 1975. Establishment of tight junctions between epithelial cells. *Proc. Natl. Acad. Sci. USA* 72:2711–13.

Jaffe, L. F. 1985. Extracellular current measurements with a vibrating probe. *Trends Neurosci.* 8:517–21.

Josephson, R. K., and Macklin, M. 1967. Transepithelial potentials in *Hydra*. *Science* 156:1629–31.

Klyce, S. D. 1975. Transport of Na, Cl, and water by the rabbit corneal epithelium at resting potential. *Am. J. Physiol.* 228:1446–52.

Klyce, S. D., and Marshall, W. S. 1982. Effects of Ag+ on ion transport by the corneal epithelium of the rabbit. *J. Membr. Biol.* 66:133–44.

Koefoed-Johnsen, V., and Ussing, H. H. 1958. The nature of the frog skin potential. *Acta Physiol Scand.* 42:298–308.

Matsuda, H., Koyama, H., Sato, H., Sawada, J., Itakura, A., Tanaka, A., Matsumoto, M., Konno, K., Ushio, H., and Matsuda, K. 1998. Role of nerve growth factor in cutaneous wound healing: Accelerating effects in normal and healing-impaired diabetic mice. *J. Exp. Med.* 187:297–306.

McCaig, C. D., and Dover, P. J. 1991. Factors influencing perpendicular elongation of embryonic frog muscle cells in a small applied electric field. *J. Cell Sci.* 98:497–506.

McCaig, C. D., Rajnicek, A. M., Song, B., and Zhao, M. 2005. Controlling cell behaviour electrically: Current views and future potential. *Physiol. Rev.* 85:943–78.

McCaig, C. D., and Robinson, K. R. 1981. The ontogeny of the transepidermal potential difference in frog embryos. *Dev. Biol.* 90:335–39.

Messerli, M., and Robinson, K. R. 1997. Endogenous electrical fields affect the distribution of extracellular protein in *Xenopus* embryos. *Mol. Biol. Cell.* 8:1296A.

Metcalf, M. E. M., and Borgens, R. B. 1994. Weak applied voltages interfere with amphibian morphogenesis and pattern. *J. Exp. Zool.* 268:322–38.

Nuccitelli, R., Nuccitelli, P., Ramlatchan, S., Sanger, R., and Smith, P. J. 2008. Imaging the electric field associated with mouse and human skin wounds. *Wound Repair Regen.* 16:432–41.

Pullar, C. E., Baier, B. S., Kariya, Y., Russell, A. J., Horst, B. A. J., Marinkovitch, M. P., and Isseroff, R. R. 2006. β4 integrin and epidermal growth factor co-ordinately regulate electric field-mediated directional migration via Rac1. *Mol. Biol. Cell.* 17:4925–35.

Pullar, C. E., Zhao, M., Song, B., Pu, J., Reid, B., Ghoghawala, S., McCaig, C. D., and Isseroff, R. R. 2007. β-Adrenergic receptor agonists delay while antagonists accelerate epithelial wound healing: Evidence of an endogenous adrenergic network within corneal epithelium. *J. Cell. Physiol.* 211:261–72.

Rajnicek, A. M., Stump, R., and Robinson, K. R. 1988. An endogenous sodium current may mediate wound healing in *Xenopus neurulae. Dev. Biol.* 128:290–99.

Robinson, K. R. 1983. Endogenous electrical current leaves the limb and pre-limb region of the *Xenopus* embryo. *Dev. Biol.* 97:203–11.

Rosza, A. J., Guss, R. B., and Beuerman, R. W. 1983. Neural remodelling following experimental surgery of the rabbit cornea. *Invest. Ophthalmol. Vis. Sci.* 24 :1033–51.

Scott, W. N., and Cooperstein, D. F. 1975. Ascorbic acid stimulates chloride transport in the amphibian cornea. *Invest. Ophthalmol.* 14:763–66.

Shen, L., Weber, C. R., and Turner, J. R. 2008. The tight junction protein complex undergoes rapid and continuous molecular remodelling at steady state. *J. Cell Biol.* 181:683–95.

Shi, R., and Borgens, R. B. 1994. Embryonic neuroepithelium sodium transport, the resulting physiological potential, and cranial development. *Dev. Biol.* 165:105–16.

Shi, R., and Borgens, R. B. 1995. Three dimensional gradients of voltage during development of the nervous system as invisible coordinates for the establishment of the embryonic pattern. *Dev. Dyn.* 202:101–14.

Song, B., Zhao, M., Forrester, J. V., and McCaig, C. D. 2002. Electrical cues regulate the orientation and frequency of cell division and the rate of wound healing *in vivo. Proc. Natl. Acad. Sci. USA* 99:13577–82.

Song, B., Zhao, M., Forrester, J. V., and McCaig, C. D. 2004. Nerves are guided and nerve sprouting is stimulated by a naturally occurring electrical field *in vivo. J. Cell Sci.* 117:4681–90.

Song, H. J., Ming, G. L., and Poo, M.-M. 1997. cAMP-induced switching in turning direction of nerve growth cones. *Nature* 388:275–79.

Sta. Iglesia, D. D., and Vanable, J. W., Jr. 1998. Endogenous lateral electric fields around bovine corneal lesions are necessary for and can enhance normal rates of wound healing. *Wound Rep. Reg.* 6:531–42.

Terasaki, M., Miyake, K., and McNeil, P. L. 1997. Large plasma membrane disruptions are rapidly resealed by Ca^{2+}-dependent vesicle–vesicle fusion events. *J. Cell Biol.* 139:63–74.

Vanable, J. W., Jr. 1989. Integumentary potentials and wound healing. In *Electric fields in vertebrate repair*, ed. R. B. Borgens, K. R Robinson., J. W. Vanable Jr., and M. E. McGinnis, 171–224. New York: Alan R. Liss.

Zhao, M., Agius-Fernandez, A., Forrester, J. V., and McCaig, C. D. 1996. Orientation and directed migration of cultured corneal epithelial cells in small electric fields are serum dependent. *J. Cell Sci.* 109:1405–14.

Zhao, M., Dick, A., Forrester, J. V., and McCaig, C. D. 1999. Fibronectin and laminin enhance electric field-directed migration of corneal epithelial cells by a mechanism involving upregulation and cathodal redistribution of EGF receptors. *Mol. Biol. Cell* 10:1259–76.

Zhao, M., Forrester, J. V., and McCaig, C. D. 1999. Physiological electric fields orient the axis of cell division. *Proc. Natl. Acad. Sci. USA* 96:4942–46.

Zhao, M., Pu, J., Forrester, J. V., and McCaig, C. D. 2002. Membrane lipids, EGF receptors and intracellular signals co-localize and are polarized in epithelial cells moving directionally in a physiological electric field. *FASEB J.* 16:857–59 and 10.1096/fj.01–0811fje.

Zhao, M., Song, B., Pu, J., Wada, T., Reid, B., Tai, G., Wang, F., Guo, A., Walczysko, P., Gu, Y., Sasaki, T., Suzuki, A., Forrester, J. V., Bourne, H. R., Devreotes, P. N., McCaig, C. D., and Penninger, J. M. 2006. Electrical signals control wound healing through phosphatidylinositol-3-OH kinase-gamma and PTEN. *Nature* 442(7101):457–60.

6 Physiological Electric Fields Can Direct Keratinocyte Migration and Promote Healing in Chronic Wounds

Christine E. Pullar
Department of Cell Physiology and Pharmacology
University of Leicester
Leicester, United Kingdom

CONTENTS

INTRODUCTION

The skin is the largest organ in the body, and its outer layer, the epidermis, forms a protective barrier against the external environment, protecting us from a myriad of potentially harmful agents, such as heat and bacterial infections. It is therefore imperative that any damage to the epidermis is repaired as quickly as possible. Consequently, evolution has primed our healing responses to repair wounds rapidly. Inflammatory cells arrive at a wound site within minutes to clear the wound of any infection (see Chapter 8), and sedentary keratinocytes at the wound edge become migratory within a few hours to begin the process of reepithelialization. The exact cues that initiate the changes in keratinocyte morphology and motility are unknown, but chemical, physical, and electrical cues are present at the wound edge.

An electric field gradient is generated immediately upon wounding skin, and the center of the wound quickly becomes more negative than the surrounding tissue; it is therefore the cathode of this endogenous field. This can be measured with the Dermacorder and is highest at the wound edge (see Chapter 1). Measurements at the edge of a human skin wound are generally in the range of 100 to 150 mV/mm. In the mid-1990s it was demonstrated that human keratinocytes could sense and respond to an applied direct current electric field of physiological strength (100 mV/mm). Within a few minutes of field application, cells would turn and migrate directionally toward the cathode, a process known as galvanotaxis or electrotaxis. Over the past fifteen years a number of studies have revealed various proteins that play a role in the mechanism of keratinocyte galvanotaxis, and these studies are described within this chapter. In addition, applied electric fields have recently been shown to alter the transcription of various inflammatory genes in keratinocytes.

Unfortunately, wounds do not always heal in a timely manner. Chronic wounds are very prevalent in the elderly and are extremely debilitating for the patient and costly for health care systems worldwide. Impairments in healing seem to coincide with alterations in keratinocyte morphology and function at the wound edge. Electrical stimulation is currently successful in enhancing the healing of chronic wounds. Hopefully, the knowledge gained from continued research on how keratinocytes and other cell types respond to applied electric fields will allow the optimization of current electrical stimulation protocols to further enhance chronic wound repair.

THE SKIN

The skin consists of two main layers, the epidermis and the dermis. The epidermis consists of consecutive layers of keratinocytes. The deepest layer of the epidermis, the stratum basale, contains the stem cells that undergo mitotic divisions to produce cells that regenerate the upper epidermal layers. Each epidermal layer differs in its state of differentiation. The basal layer consists of undifferentiated cells, and the level of differentiation increases toward the stratum corneum (Boukamp et al., 1988). As described in Chapter 1, keratinocytes are highly polarized, and the asymmetric distribution of ion channels on their apical and basolateral membranes concentrates sodium ions in the deeper epidermal layers, generating the transepithelial potential (TEP).

The dermal layer of skin lies beneath the epidermis, and consists of cellular components such as fibroblasts, connective tissue, and collagen. The dermis can be further divided into two layers, the papillary and the reticular dermis. The papillary dermis has a depth ranging from 300 to 400 μm (Sorrell and Caplan, 2004), depending on the anatomical location. The papillary dermis is characterized by poorly organized collagen fiber bundles that consist of type I and III collagen in comparison to the reticular dermis, which is comprised of thick, well-organized fiber bundles aligned parallel to the surface of the skin. The papillary dermis also contains more collagen III than the reticular dermis (Sorrell and Caplan, 2004).

AN OVERVIEW OF SKIN WOUND REPAIR

Wound healing requires the complex orchestration of a number of physiological processes that overlap in time. Adult wounds always heal with a scar, as healing responses appear primed to repair a wound quickly with little regard to the quality of the repaired skin (Shaw and Martin, 2009).

In order to orchestrate this complex sequence of overlapping physiological processes, performed by numerous cell types, guidance cues exist within the wound. As previously described in Chapter 1, a direct current (DC) electric field gradient is generated immediately upon wounding skin; therefore, all wound repair processes take place in the presence of this steady, long-lasting voltage gradient. In addition, chemical and physical cues are present in the wound environment, and presumably, the different wound guidance cues interact with each other to direct wound repair processes.

In general, wound repair can be divided into three phases: a hemostasis/inflammatory phase, a proliferative phase, and a remodeling phase. During the initial hemostasis/inflammatory phase, platelets degranulate to achieve hemostasis and prevent excessive blood loss, and the clot forms a provisional matrix for cells to migrate through. The first inflammatory cells to arrive at the wound site are neutrophils. They migrate to the wound within minutes of injury and remove any bacteria and foreign material from the site of injury. Macrophages arrive at the wound site around three days postinjury, while T cells and mast cells appear to play a role in the later phases of wound repair (Stramer et al., 2007).

It is essential that the epidermal barrier be repaired quickly to prevent infection. Within a few hours of injury, keratinocytes at the wound edge begin to migrate over the provisional matrix into the wound bed, to initiate reepithelialization. The cues that initiate this migratory response are thought to include mechanical and chemical signals in the local epidermal environment, but one important guidance cue is often overlooked, the wound electric field or injury potential. Several hours postwounding, keratinocytes become activated and alter their integrin expression profile to allow them to detach from the basement membrane and migrate over the provisional matrix in the wound bed (Martin 1997).

Endothelial cells and dermal fibroblasts take center stage in the proliferative phase. Dermal fibroblasts migrate into the provisional wound bed formed by the blood clot and begin to deposit collagen. Initially collage III is synthesized, which is later replaced by collagen I. Endothelial cells become activated to migrate and align to form new blood vessels in the provisional wound bed, supplying the regenerating tissue with oxygen and nutrients. Finally, dermal fibroblasts remodel the provisional wound bed in the ultimate phase of wound repair to form a mature wound scar (Martin 1997).

CHRONIC WOUNDS

Unfortunately, a large number of skin wounds do not heal normally. A plethora of factors can interrupt the healing process, including chronic disease, vascular insufficiency, diabetes, nutritional deficiencies, infection, sustained inflammation, advanced age, mechanical pressure, and neuropothies, to name a few (Fonder et al. 2008). A chronic wound is defined as a break in the skin for a long duration (>6 weeks) or wound recurrence at the same site. They are extremely prevalent in the elderly and are both painful and debilitating for the patient and very costly for health care systems worldwide. Global wound care expenditures are around \$13 to 15 billion annually (Walmsley, 2002). There are three main types of chronic wounds: pressure ulcers, lower limb ulcers of vascular aetiology (arterial and venous ulcers), and diabetic foot ulcers.

Efforts to develop new therapies to heal chronic wounds are hampered by the lack of knowledge of the mechanisms responsible for the chronicity of the wounds (Stojadinovic et al., 2005). However, it is well known that impaired healing coincides with alterations in keratinocyte function at the edges of chronic wounds. Activated β-catenin and c-myc are present in the epidermis of chronic wounds and appear to inhibit keratinocyte migration and alter their differentiation (Stojadinovic et al., 2005). Indeed, keratinocytes at the wound edge of venous ulcers do not initiate activation or differentiation pathways, resulting in a thick callus-like formation at the wound edge (Stojadinovic et al. 2005). Removal of the "bad" keratinocytes at the wound edge to allow healthy keratinocytes to replace them provides the biological basis to support debridement, enhancing reepithelialization and healing in chronic wounds (Pastar et al., 2008).

THE USE OF APPLIED DIRECT CURRENT ELECTRIC FIELDS TO HEAL CHRONIC WOUNDS

Since the mid-1960s considerable research has been directed at evaluating the effects of exogenous electrical currents on the healing of chronic wounds that are frequently unresponsive to standard treatment. The treatments that are currently available to patients with chronic wounds are influenced by federal and regional insurance authorities, who base reimbursement decisions on treatment effectiveness, which they in turn establish by determining the strength of evidence derived from basic science and clinical research trials (McCulloch and Kloth, 2010).

Electrical stimulation (ES) used to treat chronic wounds has been shown in numerous clinical studies to enhance wound healing. In a Cochrane review that is currently submitted (Kohl and Houghton, 2010), wounds (n = 1,075) treated with low-frequency ES, in addition to the standard wound care (SWC), showed a statistically significant and clinically relevant extra 25.5% wound area reduction after four weeks (PAR4) of treatment. Different wound types-pressure ulcers, diabetic foot ulcers, and lower leg ulcers-showed similar effects. Comparing unidirectional, monophasic ES with bidirectional ES showed consistent and clear positive effects of unidirectional ES; the extra PAR4 healing rate of unidirectional ES was 34.68% (n = 493; 95% confidence interval = 26.67 to 42.68), whereas the extra PAR4 healing rate of bidirectional ES dropped to 8.51% (n = 582; 95% confidence interval = −4.44 to 21.46) (Kohl and Houghton, 2010). In conclusion, the results compiled in the Cochrane meta-analysis demonstrate that ES increased wound closure rate. This effect was evident when ES was used to treat pressure ulcers, diabetic foot ulcers, lower leg ulcers, and ulcers of mixed aetiology. ES treatments that employed unidirectional ES produced statistically significant and clinically relevant improved outcomes (Kohl and Houghton, 2010).

WOUND CELL PHYSIOLOGY CAN BE DIRECTED BY PHYSIOLOGICAL EFS *IN VITRO*

In order to understand how electrical stimulation can enhance the healing of chronic wounds, a number of labs around the world are studying how the application of direct current EFs can alter cell physiology to enhance chronic wound repair.

AN APPLIED DC EF CAN ALTER THE DIRECTION OF CELL MIGRATION

The majority of the cell types that play a role in wound repair are capable of responding by migrating directionally to applied electrical guidance cues *in vitro*, as described in numerous chapters within this book. Cells can be plated onto extracellular matrix-coated glass in custom chambers where the volume of the cell chamber is limited to restrict Joule heating (Figure 2.5). An EF of physiological strength (100 mV/mm) can be applied across the chamber, which rests in a heating portal on the stage of an inverted microscope, and images can be automatically captured at defined intervals over a one- to two-hour period, controlled by custom automation of Improvision software. Using this setup, researchers have demonstrated that the majority of cells in the skin can sense and respond to an applied EF within minutes of application. The EF can direct cell migration, alter cell proliferation rate, align the axis of division, elongate cells, increase the level of growth factors secreted from cells, and direct nerve growth, as described in Chapters 4 through 11. The apparatus design for studying EF-mediated directional migration, galvanotaxis, or electrotaxis is covered in Chapter 2, and the experimental conditions are described below.

PRIMARY HUMAN KERATINOCYTES

In order to study the role of endogenous EFs in guiding keratinocyte migration, human keratinocytes were isolated directly from human neonatal foreskin and used at low passage (Rheinwald and Green, 1975). The initial isolation from the epidermis is termed passage 1. To expand the cells, they are trypsinized from the original culture and transferred to larger dishes, grown to almost confluence and frozen at passage 2. Each time a cell population is trypsinized, the passage number is increased by 1. We routinely use cells for galvanotaxis experiments between passages 3 and 7 (Pullar, personal communication).

EXPERIMENTAL CONDITIONS

Human neonatal keratinocytes are routinely plated as single cells onto collagen I–coated glass (60 µg/ml), in custom chambers for two to six hours, and the chambers are then placed on a heated stage (37°C), in the presence or absence of an applied electric field of 100 mV/mm (Pullar et al. 2001, 2006; Pullar and Isseroff, 2005). Experiments have been performed in the presence of a continuous cross-flow of medium (0.77 ml/s) to ensure that the cells were responding directly to the EF and not to an established chemical gradient (Nishimura et al. 1996).

Images of the cells are recorded at time 0 and then every ten minutes for one hour using an automation in Openlab software from Improvision (PerkinElmer). Cells are manually tracked, and the software is used to calculate the true speed (µm/min) and directionality (cosine of the angle of migration (θ)) for each cell and the average values for each experiment (n > 50). True distance indicates the actual distance that each cell traveled in the one-hour period, rather than the straight-line distance between the starting point and end point of each cell (net distance). The cosine θ describes the direction of migration and is a measure of the persistence of directionality, where θ is the angle between the field axis and the cells' path of migration during the last ten minutes of the experiment, i.e., the angle between the penultimate and final cell position (Figure 6.1). The field axis is established with an angle of 0° assigned to the cathode and an angle of 180° assigned to the anode. The cosine of a cell moving directly toward the cathode would therefore be 1 (cosine of 0 = 1), and the cosine of a cell moving directly toward the anode would be –1 (cosine of 180° = –1). A population of cells migrating in the absence of a directional cue will migrate randomly with an average cosine of 0 (Figure 6.2). The cosine θ will give a number between –1 and 1 for each cell, and the average of the cosine θ for all the cells in an experiment will provide an index of directional migration. All experiments are repeated at least

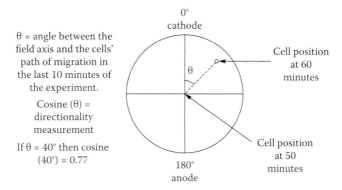

FIGURE 6.1 Calculating the directionality of migration of each single cell. An automation in Openlab (Improvision, PerkinElmer) software controls a camera on an inverted Nikon microscope to take pictures every ten minutes over a one-hour period. The electric field (100 mV/mm) is either on (field) or off (nonfield). To determine how directionally each cell is moving we calculate the angle (θ) between the cell trajectory in the last ten minutes of the experiment and the field axis that runs directly from the cathode (0°) to the anode (180°). We then calculate the cosine (θ). A cell moving directly toward the cathode will have an angle of 0° and the cosine of 0° = 1. A cell moving directly toward the anode will have an angle of 180° and the cosine of 180° = –1. In the absence of an EF, cells will move randomly in all directions and the average of all cosine (θ) = 0.

three times and the Student's t-test and analysis of variance (ANOVA) are used to test the significance of differences between data sets.

KERATINOCYTES MIGRATE TOWARD THE CATHODE OF AN APPLIED ELECTRIC FIELD OF 100 MV/MM

Initial studies demonstrated that keratinocytes would begin to respond to an applied EF above 5 mV/mm; below this field strength, cells would migrate randomly indistinguishable from cells migrating in the absence of an EF (Figure 6.2). When an EF of 10 mV/mm was applied to the keratinocytes, they began to migrate directionally toward the cathode (average cosine (θ) = 0.26 ± 0.08), and the maximum cathodal migration was achieved between 100 and 400 mV/mm (average cosine (θ) = 0.79 ± 0.05 to 0.81 ± 0.03). There was no alteration in speed in the absence or presence of any field strength (Nishimura et al., 1996). An example of keratinocyte galvanotaxis in the presence of a 100 mV/mm EF is shown in Figure 6.3. Cells are migrating at an average speed of 0.8 μm/min with a cosine of 0.72.

   In conclusion, primary human keratinocytes migrated robustly *in vitro*, directionally toward the cathode of applied physiological EFs, appear to sense and respond to the applied EF within ten to fifteen minutes of field application and maintain this directionality until the field is removed (Nishimura et al., 1996). There appears to be no direct relationship between the presence of an EF and cell

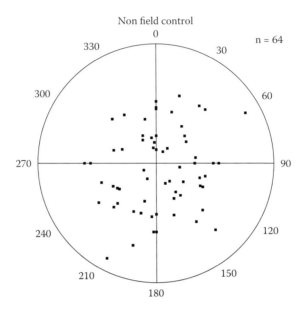

FIGURE 6.2   A circle graph of a population of human keratinocytes migrating randomly for one hour in the absence of an applied EF. Each cell's position at time t = 0 minutes is at the origin of the graph (0,0), and its final position at the end of the experiment is plotted as a single point on the graph. The radius of the circle graph represents 100 μM of translocation distance. The data shown are combined from three independent experiments on two separate cell strains. N = 64. Cosine (θ) = 0 ± 0.04.

migration speed. In addition, it was noted that single cells (mean cosine = 0.8 ± 0.03) respond more robustly to the applied EF than paired cells (0.74 ± 0.05) and grouped cells (0.57 ± 0.09) (Nishimura et al., 1996).

DOES THE DIFFERENTIATION STATE OF THE KERATINOCYTE ALTER GALVANOTAXIS?

The epidermis is composed of multiple layers of keratinocytes that differ in their state of differentiation. The basal layer consists of undifferentiated cells, and the level of differentiation increases toward the stratum corneum, as previously described.

It was important to determine if all cells in the epidermis could respond to an applied EF, and therefore possibly participate in the reepithelialization process *in vivo*. Involucrin is a keratinocyte differentiation marker expressed by keratinocytes in the suprabasilar layers of the epidermis, but not in the basal layer. Involucrin negative keratinocytes migrated almost twice as fast as involucrin positive cells (46.49 ± 2.49 μm/min vs. 23.60 ± 1.16 μm/min) and were also able to sense and respond to an applied directional cue more robustly (0.56 ± 0.095 vs. 0.42 ± 0.042). However, involucrin positive cells could respond fairly well

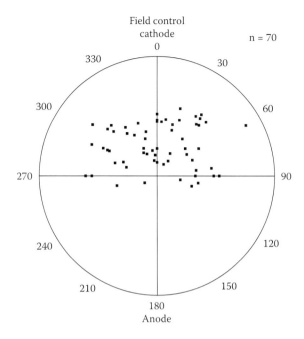

FIGURE 6.3   A circle graph of a population of human keratinocytes migrating direction-ally for one hour in the presence of an applied EF (100 mV/mm). The cathode (0°) is at the top of the circle graph and the anode (180°) is at the bottom. Each cell's position at time $t = 0$ minutes is at the origin of the graph (0,0), and its final position at the end of the exper-iment is plotted as a single point on the graph. The radius of the circle graph represents 100 μM of translocation distance. The data shown are combined from three independent experiments on two separate cell strains. N = 70. Cosine (θ) = 0.72 ± 0.06.

to applied EFs, suggesting that both undifferentiated and partially differentiated keratinocytes could respond to the physiological EF to initiate reepithelialization (Obedencio et al. 1999).

THE INFLUENCE OF EXTRACELLULAR MATRIX ON KERATINOCYTE GALVANOTAXIS

In order to migrate, cells extend membrane protrusions (lamellipodia or filipo-dia) and make contacts with the extracellular matrix (ECM) via focal adhesion complexes containing integrins. Focal adhesions serve as points of traction that the cell uses to pull its cell body forward. To complete a cycle of motion, the attachments at the rear of the cell are released (Chicurel, 2002). In a wound, keratinocytes at the wound edge will come into contact with different extracel-lular matrix proteins in the provisional wound bed. Presumably, these wound ECMs can provide physical guidance cues that could alter the ability of wound edge keratinocytes to respond to electrical guidance cues. Indeed, the surface that keratinocytes are migrating over has a direct effect on their ability to sense and respond to applied EFs. Human neonatal keratinocytes were plated on collagens I and IV, fibronectin, laminin, and cell culture plastic. The cells were capable of sensing and responding to an applied EF on all surfaces, but maximal galvanotaxis was observed when cells were plated on collagens I and IV and cell culture plastic. There was a 50% decrease in galvanotaxis on a laminin substrate, whereas cells plated on fibronectin could sense and respond to the applied EF robustly, but directionality was reduced by 15% compared to cells plated on collagen I. This suggests that the underlying matrix may influence the ability of EFs to direct keratinocyte migration (Sheridan et al., 1996).

THE KERATINOCYTE ELECTRICAL COMPASS

Studies to dissect the mechanisms underpinning the electrical compass began about ten years ago. Their ability to migrate requires external growth factors, particularly epidermal growth factor (EGF) and insulin, but cells were able to sense and respond to applied EFs even in the absence of added growth factor (Fang et al., 1998), perhaps because they have the ability to synthesize growth factors in sufficient quantities to sustain directional migration.

THE ROLE OF THE EGFR IN KERATINOCYTE GALVANOTAXIS

Protein phosphorylation by various protein kinases was initially examined as a posttranslational modification that could play a role in directional sensing. In order to tease out the signaling pathways that play a role in electrical guidance, experiments were performed in the presence of a number of kinase inhibitors. One kinase inhibitor in particular, PD158780, a specific inhibitor of the epidermal growth factor receptor (EGFR) kinase, inhibited directionality at 0.5 μM, while having no effect on migration rate. Other kinase inhibitors, including genistein, tyrophostin B46, and lavendustin A, had a more significant effect on migration speed and only modest effects on directionality. Immunostaining with anti-EGFR antibodies suggested stronger staining for the EGFR on the cathodal-facing side of the plasma membrane, as early as five minutes post-EF application, and was maintained for the duration of the one-hour experiment. In contrast, cells not exposed to EFs exhibited a relatively uniform distribution of EGFRs on the plasma membrane (Fang et al., 1999). Another study revealed a role for cyclic AMP-dependent protein kinase A (PKA) by demonstrating a 50% decrease in directionality in the presence of an inhibitor (KT5720) with no change in migration rate. While an inhibitor of myosin light chain kinase (ML-7) reduced migration speed, inhibitors of PKC and CaM kinase had no effect on either motility or directionality (Pullar et al., 2001).

More recently, exposure of both keratinocytes and neutrophils to EFs in serum-free medium induced the rapid and sustained phosphorylation of extracellular-signal-regulated kinase (ERK), p38 mitogen-activated kinase (MAPK), Src, and Akt, but not the janus kinase JAK1, demonstrating that the EF activates selective signaling pathways (Zhao et al., 2006). Indeed, phosphorylated src kinase and the product of phosphatidylinositol 3 kinase (PI3K), phosphatidylinositol-3,4,5-tris phosphate, became polarized toward the cathode of an applied EF of 150 mV/mm in mouse keratinocytes and HL60 cells, respectively. Finally, an essential role for phosphatidylinositol 3 kinase (PI3K) in EF sensing was confirmed in keratinocytes, neutrophils, and dermal fibroblasts isolated from mutant mice where the PI3K p110γ catalytic subunit was disrupted. EF-mediated directional migration was impaired and cells showed both reduced activation of Akt and reduced phosphorylation of src, p38, and ERK (Zhao et al., 2006).

It is highly likely that as kinase-mediated phosphorylation appears important for keratinocytes to sense and respond to applied electric fields, phosphatases will be equally important. Indeed, the tumor surpressor PTEN negatively regulates keratinocyte galvanotaxis (Zhao et al., 2006). Loss of the phosphatase PTEN, which negatively regulates the PI3K/Akt pathway by dephosphorylating phosphatidylinositol-3,4,5-tris phosphate, increased keratinocyte galvanotaxis (Zhao et al. 2006).

THE ROLE OF CAMP IN KERATINOCYTE GALVANOTAXIS

Keratinocytes express a high number of β_2-adrenergic receptors (Steinkraus et al. 1992) and can synthesize (Schallreuter et al. 1992) and secrete (Pullar et al. 2006c) the catecholamine adrenaline, the natural ligand for β_2-adrenergic receptors. We have previously shown that β_2-adrenergic receptor agonists can decrease keratinocyte migration by the PP2A-mediated dephosphorylation of ERK (Pullar et al., 2003).

β_2-adrenergic receptors are known to couple to the G protein $G_{\alpha s}$, activating adenylyl cyclase and increasing intracellular cyclic AMP (Lefkowitz et al., 2002). We explored the role of cAMP in galvanotaxis and discovered that very low concentrations of β-adrenergic receptor agonists, 0.1 pM and 0.1 nM, decreased directionality by 52 and 75% respectively, while having no effect on migration rate (Pullar and Isseroff, 2005). Higher concentrations of β-adrenergic receptor agonists (0.1 µM) decreased directionality, but also decreased migration rate, as expected (Pullar et al., 2003; Pullar et al., 2006b). Interestingly, by presumably blocking the endogenous ligand-mediated decrease in galvanotaxis, β_2-adrenergic receptor antagonists improve the ability of keratinocytes to sense and respond to EFs, increasing galvanotaxis by 26%.

Increasing intracellular cAMP levels with the active cAMP analog, sp-cAMP, also decreased galvanotaxis by 64%, while having no effect on migration rate. Forskolin, a well-known activator of adenylyl cyclase (Insel and Ostrom, 2003), and pertussis toxin, an inhibitor of $G_{\alpha i}$ (Choi and Toscano, 1988), which normally inhibits adenylyl cyclase, both increase cAMP and both decrease EF-mediated directional migration by 75 and 58%, respectively, while having no effect on migration rate. EF-mediated changes to the levels of intracellular cAMP appear to play a major role in the ability of keratinocytes to sense and respond to applied EFs.

Finally, pretreating keratinocytes with the inactive cAMP analog, rp-cAMP, which inhibits PKA, increased keratinocyte directional migration by 19% and prevented any β-adrenoceptor-mediated decrease in keratinocyte galvanotaxis. It therefore appears that cAMP-dependent downstream signaling negatively regulates keratinocyte galvanotaxis (Pullar and Isseroff, 2005).

THE ROLE OF ION CHANNELS IN KERATINOCYTE GALVANOTAXIS

In a further study, the role of ion channels in keratinocyte galvanotaxis was investigated. Amiloride, the epithelial sodium channel blocker, had no effect on either speed or directionality of migration (Trollinger et al., 2002), a finding that we have repeated in our laboratory (Pullar, unpublished data). Verapamil, a voltage-gated calcium channel blocker, had no effect on speed or directionality of migration at 10 μM (Trollinger et al., 2002), but we found a 50% reduction in directionality at 100 μM, with no effect on speed (Pullar, unpublished data) (Figure 6.4). In addition, we have discovered that two other calcium channel blockers, diltiazem (10 μM) and nifedipine (1 mM), also decrease directionality by 29 and 38%, respectively, while migration speed was unaffected (Pullar, unpublished data).

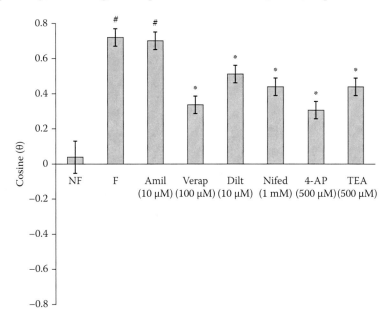

FIGURE 6.4 The role of ion channels in keratinocyte migration. Human keratinocytes were plated onto collagen I–coated glass (60 μg/ml) for three to five hours and their migration in the absence (NF) or presence of an EF (100 mV/mm) alone (F), or in the presence of a series of ion channel inhibitors (amiloride (10 μM), verapamil (100 μM), diltiazem (10 μM), nifedipine (1 mM), 4-AP (500 μM), TEA (500 μM)), was monitored every ten minutes over a one-hour period. Ion channel inhibitors were added at time 0. The data shown are combined from three independent experiments on two separate cell strains (N = 64 – 192). Error bars indicate s.e.m. *, p < 0.01 from F; #, p < 0.01 from NF.

We also saw a 40 to 50% decrease in directionality in the presence of a number of potassium channel blockers, including 4-AP (500 μM) and TEA (500 μM), with no effect on general speed of migration (Pullar, unpublished data) (Figure 6.4).

Perhaps EF-mediated polarized changes in membrane potential play a role in EF sensing (see Chapter 1); therefore, activating or inhibiting any ion channel that would alter the membrane potential could alter the ability of the cell to respond to applied EFs.

To determine if Ca^{2+} influx via a non-voltage-gated ion channel might play a role in EF sensing, inorganic ions were used. Nickel (1 mM) reduced directionality by 53% while having no effect on migration speed. At higher concentrations (5 mM) nickel inhibited both directionality and speed of migration, as did gadalinium, a well-known calcium channel blocker. In contrast, strontium (5 mM), which passes through calcium channels more readily than calcium itself and is often used to increase the current when studying calcium channel conductance, blinded the cells to the applied EF, decreasing directionality by 94%, while only decreasing speed of movement by 33%. Perhaps strontium can substitute for calcium in the mechanisms underpinning motility, but not in the mechanisms involved in the electrical compass. In conclusion, it appeared that calcium influx was required for keratinocytes to sense and respond to an applied EF, although the method of entry and the specific ion channels involved remained elusive (Trollinger et al., 2002).

Finally, a role for the ion transporter Na^+/H^+ exchanger 1 (NHE1) has been implicated in migration (Denker and Barber, 2002), and NHE1 inhibitors abrogated EF-induced Akt activation and decreased galvanotaxis of fibroblasts (Zhao et al., 2006).

THE ROLE OF INTEGRINS IN KERATINOCYTE GALVANOTAXIS

Integrins are heterodimeric transmembrane receptors that are composed of two subunits, which together form the binding site for a range of extracellular matrix proteins. Keratinocytes express a number of different integrins constitutively, including the collagen receptor $\alpha_2\beta_1$, the laminin 5 receptor $\alpha_3\beta_1$, and the laminin receptor $\alpha_6\beta_4$. $\alpha_6\beta_4$ is a component of hemidesmosomes and is primarily concentrated in the basement membrane zone in normal, undamaged epidermis. In patients expressing a null mutation of the β_4-integrin, epidermal blistering occurs, due to the poor adhesion of the epidermis to the basement membrane. Upon wounding the epidermis, the hemidesmosomes are disassembled and $\alpha_6\beta_4$ is delocalized to the lemellipodia of migrating cells. The ligand for $\alpha_6\beta_4$-integrin, laminin 5, is secreted both by keratinocytes in culture and by the leading edge of the epidermal outgrowth into the wound bed, providing an adhesive binding partner for the keratinocyte as it migrates. Previously, the $\alpha_6\beta_4$-integrin has been implicated in keratinocyte chemotaxis, where it plays a role in EGF-mediated migration through the sustained activation of Rac1 (Russell et al., 2003).

We explored the possibility that cooperation between $\alpha_6\beta_4$-integrin and the EGFR also played a role in keratinocyte galvanotaxis. β_4-integrin null human keratinocytes (β_4-), obtained from a patient with epidermolysis bullosa with

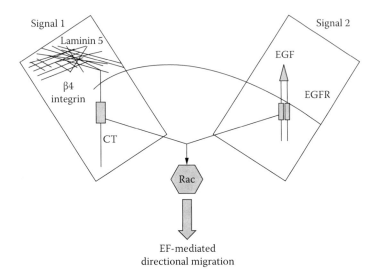

FIGURE 6.5   A diagrammatic illustration is shown outlining the synergistic integrin and growth factor pathways required for keratinocytes to sense and respond to an applied EF by migrating directionally toward the cathode.

pyloric atresia (EB-PA), and keratinocytes in which β_4-integrin was reexpressed (β_4+) (Russell et al. 2003) migrated at the same rate in the presence or absence of EGF, although the rate of migration was 60% slower in the absence of EGF. However, in the absence of EGF, β_4-keratinocytes were completely blinded to the applied EF and migrated randomly. The application of EGF could partially restore their ability to sense and respond to applied EFs. This suggested that at least two distinct but synergistic signaling pathways can coordinate keratinocyte galvanotaxis. Removing the ability of the β_4-integrin to either ligate to laminin 5, using an adhesion defective mutant (β_4+AD), or signal intracellularly, by removing the cytoplasmic tail (β_4+CT), resulted in loss of galvanotaxis in the absence of EGF. However, inhibition of Rac1 inhibited galvanotaxis in both the presence and absence of EGF, suggesting that the two signaling pathways via β_4-integrin and EGFR both require Rac1 activation to transduce the signals required for keratinocytes to sense and respond to an applied EF (Figure 6.5) (Pullar et al., 2006a).

GALVANOTAXIS OF MONOLAYERS OF KERATINOCYTES

The experiments described above are all performed on single cells, mainly isolated from human neonatal skin, but wounded keratinocyte monolayers can also respond to applied EFs (Zhao et al., 2006). When the cathode of a 200 mV/mm EF was placed in the center of a scratch wound within a keratinocyte monolayer, keratinocytes migrated into the scratch to heal the defect. However, when the anode was placed in the center of the scratch wound, keratinocytes migrated in the opposite direction and the wound opened up. In the absence of the lipid phosphatase

PTEN, which negatively regulates PI3K, keratinocytes migrated more directionally toward the cathode of the externally applied EF, irrespective of whether it was located within the scratch wound or outside of it (Zhao et al., 2006).

EF-MEDIATED CHANGES TO KERATINOCYTE GENE TRANSCRIPTION

Recently, the early transcriptional response of human adult keratinocytes to applied EFs of 100 mV/mm has been studied using microarrays. An increase in expression of a number of cytokines, interleukins, and other inflammatory response genes indicated that EFs can upregulate the expression of genes that play a role in the cell's ability to regulate inflammation in a paracrine manner. CCL20 expression was increased sevenfold after exposure to an EF for eight hours (Jennings et al. 2009).

CONCLUSIONS

In conclusion, it is clear that electrical stimulation enhances the healing of chronic wounds, and the research described above suggests that it may act as an initial signal to guide the migration of keratinocytes from the wound edge into the wound, as well as alter the level of transcription of inflammatory response genes. It is possible that the application of a direct current electric field to chronic wounds acts to perhaps activate signaling pathways that promote keratinocyte migration into the wound and alter keratinocyte differentiation at the chronic wound edge. As we deepen our understanding of how the behavior of keratinocytes can be modified by EFs, we will be able to optimize the current protocols used for application of ES to chronic wounds and perhaps even design particular protocols for each chronic wound type.

ACKNOWLEDGMENTS

I would like to gratefully acknowledge research support from the Wellcome Trust, the British Skin Foundation, The Medical Research Council, and Medisearch.

REFERENCES

Boukamp, P., R. T. Petrussevska, D. Breitkreutz, J. Hornung, A. Markham, and N. E. Fusenig. 1988. Normal keratinization in a spontaneously immortalized aneuploid human keratinocyte cell line. *J Cell Biol* 106(3):761–71.

Chicurel, M. 2002. Cell biology. Cell migration research is on the move. *Science* 295(5555):606–9.

Choi, E. J., and W. A. Toscano Jr. 1988. Modulation of adenylate cyclase in human keratinocytes by protein kinase C. *J Biol Chem* 263(32):17167–72.

Denker, S. P., and D. L. Barber. 2002. Cell migration requires both ion translocation and cytoskeletal anchoring by the Na-H exchanger NHE1. *J Cell Biol* 159(6):1087–96.

Fang, K. S., B. Farboud, R. Nuccitelli, and R. R. Isseroff. 1998. Migration of human kera-tinocytes in electric fields requires growth factors and extracellular calcium. *J Invest Dermatol* 111(5):751–56.

Fang, K. S., E. Ionides, G. Oster, R. Nuccitelli, and R. R. Isseroff. 1999. Epidermal growth factor receptor relocalization and kinase activity are necessary for directional migra-tion of keratinocytes in DC electric fields. *J Cell Sci* 112(Pt 12):1967–78.

Fonder, M. A., G. S. Lazarus, D. A. Cowan, B. Aronson-Cook, A. R. Kohli, and A. J. Mamelak. 2008. Treating the chronic wound: A practical approach to the care of non-healing wounds and wound care dressings. *J Am Acad Dermatol* 58(2):185–206.

Insel, P. A., and R. S. Ostrom. 2003. Forskolin as a tool for examining adenylyl cyclase expression, regulation, and G protein signaling. *Cell Mol Neurobiol* 23(3):305–14.

Jennings, J. A., D. Chen, and D. S. Feldman. 2009. Upregulation of chemokine (C-C motif) ligand 20 in adult epidermal keratinocytes in direct current electric fields. *Arch Dermatol Res*.

Kohl, G., and P. E. Houghton. 2010. Electrical stimulation for chronic wounds. *Cochrane Rev* (submitted and under review).

Lefkowitz, R. J., K. L. Pierce, and L. M. Luttrell. 2002. Dancing with different partners: Protein kinase a phosphorylation of seven membrane-spanning receptors regulates their G protein-coupling specificity. *Mol Pharmacol* 62(5):971–74.

Martin, P. 1997. Wound healing—Aiming for perfect skin regeneration. *Science* 276(5309):75–81.

McCulloch, J. M., and L. Kloth. 2010. *Wound healing: Evidence-based management.* 4th ed. F. A. Davis Company: Philadelphia, PA.

Nishimura, K. Y., R. R. Isseroff, and R. Nuccitelli. 1996. Human keratinocytes migrate to the negative pole in direct current electric fields comparable to those measured in mammalian wounds. *J Cell Sci* 109(Pt 1):199–207.

Obedencio, G. P., R. Nuccitelli, and R. R. Isseroff. 1999. Involucrin-positive keratinocytes demonstrate decreased migration speed but sustained directional migration in a DC electric field. *J Invest Dermatol* 113(5):851–55.

Pullar, C. E., B. S. Baier, Y. Kariya, A. J. Russell, B. A. Horst, M. P. Marinkovich, and R. R. Isseroff. 2006a. beta4 integrin and epidermal growth factor coordinately regulate electric field-mediated directional migration via Rac1. *Mol Biol Cell* 17(11):4925–35.

Pullar, C. E., J. Chen, and R. R. Isseroff. 2003. PP2A activation by beta2-adrenergic recep-tor agonists: Novel regulatory mechanism of keratinocyte migration. *J Biol Chem* 278(25):22555–62.

Pullar, C. E., J. C. Grahn, W. Liu, and R. R. Isseroff. 2006b. Beta2-adrenergic receptor activation delays wound healing. *FASEB J* 20(1):76–86.

Pullar, C. E., and R. R. Isseroff. 2005. Cyclic AMP mediates keratinocyte directional migration in an electric field. *J Cell Sci* 118 Pt 9):2023–34.

Pullar, C. E., R. R. Isseroff, and R. Nuccitelli. 2001. Cyclic AMP-dependent protein kinase A plays a role in the directed migration of human keratinocytes in a DC electric field. *Cell Motil Cytoskel* 50(4):207–17.

Pullar, C. E., A. Rizzo, and R. R. Isseroff. 2006c. beta-Adrenergic receptor antagonists accelerate skin wound healing: Evidence for a catecholamine synthesis network in the epidermis. *J Biol Chem* 281:21225–35.

Rheinwald, J. G., and H. Green. 1975. Serial cultivation of strains of human epider-mal keratinocytes: The formation of keratinizing colonies from single cells. *Cell* 6(3):331–43.

Russell, A. J., E. F. Fincher, L. Millman, R. Smith, V. Vela, E. A. Waterman, C. N. Dey, S. Guide, V. M. Weaver, and M. P. Marinkovich. 2003. Alpha 6 beta 4 integrin regulates keratinocyte chemotaxis through differential GTPase activation and antagonism of alpha 3 beta 1 integrin. *J Cell Sci* 116(Pt 17):3543–56.

Schallreuter, K. U., J. M. Wood, R. Lemke, C. LePoole, P. Das, W. Westerhof, M. R. Pittelkow, and A. J. Thody. 1992. Production of catecholamines in the human epidermis. *Biochem Biophys Res Commun* 189(1):72–78.

Shaw, T. J., and P. Martin. 2009. Wound repair at a glance. *J Cell Sci* 122(Pt 18):3209–13.

Sheridan, D. M., R. R. Isseroff, and R. Nuccitelli. 1996. Imposition of a physiologic DC electric field alters the migratory response of human keratinocytes on extracellular matrix molecules. *J Invest Dermatol* 106(4):642–46.

Sorrell, J. M., and A. I. Caplan. 2004. Fibroblast heterogeneity: More than skin deep. *J Cell Sci* 117(Pt 5):667–75.

Steinkraus, V., M. Steinfath, C. Korner, and H. Mensing. 1992. Binding of beta-adrenergic receptors in human skin. *J Invest Dermatol* 98(4):475–80.

Stojadinovic, O., H. Brem, C. Vouthounis, B. Lee, J. Fallon, M. Stallcup, A. Merchant, R. D. Galiano, and M. Tomic-Canic. 2005. Molecular pathogenesis of chronic wounds: The role of beta-catenin and c-myc in the inhibition of epithelialization and wound healing. *Am J Pathol* 167(1):59–69.

Stramer, B. M., R. Mori, and P. Martin. 2007. The inflammation-fibrosis link? A Jekyll and Hyde role for blood cells during wound repair. *J Invest Dermatol* 127(5):1009–17.

Trollinger, D. R., R. R. Isseroff, and R. Nuccitelli. 2002. Calcium channel blockers inhibit galvanotaxis in human keratinocytes. *J Cell Physiol* 193(1):1–9.

Walmsley, S. 2002. *Advances in wound management: Executive summary, clinical reports.* London: PJB Publications Ltd.

Zhao, M., B. Song, J. Pu, T. Wada, B. Reid, G. Tai, F. Wang, A. Guo, P. Walczysko, Y. Gu, T. Sasaki, A. Suzuki, J. V. Forrester, H. R. Bourne, P. N. Devreotes, C. D. McCaig, and J. M. Penninger. 2006. Electrical signals control wound healing through phosphatidylinositol-3-OH kinase-gamma and PTEN. *Nature* 442(7101):457–60.

7 Electrical Control of Angiogenesis

Entong Wang
Department of Otolaryngology-Head and Neck Surgery
General Hospital of Air Force
Beijing, People's Republic of China

Yili Yin
Wellcome Trust Centre for Gene Regulation
and Expression, College of Life Sciences
University of Dundee
Dundee, Scotland

Huai Bai
Unit of Laboratory Medicine, West China
Second Hospital of Sichuan University
Chengdu, People's Republic of China

Brian Reid
Department of Dermatology, School of
Medicine, Center for Neurosciences
University of California, Davis
Davis, California

Zhiqiang Zhao
Institute of Medical Sciences
University of Aberdeen
Aberdeen, Scotland

Min Zhao
Department of Dermatology, School of
Medicine, Center for Neurosciences
University of California, Davis
Davis, California

CONTENTS

INTRODUCTION

Many extracellular biochemical signals control angiogenesis. In addition, biophysical factors are being recognized that affect endothelial behaviors and angiogenesis signaling pathways. Here we review the experimental evidence for electrical signals as a controller of angiogenesis. *In vivo* electrical stimulation enhances angiogenesis in muscle and brain tissues. Electric fields (EFs) regulate expression of vascular endothelial growth factor (VEGF) and interleukin 8 (IL-8) in muscle cells and vascular endothelial cells. Electrical stimulation activates multiple signaling pathways that control endothelial cells and are important for angiogenesis. In *ex vivo* models, sprouting of endothelial cells and vessel formation from explants can be reorientated by static electric charges or applied EFs. While the role of endogenous EFs in angiogenesis needs further investigation, EFs represent a novel type of signaling paradigm for control of angiogenesis. The combination of electrical stimulation and other regulatory mechanisms may offer an effective technique to modulate angiogenesis, which may lead to therapies for angiogenesis-related diseases.

ANGIOGENESIS

Angiogenesis is a process whereby new blood vessels form from existing vasculature and is critical in many physiological and pathophysiological processes. It is essential for embryonic development, injury repair, and tissue regeneration, and is involved in tumor growth, arthritis, and retinopathy. Angiogenesis is a complex process that requires a finely tuned balance between numerous stimulatory (pro-angiogenic) and inhibitory (anti-angiogenic) signals (Risau, 1997; Buschmann and Schaper, 1999; Carmeliet and Jain, 2000; Liekens et al., 2001; Carmeliet, 2005; Folkman, 2007a, 2007b; Jain, 2008, 2009). Normally, angiogenesis is tightly controlled through the balance between pro-angiogenic factors and anti-angiogenic factors. Imbalance of the pro- and anti-angiogenic mechanisms results in abnormal angiogenesis, which can result in insufficient or excessive blood vessel growth. Insufficient blood vessel growth leads to ischemic conditions, slow wound healing, infections, and immune disorders. Excessive angiogenesis contributes to the pathogenesis of numerous angiogenesis-dependent disorders, such as inflammation, tumor growth, arthritis, and proliferative retinopathy. Angiogenesis is closely related to tumor development, including growth, invasion, and metastasis of cancer cells. Elucidating the mechanisms of angiogenesis therefore has wide clinical implications.

Research over the past few decades has provided significant understanding of the mechanisms of angiogenesis (Folkman and Klagsbrun, 1987; Nelson et al., 2000; Carmeliet and Jain, 2000; Carmeliet, 2005; Ferrara and Kerbel, 2005; Jain, 2005; 2008, 2009; Graupera et al., 2008; Fukumura and Jain, 2008). Angiogenic factors, including VEGF, fibroblast growth factors (FGFs), platelet-derived growth factor (PDGF), and IL-8, stimulate endothelial cells and initiate multiple tightly controlled steps to form new blood vessels. These steps are degradation of basement membrane, migration of endothelial cells into stroma, proliferation of endothelial cells, alignment, lumen formation, and finally, connection establishment. In normal conditions, these steps are balanced by anti-angiogenic factors such as angiostatin, endostatin, and interleukin 4 (IL-4). Angiogenesis in the female reproductive cycle and wound healing are two examples where new blood vessel formation is tightly controlled. Many extracellular stimuli may affect this pro- and anti-angiogenic balance and influence angiogenesis (Risau, 1997; Carmeliet, 2005; Folkman, 2007a, 2007b; Matsunaga et al., 2008).

Recent studies have demonstrated the important roles of physical factors. For example, mechanical forces regulate angiogenesis through the transcriptional regulation of VEGF receptor 2 (Brown et al., 1996; Brown and Hudlicka, 2003; Mammoto et al., 2009). EFs are another physical factor that participates in the control of angiogenesis. Accumulating experimental evidence suggests that electricity affects many important aspects of angiogenesis. Electrical stimulation can regulate the expression of VEGF and angiogenesis *in vivo*. *In vitro* experiments demonstrated EF-mediated upregulation of pro-angiogenic factors. EFs also activate multiple intracellular signaling pathways in endothelial cells and control endothelial behaviors. Direct current (DC) EFs guide endothelial sprouting and the direction of new vessel formation. EFs may therefore offer a new technique for the modulation of angiogenesis.

ELECTRICAL STIMULATION ENHANCES ANGIOGENESIS *IN VIVO*

Increasing experimental evidence suggests that electrical stimulation is able to significantly alter the pro- and anti-angiogenic balance. Experiments have used either DC or alternating current (AC) at certain frequencies. The majority of published experiments with animal models use AC electrical stimulation. Overall, electrical stimulation enhances angiogenesis *in vivo* (Hudlicka and Brown, 2009). The direct electrical stimulation of muscles or nerves with alternating currents induces muscle contraction. The enhanced contractile activity requires increased capillary density and blood supply, which occur through several mechanisms, including increased expression of some immediate early genes, c-fos, c-jun, and egr-1, and importantly, increased transcription and translation of VEGF (Michel et al., 1994; Annex et al., 1998). The increase in VEGF secretion is responsible for the EF-mediated increase in angiogenesis (Hudlicka and Tyler, 1984; Hang et al., 1995). In other experimental models, electrical stimulation increased capillary growth in ischemic muscle tissues (Egginton and Hudlicka, 2000; Chekanov et al., 2002; Hudlicka and Brown, 2009) and opened pre-existing collaterals (arteriogenesis) (Sheikh et al., 2005).

A surprising finding was that muscle treated with lower levels of electrical stimulation, which did not cause contraction, was also very effective in inducing significant angiogenesis (Kanno et al., 1999; Linderman et al., 2000). Kanno et al. (1999) used AC stimulation of 50 Hz and 0.1 V, which was far below the threshold for muscle contraction. This AC electrical stimulation activated the transcription of VEGF mRNA and significantly increased the level of VEGF protein sixfold. Capillary density and blood flow increased significantly after electrical stimulation. The increase in VEGF, blood flow, and capillary density in the ipsilateral, electrically stimulated limb, but not in the contralateral limb, suggests that angiogenesis was indeed enhanced by the muscle subcontraction induced by electrical stimulation.

Electrical stimulation also enhanced angiogenesis in nonmuscle tissues. Most recently, a Japanese group succeeded in enhancing angiogenesis in the rat ischemic cortex with cortical electrical stimulation (Baba et al., 2009). In this study, the middle cerebral artery was occluded for ninety minutes. One hour after reperfusion, the anesthetized rats were implanted with electrodes in the right frontal epidural space. Square-wave pulses of 1 ms/100 μA at 2 Hz were used to stimulate for one week. The stimulation significantly increased neurotrophic factors (glial cell line-derived neurotrophic factor, brain-derived neurotrophic factor) and VEGF in the brain tissues. They used laminin staining to show that angiogenesis was significantly enhanced after electrical stimulation in the ischemic cortex compared with the nonstimulated cortex and the intact cortex.

ELECTRICAL STIMULATION ACTIVATES PRO-ANGIOGENIC SIGNALING PATHWAYS

ELECTRICAL STIMULATION UPREGULATES EXPRESSION OF VEGF AND HEPATOCYTE GROWTH FACTOR (HGF)

VEGF is one of the most potent angiogenic factors, inducing angiogenesis by activating VEGF receptors (Risau, 1997; Folkman, 2007a, 2007b; Matsunaga et al., 2008), initiating a central pathway of angiogenesis (Hang et al., 1995; Kanno et al., 1999; Patterson and Runge, 1999; Ferrara, 2000; Linderman et al., 2000; Nagasaka et al., 2006; Matsunaga et al., 2008). VEGF promotes many of the events necessary for angiogenesis, such as proliferation and migration of vascular endothelial cells, remodeling of the extracellular matrix, and the formation of capillary tubules both *in vitro* and *in vivo* (Ferrara, 1995; Matsunaga et al., 2008). Application of antibodies against VEGF and expression of antisense VEGF inhibited angiogenesis (Kim et al., 1993; Kendall and Thomas, 1993; Saleh et al., 1996). Avastin (bevacizumab), a humanized monoclonal neutralizing antibody against VEGF, blocks the binding of VEGF to its receptor and neutralizes all of the isoforms of human VEGF (Presta et al., 1997). Avastin inhibited VEGF-induced proliferation of human vascular endothelial cells in a dose-dependent manner (Li et al., 2007).

Electrical stimulation induces VEGF expression in muscle cells, endothelial cells, and osteoblasts (Baba et al., 2009; Kim et al., 2006; Zhao et al., 2004). The electrical stimulation-mediated increase in VEGF expression in muscle cells is induced in the presence or absence of muscle contraction both *in vivo* and *in vitro* (Kanno et al., 1999; see "Electrical Stimulation Enhances Angiogenesis *In Vivo*" section). In excised muscle, pulsed stimulation significantly increased expression of VEGF and HGF in the muscle, while FGF, interleukin-6, and HIP-1alpha remained unchanged (Nagasaka et al., 2006). This selective upregulation suggests that specific signaling pathways are activated upon electrical stimulation.

We discovered that electrical stimulation induced significant secretion of VEGF from endothelial cells in the absence of any other cell types. DC EFs (150 mV/mm), applied to cultured human umbilical vein endothelial cells (HUVECs), significantly increased the production of VEGF compared to control cultures that were not exposed to an EF. The increase in VEGF production was biphasic, with the first peak at thirty minutes and the second peak four hours after stimulation, which persisted at twenty-four hours (Figure 7.1) (Zhao et al., 2004). Pulsed electrical stimulation also upregulated VEGF expression in embryonic bodies derived from mouse embryonic stem cells. Treatment with pulses ranging from 250 to 750 mV/mm for 60 s significantly increased the expression of VEGF and platelet endothelial cell adhesion molecule-1 (PECAM-1) in embryonic bodies (Sauer et al., 2005). Increased VEGF secretion may exert pro-angiogenic effects through both autocrine and paracrine mechanisms by binding to VEGF receptors on the same cell or adjacent cells to activate VEGF receptor signaling

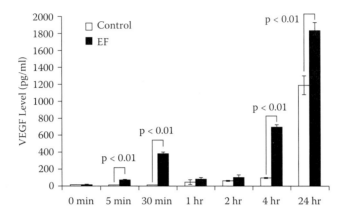

FIGURE 7.1 Electrical stimulation significantly increases VEGF expression in vascular endothelial cells. There is a biphasic increase in VEGF in the culture medium after exposure of human umbilical vein endothelial cells (HUVECs) in a DC EF = 200 mV/mm. VEGF in the medium was quantified by ELISA. (Reprinted from Zhao et al., *J. Cell Sci.*, 117(Pt 3), 397–405, 2004.)

pathways (Rousseau et al., 1997; Wu et al., 2000; Shiojima and Walsh, 2002; Matsunaga et al., 2008).

The mechanisms by which electrical stimulation enhances the expression of VEGF are not understood. Reactive oxygen species (ROS) may mediate VEGF increase as the ROS scavenger, vitamin E, significantly decreases electrically induced VEGF expression (Sauer et al., 2005). We have started to investigate the transcriptional changes of VEGF and IL-8 induced by DC electrical stimulation. VEGF and IL-8 mRNA were significantly upregulated following electrical stimulation (our unpublished data); therefore, electrical stimulation increases both the transcription and the translation of these pro-angiogenic factors. It is possible that a positive feedback loop, via VEGF receptor signaling, may be involved in the upregulation of VEGF translation. Indeed, the inhibition of VEGF receptors prevented the electrical stimulation-induced increase in VEGF. Intriguingly, stimulation frequency has a significant effect on the production of VEGF and HIF-1alpha in rabbit skeletal muscle (Shen et al., 2009).

ELECTRICAL STIMULATION RAISES REACTIVE OXYGEN SPECIES (ROS) AND HYPOXIA-INDUCING FACTOR (HIF)

The electrical stimulation of embryonic bodies derived from mouse embryonic cells activated ROS-HIF signaling (Sauer et al., 2005). HIF signaling is a key element in angiogenesis. HIF is a transcription factor that normally responds to changing intracellular oxygen concentration. Under typical oxygen levels (normoxia), HIF is hydroxylated and acetylated, and then degraded through ubiquitination. Hypoxia results in HIF accumulation and transportation to the nucleus, where HIF induces expression of a wide variety of target gene products, including

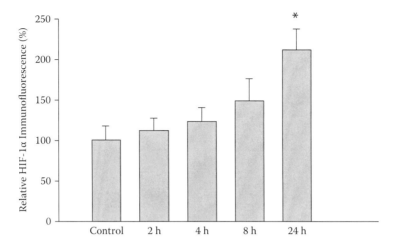

FIGURE 7.2 DC electrical stimulation significantly increases HIF1 expression in embryonic bodies derived from mouse embryonic stem cells. Four-day-old embryoid bodies were treated with an EF (500 V/m) for 60 s, and protein expression was quantified in individual embryoid bodies at times indicated by confocal laser scanning microscopy. *, $p < 0.05$, significantly different from the untreated control. (Reprinted from Sauer et al., *Exp. Cell. Res.*, 304, 380–90, 2005.)

growth factors (such as VEGF, FGF, and TGF). Subsequent activation of signaling pathways (including PLCγ, PI3K, Src, Smad signaling) would instigate proliferation and migration of endothelial cells (Adams and Alitalo, 2007). Pulsed electrical stimulation raised the levels of HIF-1alpha in embryonic bodies derived from mouse embryonic stem cells (Figure 7.2) (Sauer et al., 2005). This increase was dependent on EF-induced elevation of ROS. Inhibition of NAD(P)H oxidase, or using antioxidant, abolished the EF-induced HIF increase.

ELECTRICAL ACTIVATION OF MAP KINASE AND PI3K/Akt SIGNALING PATHWAYS

EFs activate MAP kinase and PI3K signaling pathways, which play important roles in proliferation and migration of endothelial cells, and also in angiogenesis. Indeed, knockout studies revealed that ERK2 is required for placental angiogenesis (Kuida and Boucher, 2004). Elevated HIF induces VEGF expression initiating VEGF receptor signaling and activation of both MAP kinase and PI3K signaling pathways. Activation of PI3K/Akt signaling is known to be an important step in the VEGF-induced survival, proliferation, and migration of vascular endothelial cells (Morales-Ruiz et al., 2000; Kureishi et al., 2000; Dimmeler et al., 2000; Foster et al., 2003). The PI3K inhibitor LY294002 inhibited the VEGF-induced migration of endothelial cells (Ali et al., 2005; Matsunaga et al., 2008). EFs induce activation of MAP kinases and PI3K in corneal epithelial cells and keratinocytes within a few minutes of exposure in a polarized manner (Zhao et al., 2003, 2006).

EFFECTS OF ELECTRIC FIELDS (EFs) ON MIGRATION, ELONGATION, AND ORIENTATION OF ENDOTHELIAL CELLS

Applied EFs induce multiple cellular responses in endothelial cells: migration, orientation, survival, and proliferation, which are essential steps in angiogenesis and vascular remodeling *in vivo* (Risau, 1997; Jain, 1997; Buschmann and Schaper, 1999; Auerbach et al., 2003; Goodwin, 2007). Endothelial cells migrate directionally, elongate, and align themselves strikingly in a physiological strength EF (Figure 7.3). As the mechanisms that cells use to sense EFs are still unclear, it is difficult to separate the effects that are induced directly by the EFs or are secondary to the increase of angiogenic factors (VEGF, IL-8, etc.) and the activation of

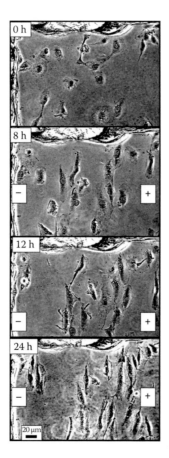

FIGURE 7.3 Endothelial cells reoriented, elongated, and migrated directionally in a small physiological EF. Obvious elongation and orientation of HUVECs can be seen after eight hours in an EF of 150 mV/mm. Directional lamellipodial extension and cell migration were evident at twelve and twenty-four hours after EF exposure. Scratches made on the culture dish as static reference points can be seen to the left and top of each frame. (Reprinted from Zhao et al., *J. Cell Sci.*, 117(Pt 3), 397–405, 2004.)

their downstream signaling pathways, as described in the "Electrical Stimulation Activates Pro-angiogenic Signaling Pathways" section. We will discuss the effects on cellular behaviors without distinguishing the direct and indirect effects.

APPLIED EFs DIRECT THE MIGRATION OF ENDOTHELIAL CELLS

One of the most noticeable cellular responses to applied EFs is directed cell migration, termed galvanotaxis or electrotaxis, which has been described for many types of cells (Robinson, 1985; Nuccitelli, 1988; McCaig et al., 2005; Zhao et al., 2006). Cell migration is a key behavior in many biological processes, such as embryonic development, tissue repair and regeneration, tumor metastasis, and angiogenesis. In addition to directing migration, applied EFs can also accelerate migration rates. This response is nonlinear and varies with cell type (Nuccitelli et al., 1993; Farboud et al., 2000; Wang et al., 2000, 2003b).

Vascular endothelial cells respond to applied EFs by directional migration (Li and Kolega, 2002; Zhao et al., 2004; Bai et al., 2004) (Figure 7.4). Most intriguingly, endothelial cells isolated from large and small vessels migrate in applied EFs in different directions. Bovine aortic endothelial cells (BAECs) migrated

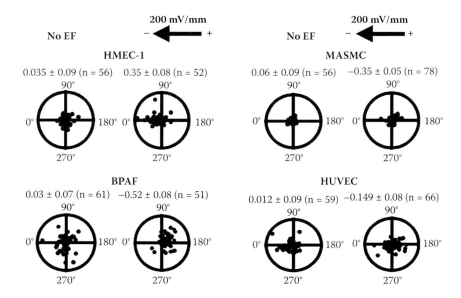

FIGURE 7.4 Scatter plots show cumulated cell migration of vascular cells in applied EFs. Each point represents a single cell, located initially at the center of the circular graph (zero hour) and plotted using their final location after five hours. The radius of each circle is 150 μm, and the directedness is indicated. Human microvascular endothelial cells (HMEC-1) migrated toward the cathode (left), whereas vascular fibroblasts (bovine pulmonary artery fibroblasts (BPAFs)), murine aorta smooth muscle cells (MASMCs), and human umbilical vein endothelial cells (HUVECs) moved anodally (to the right). (Reprinted from Bai et al., *Arterioscler. Thromb. Vasc. Biol.*, 24, 1234–39, 2004.)

toward the cathode (Li and Kolega, 2002), while HUVECs respond to applied EFs by migration to the anode (Zhao et al., 2004). An immortalized human dermal microvascular endothelial cell line (hMEC-1), however, migrated to the cathode (Bai et al., 2004) (Figure 7.4). The physiological significance of such differences is not known but may be significant considering that the blood vessels are complex and highly organized structures.

EF-directed migration of endothelial cells is field strength dependent, and the response thresholds are dependent on cell type and experimental conditions. Cultured HUVECs show an obvious directional migration response in an applied EF of 100 mV/mm. The threshold that could induce directional migration should therefore be below 100 mV/mm (Zhao et al., 2004). Li and Kolega (2002) showed that applied EFs less than 100 mV/mm did not induce the directed migration of bovine aortic endothelial cells. Most bovine aortic endothelial cells showed directional migration when the field was increased to 200 mV/mm, and all cells showed directional migration at 500 mV/mm (Li and Kolega, 2002). The directed migration of several types of vascular endothelial cells was most obvious at 150 to 200 mV/mm; those include cells from different origins: the human microvascular endothelial cell line (hMEC-1), bovine pulmonary artery fibroblasts (BPAFs), murine aorta smooth muscle cells (MASMCs), and HUVECs. Additional increases of EF strength did not increase the directedness (Bai et al., 2004) (Figure 7.4). In addition, EFs significantly increased the migration rate of vascular endothelial cells. Migration rates of MASMCs, BPAFs, HUVECs, and HMECs increased by 25, 35, 55, and 90%, respectively, in an EF of 200 mV/mm, compared to cells cultured without EF exposure (Bai et al., 2004).

Applied EFs Elongate and Align Endothelial Cells

Many types of cells reorientate and change their shape in applied EFs. Cells usually orient their long axis perpendicular to the EF vector (see Chapter 2). The perpendicular orientation of cells induced by applied EFs is often accompanied by elongation of the cells (Nuccitelli, 1988). Changes in cell shape and alignment in a certain direction in relation to each other may be important for vessel formation. In sprouting angiogenesis, vascular endothelial cells must orientate in order to invade tissues and form vascular patterns (Gerhardt and Betsholtz, 2005). As angiogenesis progresses, endothelial cells actively change shape and polarity for vessel remolding (Risau, 1997; Jain, 1997; Auerbach et al., 2003; Goodwin, 2007).

Vascular endothelial cells align perpendicularly and elongate in EFs with their long axis perpendicular to the EF vector, resembling their response to fluid shear stress (Figures 7.3 and 7.5) (Zhao et al., 2004; Bai et al., 2004). HUVECs cultured without exposure to the EF had a typical cobblestone-like appearance, with the long axes of the cell bodies oriented randomly, while cells exposed to applied DC EFs underwent a striking reorientation, with their long axes aligning perpendicular to the vector of the applied EF and elongating significantly (Figure 7.5). The orientation and elongation responses of vascular endothelial cells in EFs were voltage dependent. The threshold field strength for inducing their perpendicular

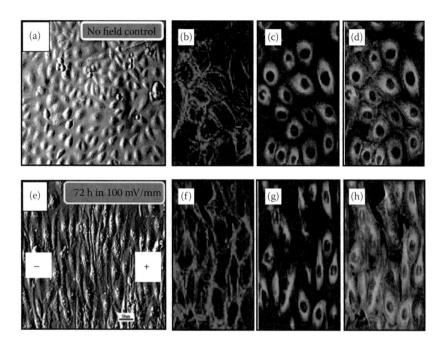

FIGURE 7.5   See color insert. Perpendicular orientation and elongation of endothelial cells in a small physiological EF. (a) HUVEC cultured in the absence of an EF showed a typical cobblestone morphology and random orientation. (b–d) No-field controls showed no obvious alignment and cell elongation. (e) Cells exposed to a small applied EF showed dramatic elongation and perpendicular orientation in the EF (seventy-two hours, 100 mV/mm). (f–h) Most actin filaments (red) and microtubules (green) became aligned along the long axes of the cells (twelve hours at 150 mV/mm). (a,e) Images taken with Hoffman modulation optics. (d,h) Merged images scale bar = 50 μm. (Modified from Zhao et al., *J. Cell Sci.*, 117(Pt 3), 397–405, 2004.)

orientation and elongation was between 50 and 150 mV/mm. Both orientation and elongation of vascular endothelial cells in EFs also showed time dependency. After twenty-four hours at 300 mV/mm, almost all the cells showed an alignment perpendicular to EF vector (Zhao et al., 2004; Bai et al., 2004).

SIGNALING MECHANISMS OF EF-INDUCED ELONGATION, ORIENTATION, AND MIGRATION OF ENDOTHELIAL CELLS

As previously described, VEGF receptors are the proximal element to transduce and induce EF-mediated pro-angiogenic responses through downstream signaling elements involving PI3K-Akt, Rho-ROCK, and the F-actin cytoskeleton, as inhibition of the VEGF receptors abolished both the EF-induced reorientation and the EF-induced decrease of the long:short axis ratio of endothelial cells (Zhao et al., 2004). In addition, mechanical force-induced angiogenesis also depends on soluble VEGF (Mammoto et al., 2009). The exact mechanisms of EF-induced activation of VEGF receptors may involve the direct effects of the EFs and indirect effect through increased expression of VEGF.

PI3K-Akt also mediates EF-induced endothelial responses (Zhao et al., 2004). The PI3K inhibitor LY294002, an Akt inhibitor, and the Rho kinase inhibitor Y27632 all significantly reversed the EF-mediated changes to the long:short axis ratio and the elongation of endothelial cells. Individually, the Akt and Rho kinase inhibitors partially inhibited the EF-mediated vascular endothelial cell orientation response, but it was completely abolished when a combination of the two inhibitors was used, indicating the two enzymes function in two different pathways that converge to induce cell reorientation.

The Rho family GTPases regulate VEGF-stimulated endothelial cell motility and reorganization of the actin cytoskeleton, which are important in endothelial cell retraction and in the formation of intercellular gaps (Wojciak-Stothard et al., 1998). A Rho small GTPase kinase, Rho-p160ROCK, mediates shear stress-induced cell alignment (Li et al., 1999). In addition, a Rho inhibiting molecule, p190RhoGAP (also known as GRLF1), controls capillary network formation *in vitro* in human microvascular endothelial cells and retinal angiogenesis *in vivo* by modulating the balance of activities between two antagonistic transcription factors, TFII-I (also known as GTF2I) and GATA2. A mechanosensitive mechanism regulates the balance of these two transcription factors and regulates gene expression of the VEGF receptor-2 (Mammoto et al., 2009). Inhibition of ROCK also significantly decreased EF-induced orientation of HUVECs (Zhao et al., 2004). A small DC EF might activate the Rho-p160ROCK pathway, which in turn modulates myosin light-chain phosphorylation, or intracellular Ca^{2+}, to regulate the cell alignment. Both actin filaments and microtubules were aligned in the direction of cell elongation (Figure 7.5). Latrunculin A, which inhibits actin polymerization, suppressed the orientation response significantly but not fully, and completely abolished the EF-induced elongation response, suggesting differential mechanisms between orientation, migration, and elongation of endothelial cells in EFs (Zhao et al., 2004).

EFS MAY REGULATE CELL SURVIVAL AND PROLIFERATION

PI3K/Akt signaling is known to be both anti-apoptotic and pro-proliferative (Kennedy et al., 1997). EF-induced activation of PI3K signaling may increase the ability of endothelial cells to survive either through direct effects or through VEGF expression. Pulsed electrical stimulation significantly decreased the number of apoptotic cells in brain tissue after ischemic insults. Inhibition of PI3K/Akt signaling, with the Akt inhibitor LY294002, abolished the protective effect of electrical stimulation. The antiapoptotic effect is likely to be at least partially mediated by PI3K and is in keeping with activation of VEGF receptor signaling and activation of the PI3K/Akt pathway (Baba et al., 2009).

EFs can promote or inhibit cell proliferation, depending on cell type and electrical field strength (Armstrong et al., 1988; Goldman and Pollack, 1996; Binhi and

Goldman, 2000; Azadniv et al., 1993; Wang et al., 2005). Applied EFs stimulate the growth of human dermal fibroblasts within a collagen matrix (Goldman and Pollack, 1996; Binhi and Goldman, 2000). Electrical stimulation at field strengths of 150 to 300 mV/mm significantly increased proliferation of bovine growth plate chondrocytes (Armstrong et al., 1988); however, in contrast, their proliferation was significantly inhibited at a higher field strength of 450 mV/mm. Meanwhile, an applied EF of 200 mV/mm inhibited the proliferation of HUVECs by 50%, with a decrease of cell density and cell growth rate, but EFs of lower strengths (50 or 100 mV/mm) did not affect their proliferation rate significantly (Wang et al., 2003a). In addition, DC EFs of 200 to 300 mV/mm also significantly inhibited the proliferation of ocular lens epithelial cells (Wang et al., 2005). The inhibition of proliferation at higher EF strengths may be through cell cycle control by inhibiting cell cycle progression at the G1/S transition, resulting in cell cycle arrest at G1. An EF of 200 mV/mm decreased expression of cyclin E. Reduced expression of cyclin E would reduce activity of the cyclin E/Cdk2 complexes and prevent passage through G1. EF exposure also significantly increased the expression of p27[kip1], an inhibitor of the cyclin E/Cdk2 complex (Wang et al., 2003a).

Electric stimuli (DC, AC, or pulses) appear to have diverse effects on angiogenic signaling pathways and endothelial behaviors, and due to the variable nature of electrical stimulation in terms of the magnitude, frequency, pulse width, etc., further research will be needed to optimize electrical stimulation regimes to have the desired effect on angiogenesis.

APPLIED EFS REGULATE THE DIRECTION OF NEW VESSEL FORMATION *IN VITRO*

The *in vivo* angiogenesis experiments described above did not take into account the directional nature of EFs. DC EFs have a unique directional component. Application of EFs to *in vivo* angiogenesis models, e.g., the chick chorioallantoic membrane assay and corneal pocket assay, will be good approaches to answer the question. We have explored if EFs can direct vessel formation using explant cultures where vessel-like structures can be monitored continuously and EFs can be applied in a steady controllable manner. Over several days in culture, rings of mouse and rat aorta form vessel-like structures that are useful for studying angiogenesis *in vitro* (Diglio et al., 1989). We have developed a method to culture mouse and rat aortic rings directly in the electrotaxis chamber, where we can apply EFs. Our results have shown that indeed applied EFs give the vessel-like structure a direction to form and grow (Figure 7.6). Vessel outgrowth from a rat aortic ring appeared to be directed toward the anode.

POSSIBLE ROLE OF ENDOGENOUS EFS IN ANGIOGENESIS

Many aspects of angiogenesis may be affected by steady DC or AC electrical stimulation and pulsed EFs of certain frequency and magnitude. Electrical stimulation

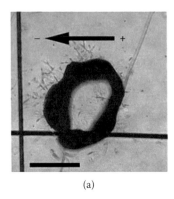
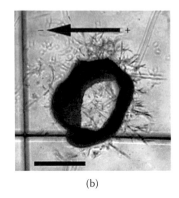

(a)   (b)

FIGURE 7.6   A small applied EF directs vessel-like structure formation. The vessel-like structures from a rat aortic ring grew toward the anode in a three-dimensional electrotaxis chamber, indicating directional angiogenesis. (a) Day 0, immediately after embedding the aorta ring. (b) Day 3, after 200 mV/mm EF treatment. Bar = 0.5 mm. (Reprinted from Song et al., *Nat. Protoc.*, 2, 1479–89, 2007.)

induces production of pro-angiogenic factors, activation of intracellular signaling pathways, and affects survival and proliferation of endothelial cells *in vitro* and *in vivo*. DC EFs guide migration and orientation of endothelial cells. *In vivo* experiments demonstrated convincing effects of electrical stimulation on angiogenesis in ischemic muscles and ischemic brain. Tantalizing preliminary results showed new vessel formation may be guided in a particular direction with applied EFs. This raises the interesting question of whether endogenous EFs contribute to the regulation of angiogenesis.

Endogenous EFs are found in areas where angiogenesis occurs during embryonic development, wound healing, regeneration, and tumor growth (reviewed by Levin, 2007; McCaig et al., 2005; Nuccitelli, 2003; Robinson and Messerli, 2003; Stewart et al., 2007; Zhao, 2009) (see Chapters 1 and 3). As described in Chapter 1, polarized epithelia transport ions and generate a transepithelial potential across the epithelial layers. Indeed, endogenous EFs have been measured in developing frog and chick embryo (Borgens et al., 1983; Hotary and Robinson, 1992; Levin, 2007; Jaffe and Stern, 1979; Nuccitelli, 2003; Robinson and Messerli, 2003; McCaig et al., 2005). Upon wounding, the loss of the epithelial barrier forms a site of short circuit that allows ions to leak through, generating the wound electric current (Chiang et al., 1992; Jaffe and Vanable, 1984; Reid et al., 2005, 2007; Zhao et al., 2006; Nuccitelli et al., 2008; and reviewed by McCaig et al., 2005; Zhao, 2009). Corneal epithelium, for example, normally maintains a transepithelial potential difference of ~25 to 40 mV, with the stroma positive relative to the tear fluid. Upon injury, the transepithelial potential collapses at the wound center. Surrounding intact epithelia maintain a transepithelial potential of ~25 to 40 mV distally (~1 mm) from the wound edge. This voltage gradient establishes an EF in the corneal tissues that has a vector parallel to the epithelial surface and the wound center as the cathode (Reid et al., 2005, 2007) (see Chapter 5) (Figure 7.7). This wound center-orientated EF

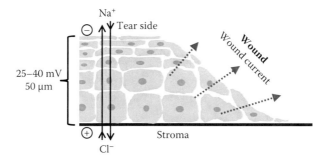

FIGURE 7.7   See color insert. Mechanisms of endogenous electric fields at a corneal epithelium wound. Active pumping of Na^+ from the apical (tear) side of the cornea across the epithelium and stroma to the basal side (aqueous humor), and of Cl^- from the basal to apical side, establishes a transepithelial potential difference of about 25 mV across 50 μm, with the inside positive. Damage to the epithelium disrupts tight junctions that maintain high resistance. This generates a short circuit at the wound site, which drains high concentrations of ions toward the wound from surrounding tissue. This generates a net flow of positive charge out of the wound into the tear solution, i.e., an endogenous wound electric current (blue dashed arrows). This wound electric current can be measured, and pharmacological manipulation by ion substitution or drug treatment to alter ion flow should enhance or decrease the capacity of the batteries, and therefore the wound electric current. (Redrawn from Reid et al., *FASEB J.*, 19, 379–86, 2005.)

may contribute to directing migration of epithelial cells and nerve growth toward the wound center (Zhao et al., 2006; Song et al., 2004). At skin wounds, the endogenous EFs can measure up to ~150 mV/mm, and appear to be required for normal wound healing (see Chapter 1). If the EFs are inhibited, wound healing is compromised (Sta Iglesia and Vanable, 1998; Song et al., 2002, 2004; Reid et al., 2005; Zhao et al., 2006).

Rapid proliferation of cells may cause significant changes in cell surface charge (Brent and Forrester, 1967; Elul et al., 1975), and tumors become polarized relative to quiescent surrounding regions, which generates an EF with the tumor more negative than the surrounding tissue (Schauble and Habal, 1969) (see Chapter 12). In breast cancer, the potential differences between proliferating and nonproliferating regions can be measured at the surface of the skin and are being used diagnostically because they correlate well with the malignancy of the neoplasm (Cuzick et al., 1998). The significance of these field-associated changes is not precisely known, but it indicates that the electric signal may have a role in tumor development, perhaps through affecting tumor angiogenesis (see Chapter 12).

Endogenous EFs also exist in and around the vasculature. Several types of electric potential difference exist around the endothelium of blood vessels (Sawyer and Pate, 1953). ζ-Potentials are created at the endothelial cell wall by the flow of blood in both the aorta and vena cava, and range from 100 to 400 mV (Sawyer et al., 1966). Injured and ischemic tissue also becomes electrically polarized relative to surrounding normal tissue because of the depolarization of cells and the buildup of extracellular K$^+$ ions in the damaged areas, and this can produce a directly measured DC EF of 5.8 mV/mm across an 8 mm zone at the boundary with undamaged tissue (Coronel et al., 1991).

Although endogenous EFs appear to exist widely, their magnitude is relatively weak, but they often persist for many hours and days. The role of endogenous EFs in angiogenesis *in vivo* remains to be demonstrated. Synergistic control of biochemical regulators and EFs as a physical regulator may be more effective in control of angiogenesis.

CONCLUSIONS AND PROSPECTS

Angiogenesis is a complex process that involves multiple pro-angiogenic and anti-angiogenic factors, intracellular signaling pathways, endothelial migration, proliferation, and cell alignment. Electrical stimulation with either alternating or direct current may affect many steps that govern and regulate angiogenesis. We summarize the stages where EFs may affect angiogenesis in Figure 7.8.

First, electrical stimulation may increase the secretion of pro-angiogenic factors (VEGF, IL-8) from endothelial cells, or other types of cells (muscle cells and osteoblasts). Second, electrical stimulation may affect many intracellular signaling pathways that govern endothelial survival, apoptosis, and proliferation. Third,

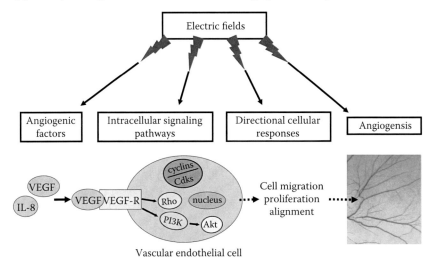

FIGURE 7.8   See color insert. Schematic diagram shows the electrical control of angiogenesis. Electrical stimulation may affect many molecular and cellular components of angiogenesis. Application of EFs upregulates production of pro-angiogenic factors, notably VEGF, in endothelial cells and many other types of cells. Important intracellular signaling pathways PI3K and ERK may be activated, and in a directional manner in direct current EFs, which results in directional organization of the cytoskeleton and directional cell migration and alignment. When multiple cells organize into vessels, the organized directional behaviors may orientate the formation of blood vessels.

DC EFs give endothelial cells a powerful directional cue to guide cell migration, elongation, and alignment. These are important angiogenic responses and may subsequently lead to organized vessel formation (Figure 7.8).

To electrically control angiogenesis is an exciting research topic. Many aspects are yet to be explored. Definitive effects on guiding angiogenesis with EFs *in vivo* need to be demonstrated. This will involve development of new techniques to deliver electric currents. How EFs mediate endothelial cell responses will be a continuing question for the research community. The therapeutic use of electrical stimulation to modulate angiogenesis may prove very practical. Developing techniques to exploit electrical signaling and to control angiogenesis will lead to new therapies to treat diseases or conditions in which angiogenesis is abnormal.

ACKNOWLEDGMENTS

The authors gratefully acknowledge research support to their labs from the Wellcome Trust (058551, 068012), the British Heart Foundation, the UC Davis Dermatology Development Fund, National Eye Institute NIH1RO1EY19101, California Institute of Regenerative Medicine; RB1-01417, NSF MCB-0951199, Research to Prevent Blindness, the Beijing Natural Science Foundation, and the National Natural Science Foundation of China.

REFERENCES

Adams RH, Alitalo K. 2007. Molecular regulation of angiogenesis and lymphangiogenesis. *Nat Rev Mol Cell Biol* 8:464–78.

Ali N, Yoshizumi M, Fujita Y, et al. 2005. A novel Src kinase inhibitor, M475271, inhibits VEGF-induced human umbilical vein endothelial cell proliferation and migration. *J Phamacol Sci* 98:130–41.

Annex BH, Torgan CE, Lin P, et al. 1998. Induction and maintenance of increased VEGF protein by chronic motor nerve stimulation in skeletal muscle. *Am J Physiol* 274:H860–67.

Armstrong PF, Brighton CT, Star AM. 1988. Capacitively coupled electrical stimulation of bovine growth plate chondrocytes grown in pellet form. *J Orthop Res* 6:265–71.

Auerbach R, Lewis R, Shinners B, et al. 2003. Angiogenesis assays: A critical overview. *Clin Chem* 49:32–40.

Azadniv M, Miller MW, Cox C, Valentine F. 1993. On the mechanism of a 60-Hz electric field induced growth reduction of mammalian cells *in vitro*. *Radiat Environ Biophys* 32:73.

Baba T, Kameda M, Yasuhara T, et al. 2009. Electrical stimulation of the cerebral cortex exerts antiapoptotic, angiogenic, and anti-inflammatory effects in ischemic stroke rats through phosphoinositide 3-kinase/Akt signaling pathway. *Stroke* 40:e598–605.

Bai H, Mccaig CD, Forrester JV, Zhao M. 2004. Direct current electric fields induce distinct preangiogenic responses in microvascular and macrovascular cells. *Arterioscler Thromb Vasc Biol* 24:1234–39.

Binhi VN, Goldman RJ. 2000. Ion-protein dissociation predicts 'windows' in electric field-induced wound-cell proliferation. *Biochim Biophys Acta* 1474:147–56.

Borgens RB, Rouleau MF, DeLanney LE. 1983. A steady efflux of ionic current predicts hind limb development in the axolotl. *J Exp Zool* 228:491–503.

Brent TP, Forrester JA. 1967. Changes in surface charge of HeLa cells during the cell cycle. *Nature* 215:92–93.

Brown MD, Egginton S, Hudlicka O, Zhou AL. 1996. Appearance of the capillary endothelial glycocalyx in chronically stimulated rat skeletal muscles in relation to angiogenesis. *Exp Physiol* 81:1043–46.

Brown MD, Hudlicka O. 2003. Modulation of physiological angiogenesis in skeletal muscle by mechanical forces: Involvement of VEGF and metalloproteinases. *Angiogenesis* 6:1–14.

Buschmann I, Schaper W. 1999. Arteriogenesis versus angiogenesis: Two mechanisms of vessel growth. *News Physiol Sci* 14:121–25.

Carmeliet P. 2005. Angiogenesis in life, disease and medicine. *Nature* 438:932–36.

Carmeliet P, Jain RK. 2000. Angiogenesis in cancer and other diseases. *Nature* 407:249–57.

Chekanov V, Rayel R, Krum D, et al. 2002. Electrical stimulation promotes angiogenesis in a rabbit hind-limb ischemia model. *Vasc Endovasc Surg* 36:357–66.

Chiang M, Robinson KR, Vanable JW Jr. 1992. Electrical fields in the vicinity of epithelial wounds in the isolated bovine eye. *Exp Eye Res* 54:999–1003.

Coronel R, Wilms-Schopman FJG, Opthof T, et al. 1991. Injury current and gradient of diastolic stimulation threshold, TQ potential and extracellular potassium concentration during acute regional ischemia in the isolated perfused pig heart. *Circ Res* 68:1241–49.

Cuzick J, Holland R, Barth V, et al. 1998. Electropotential measurements as a new diagnostic modality for breast cancer. *Lancet* 352:359–63.

Diglio CA, Grammas P, Giacomelli F, Wiener J. 1989. Angiogenesis in rat aorta ring explant cultures. *Lab Invest* 60:523–31.

Dimmeler S, Dernbach E, Zeiher AM. 2000. Phosphorylation of the endothelial nitric oxide synthase at Ser-1177 is required for VEGF-induced endothelial cell migration. *FEBS Lett* 477:258–62.

Egginton S, Hudlicka O. 2000. Selective long-term electrical stimulation of fast glycolytic fibres increases capillary supply but not oxidative enzyme activity in rat skeletal muscles. *Exp Physiol* 85:567–73.

Elul R, Brons J, Kravitz K. 1975. Surface charge modifications associated with proliferation and differentiation in neuroblastoma cultures. *Nature* 258:616–17.

Farboud B, Nuccitelli R, Schwab IR, et al. 2000. Direct current electric fields induce rapid directional migration in cultured human corneal epithelial cells. *Exp Eye Res* 70:667–73.

Ferrara N. 1995. The role of vascular endothelial growth factor in pathological angiogenesis. *Breast Cancer Res Treat* 36:127–37.

Ferrara N. 2000. VEGF: An update on biological and therapeutic aspects. *Curr Opin Biotechnol* 11:617–24.

Ferrara N, Kerbel RS. 2005. Angiogenesis as a therapeutic target. *Nature* 438:967–74.

Folkman J. 2007a. Is angiogenesis an organizing principle in biology and medicine? *J Pediatr Surg* 42:1–11.

Folkman J. 2007b. Angiogenesis: An organizing principle for drug discovery? *Nat Rev Drug Discov* 6:273–86.

Folkman J, Klagsbrun M. 1987. Angiogenic factors. *Science* 235:442–47.

Foster FM, Traer CJ, Abraham SM, Fry MJ. 2003. The phospho-inositide (PI) 3-kinase family. *J Cell Sci* 116:3037–40.

Fukumura D, Jain RK. 2008. Imaging angiogenesis and the microenvironment. *APMIS* 116:695–715.

Gerhardt H, Betsholtz C. 2005. How do endothelial cells orientate? *EXS* 94:3–15.

Goldman R, Pollack S. 1996. Electric fields and proliferation in a chronic wound model. *Bioelectromagnetics* 17:450–57.

Goodwin AM. 2007. *In vitro* assays of angiogenesis for assessment of angiogenic and anti-angiogenic agents. *Microvasc Res* 74:172–83.

Graupera M, Guillermet-Guibert J, Foukas LC, et al. 2008. Angiogenesis selectively requires the p110 alpha isoform of PI3K to control endothelial cell migration. *Nature* 453:662–66.

Hang J, Kong L, Gu JW, Adair TH. 1995. VEGF gene expression is upregulated in electrically stimulated rat skeletal muscle. *Am J Physiol* 269:H1827–31.

Hotary KB, Robinson KR. 1992. Evidence of a role for endogenous electrical fields in chick embryo development. *Development* 114:985–96.

Hudlicka O, Brown MD. 2009. Adaptation of skeletal muscle microvasculature to increased or decreased blood flow: Role of shear stress, nitric oxide and vascular endothelial growth factor. *J Vasc Res* 46:504–12.

Hudlicka O, Tyler KR. 1984. The effect of long-term intermittent high-frequency stimulation on capillary density and fibre profiles in rabbit fast muscles. *J Physiol* 353:435–45.

Jaffe LF, Stern CD. 1979. Strong electrical currents leave the primitive streak of chick embryos. *Science* 206:569–71.

Jaffe LF, Vanable JW Jr. 1984. Electric fields and wound healing. *Clin Dermatol* 2:34–44.

Jain RK. 1997. Quantitative angiogenesis assay: Progress and problems. *Nat Med* 3:1203–8.

Jain RK. 2005. Normalization of tumor vasculature: An emerging concept in antiangiogenic therapy. *Science* 307:58–62.

Jain RK. 2008. Lessons from multidisciplinary translational trials on anti-angiogenic therapy of cancer. *Nat Rev Cancer* 8:309–16.

Jain RK. 2009. A new target for tumor therapy. *N Engl J Med* 360:2669–71.

Kanno S, Oda N, Abe M, et al. 1999. Establishment of a simple and practical procedure applicable to therapeutic angiogenesis. *Circulation* 99:2682–87.

Kendall RL, Thomas KA. 1993. Inhibition of vascular endothelial cell growth factor activity by an endogenously encoded soluble receptor. *Proc Natl Acad Sci USA* 90:10705–9.

Kennedy SG, Wagner AJ, Conzen SD, et al. 1997. The PI3K/Akt signaling pathway delivers an anti-apoptotic signal. *Genes Dev* 11:701–13.

Kim IS, Song JK, Zhang YL, et al. 2006. Biphasic electric current stimulates proliferation and induces VEGF production in osteoblasts. *Biochim Biophys Acta* 1763:907–16.

Kim KJ, Li B, Winer J, et al. 1993. Inhibition of vascular endothelial growth factor-induced angiogenesis suppresses tumor growth *in vivo*. *Nature* 362:841–44.

Kuida K, Boucher DM. 2004. Functions of MAP kinases: Insights from gene-targeting studies. *J Biochem* 135:653–56.

Kureishi Y, Luo Z, Shiojima I, et al. 2000. The HMG-CoA reductase inhibitor simvastatin activates the protein kinase Akt and promotes angiogenesis in normocholesterolemic animals. *Nat Med* 6:1004–10.

Levin M. 2007. Large-scale biophysics: Ion flows and regeneration. *Trends Cell Biol* 17:261–70.

Li Q, Yano S, Ogino H, et al. 2007. The therapeutic efficacy of anti-vascular endothelial growth factor antibody, bevacizumab, and pemetrexed against orthotopically implanted human pleural mesothelioma cells in severe combined immunodeficient mice. *Clin Cancer Res* 13:5918–25.

Li S, Chen BP, Azuma N, et al. 1999. Distinct roles for the small GTPases CDC42 and Rho in endothelial responses to shear stress. *J Clin Invest* 103:1141–50.

Li X, Kolega J. 2002. Effects of direct current electric fields on cell migration and actin filament distribution in bovine vascular endothelial cells. *J Vasc Res* 39:391–404.

Liekens S, De Clercq E, Neyts J. 2001. Angiogenesis: Regulators and clinical applications. *Biochem Pharmacol* 61:253–70.

Linderman JR, Kloehn MR, Greene AS. 2000. Development of an implantable muscle stimulator: Measurement of stimulated angiogenesis and poststimulus vessel regression. *Microcirculation* 7:119–28.

Mammoto A, Connor KM, Mammoto T, et al. 2009. A mechanosensitive transcriptional mechanism that controls angiogenesis. *Nature* 457:1103–8.

Matsunaga N, Shimazawa M, Otsubo K, Hara H. 2008. Phosphatidylinositol inhibits vascular endothelial growth factor-A-induced migration of human umbilical vein endothelial cells. *J Pharmacol Sci* 106:128–35.

McCaig CD, Rajnicek AM, Song B, et al. 2005. Controlling cell behavior electrically: Current views and future potential. *Physiol Rev* 85:943–78.

Michel JB, Ordway GA, Richardson JA, Williams RS. 1994. Biphasic induction of immediate early gene expression accompanies activity-dependent angiogenesis and myofiber remodeling of rabbit skeletal muscle. *J Clin Invest* 94:277–85.

Morales-Ruiz M, Fulton D, Sowa G, et al. 2000. Vascular endothelial growth factor-stimulated actin reorganization and migration of endothelial cells is regulated via the serine/threonine kinase akt. *Circ Res* 86:892–96.

Nagasaka M, Kohzuki M, Fujii T, et al. 2006. Effect of low-voltage electrical stimulation on angiogenic growth factors in ischaemic rat skeletal muscle. *Clin Exp Pharmacol Physiol* 33:623–27.

Nelson MA, Passeri J, Frishman WH. 2000. Therapeutic angiogenesis: A new treatment modality for ischemic heart disease. *Heart Dis* 2:314–25.

Nuccitelli R. 1988. Physiological electric fields can influence cell motility, growth, and polarity. *Adv Cell Biol* 2:213–33.

Nuccitelli R. 2003. Endogenous electric fields in embryos during development, regeneration and wound healing. *Radiat Prot Dosimetry* 106:375–83.

Nuccitelli R, Nuccitelli P, Ramlatchan S, et al. 2008. Imaging the electric field associated with mouse and human skin wounds. *Wound Repair Regen* 16:432–41.

Nuccitelli R, Smart T, Ferguson J. 1993. Protein kinases are required for embryonic neural crest cell galvanotaxis. *Cell Motil Cytoskel* 24:54–66.

Patterson C, Runge MS. 1999. Therapeutic angiogenesis: The new electrophysiology? *Circulation* 99:2614–16.

Presta LG, Chen H, O'Connor SJ, et al. 1997. Humanization of an anti-vascular endothelial growth factor monoclonal antibody for the therapy of solid tumors and other disorders. *Cancer Res* 57:4593–99.

Reid B, Nuccitelli B, Zhao M. 2007. Non-invasive measurement of bioelectric currents with a vibrating probe. *Nat Protoc* 2:661–69.

Reid B, Song B, McCaig CD, et al. 2005. Wound healing in rat cornea: The role of electric currents. *FASEB J* 19:379–86.

Risau W. 1997. Mechanisms of angiogenesis. *Nature* 386:671–74.

Robinson KR. 1985. The responses of cells to electrical fields: A review. *J Cell Biol* 101:2023–27.

Robinson KR, Messerli MA. 2003. Left/right, up/down: The role of endogenous electrical fields as directional signals in development, repair and invasion (Review). *Bioessays* 25(8):759–66.

Rousseau S, Houle F, Landry J, Huot J. 1997. p38 MAP kinase activation by vascular endothelial growth factor mediates actin reorganization and cell migration in human endothelial cells. *Oncogene* 15:2169–77.

Saleh M, Stacker SA, Wilks AF. 1996. Inhibition of growth of C6 glioma cells *in vivo* by expression of antisense vascular endothelial growth factor sequence. *Cancer Res* 56:393–401.

Sauer H, Bekhite MM, Hescheler J, Wartenberg M. 2005. Redox control of angiogenic factors and CD31-positive vessel-like structures in mouse embryonic stem cells after direct current electrical field stimulation. *Exp Cell Res* 304:380–90.

Sawyer PN, Himmelfarb E, Lustrin I, Ziskind H. 1966. Measurement of streaming potentials of mammalian blood vessels, aorta and vena cava, *in vivo*. *Biophys J* 6:641–51.

Sawyer PN, Pate JW. 1953. Bio-electric phenomena as an etiologic factor in intravascular thrombosis. *Am J Physiol* 175:103–7.

Schauble MK, Habal MB. 1969. Electropotentials of tumor tissue. *J Surg Res* 9:517–30.

Sheikh I, Tchekanov G, Krum D, et al. 2005. Effect of electrical stimulation on arteriogenesis and angiogenesis after bilateral femoral artery excision in the rabbit hind-limb ischemia model. *Vasc Endovasc Surg* 39:257–65.

Shen M, Gao J, Li J, Su J. 2009. Effect of stimulation frequency on angiogenesis and gene expression in ischemic skeletal muscle of rabbit. *Can J Physiol Pharmacol* 87(5):396–401.

Shiojima I, Walsh K. 2002. Role of Akt signaling in vascular homeostasis and angiogenesis. *Circ Res* 90:1243–50.

Song B, Gu Y, Pu J, et al. 2007. Application of direct current electric fields to cells and tissues *in vitro* and modulation of wound electric field *in vivo*. *Nat Protoc* 2:1479–89.

Song B, Zhao M, Forrester JV, et al. 2002. Electrical cues regulate the orientation and frequency of cell division and the rate of wound healing *in vivo*. *Proc Natl Acad Sci USA* 99:13577–82.

Song B, Zhao M, Forrester JV, et al. 2004. Nerve regeneration and wound healing are stimulated and directed by an endogenous electrical field *in vivo*. *J Cell Sci* 117:4681–90.

Sta Iglesia DD, Vanable JW Jr. 1998. Endogenous lateral electric fields around bovine corneal lesions are necessary for and can enhance normal rates of wound healing. *Wound Rep Reg* 6:531–42.

Stewart S, Rojas-Muñoz A, Izpisúa Belmonte JC. 2007. Bioelectricity and epimorphic regeneration. *Bioessays* 29:1133–37.

Wang E, Reid B, Lois N, et al. 2005. Electrical inhibition of lens epithelial cell proliferation: An additional factor in secondary cataract? *FASEB J* 19:842–44.

Wang E, Yin Y, Zhao M, et al. 2003a. Physiological electric fields control the G1/S phase cell cycle checkpoint to inhibit endothelial cell proliferation. *FASEB J* 17:458–60.

Wang E, Zhao M, Forrester JV, McCaig CD. 2000. Re-orientation and faster, directed migration of lens epithelial cells in a physiological electric field. *Exp Eye Res* 71:91–98.

Wang E, Zhao M, Forrester JV, McCaig CD. 2003b. Bi-directional migration of lens epithelial cells in a physiological electrical field. *Exp Eye Res* 76:29–37.

Wojciak-Stothard B, Entwistle A, Garg R, Ridley AJ. 1998. Regulation of TNF-a-induced reorganization of the actin cytoskeleton and cell-cell junctions by Rho, Rac, and CDC42 in human endothelial cells. *J Cell Physiol* 176:150–65.

Wu LW, Mayo LD, Dunbar JD, et al. 2000. Utilization of distinct signaling pathways by receptors for vascular endothelial cell growth factor and other mitogens in the induction of endothelial cell proliferation. *J Biol Chem* 275:5096–103.

Zhao M, Bai H, Wang E, et al. 2004. Electrical stimulation directly induces pre-angiogenic responses in vascular endothelial cells by signaling through VEGF receptors. *J Cell Sci* 117(Pt 3):397–405.

Zhao M, Song B, Pu J, et al. 2003. Direct visualization of a stratified epithelium reveals that wounds heal by unified sliding of cell sheets. *FASEB J* 17:397–406.

Zhao M, Song B, Pu J, et al. 2006. Electrical signals control wound healing through phosphatidylinositol-3-OH kinase-gamma and PTEN. *Nature* 442:457–60.

8 Inflammatory Cell Electrotaxis

Francis Lin
Department of Physics and Astronomy
University of Manitoba
Winnipeg, Manitoba

Christine E. Pullar
Department of Cell Physiology and Pharmacology
University of Leicester
Leicester, United Kingdom

CONTENTS

INTRODUCTION

Wounding initiates a robust inflammatory response that is essential to prevent infection. Within minutes of wounding, neutrophils arrive at a wound site to remove any bacteria and other foreign microorganisms from the wound site. Neutrophils are replaced by macrophages around three days postwounding to clear the wound of matrix and cell debris. T cells and mast cells are also recruited to wounds, but slightly later, between seven and ten days postwounding. In addition to their anti-infective, phagocyte, and immune surveillance roles, all inflammatory cells release large amounts of growth factors and cytokines that can amplify the inflammatory response and also recruit and activate other cell types in the wound environment. While embryos can regenerate damaged tissue perfectly, adult skin wounds always heal with a scar, and we now know that the robust inflammatory response is partly responsible for wound fibrosis and scarring. In addition, persistent inflammation is often seen in chronic wounds, contributing to their inability to heal. Understanding how inflammatory cells are guided to the wound and how inflammation is curtailed is therefore paramount in our efforts to dampen wound inflammation, heal chronic wounds, and reduce scarring.

Recently, studies have shown that inflammatory cells can sense and migrate directionally toward the cathode of an applied electric field, the center of the wound *in vivo*. This chapter describes the current research on inflammatory cell galvanotaxis.

GALVANOTAXIS

Directional cell migration guided by direct current (DC) electric fields (i.e., electrotaxis or galvanotaxis) has been demonstrated for a variety of cell types, including neural crest cells (see Chapters 10 and 11), corneal epithelial cells (Chapter 5), keratinocytes (Chapter 6), endothelial cells (Chapter 7), fibroblasts (Chapter 9), and rat prostate cancer cells (Chapter 12) (Chang et al. 1996; Cooper and Keller 1984; Djamgoz et al. 2001; Li and Kolega 2002; Onuma and Hui 1985; Rapp et al. 1988; Sheridan et al. 1996; Zhao et al. 1996a, 1996b, 1997, 1999, 2004, 2006). Such electrotactic migration of cells has been shown to contribute to physiological processes such as wound healing (McCaig et al. 2005; Mycielska and Djamgoz 2004; Robinson and Messerli 2003). The migration of immune cells in tissues is critical for inducing immune responses, and chemotaxis provides an important guiding mechanism. However, leukocytes are also exposed to tissue-generated DC electric fields *in vivo* (such as wounded epithelium); therefore, electrotaxis may provide an additional guiding mechanism for immune cell trafficking.

LEUKOCYTE ELECTROTAXIS

In a recent study, electrotactic responses of lymphocytes and monocytes were identified and quantitatively characterized in both *in vitro* and *in vivo* settings (Lin et al. 2008). The electrotactic response of different leukocyte subsets from human peripheral blood was tested in a transwell assay that was previously used for chemotaxis studies. Human peripheral blood mononuclear cells (PBMCs) were loaded into the upper well of the transwell apparatus and were allowed to migrate to the bottom well through a membrane insert. Platinum electrodes were arranged in the top and bottom wells to allow application of an electric field of 2.5 V across the transwell. This electric field mimics the magnitude of physiologic electric fields seen in wounds. Migrating cells in the presence or absence of the electric fields were analyzed by immunofluorescence staining and flow cytometry. As shown in Figure 8.1, the migration of most lymphocytes subsets and monocytes through the membrane increased in the applied electric field when the cathode was located in the bottom well, compared to the control experiments without electric field or with the direction of the electric field reversed. These

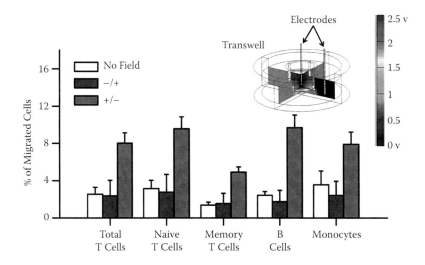

FIGURE 8.1 See color insert. Migration of human blood lymphocytes and monocytes in an applied electric field. The migration of different subsets of hPBMCs increased in transwell assay when the electric field was configured with the positive electrode in the upper well and the negative electrode in the bottom well (top(+)/bottom(−)), compared to migration in the reversed electric field (top(−)/bottom(+)) or without the field. The error bars represent the standard error of the mean (s.e.m.) of multiple independent experiments (n = 9 ~ 11). The drawing of the transwell assay with simulated electrical potential is shown.

results suggest that human blood lymphocytes and monocytes undergo electrotaxis by migrating toward the cathode of the applied electric fields.

T CELL ELECTROTAXIS

The electrotaxis results of human lymphocytes in transwell assays were further verified by analyzing the electrotactic migration of purified T cells at the single cell level in a microfluidic device. The microfluidic device was fabricated by laser cutting the microchannels in plastic sheets, and the plastic device was attached to a glass coverslide using adhesives. Two medium reservoirs were placed at both ends of a straight microchannel, and the platinum electrodes were inserted into the reservoir to apply the electric field. Such a device allowed the application of a well-defined uniform electric field across the microchannel, and the movement of individual cells could be directly monitored and quantitatively analyzed. In a 100 mV/mm electric field, purified memory T cells from human PBMCs migrated toward the cathode of the electric field. When the direction of the electric field was reversed, cells turned around and followed the cathode of the electric field (Figure 8.2). The orientation toward the electric fields and motility of cells are comparable to chemical-gradient-induced leukocyte migration (Lin et al. 2008).

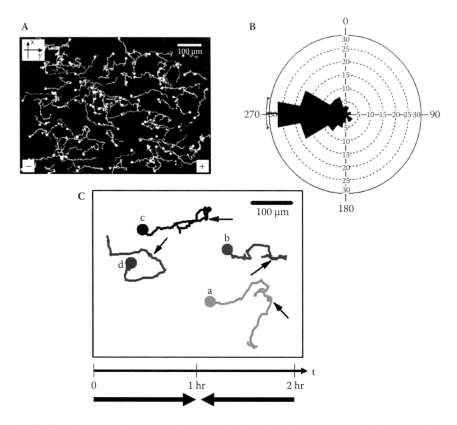

FIGURE 8.2   See color insert. Directional migration of memory T cells in an applied electric field *in vitro*. (A) Cell tracks of a representative experiment show that cells preferentially migrate toward the cathode of the electric field (to the left, 270°). An electric field (1 V cm^{-1}) was applied and the migration of cells recorded for thirty minutes at six frames per minute. The solid circles indicate the end of the tracks. (B) The rose diagram shows the distribution of migration angles (the migration angles were calculated from x-y coordinates at the beginning and end of the cell tracks, and were grouped in 20° intervals, with the radius of each wedge indicating cell number). The mean angle vector and the 95% confidence interval were also shown. Rayleigh uniformity test confirmed that the distribution of migration angles is not uniform. The modified Rayleigh test (V test) showed that the deviation of the migration angles from the direction of electric field (270°) is not significant. (C) Four representative cell tracks showing that cells migrated toward the cathode of the electric field and followed the reverse of the electric field (90° to 270° at the sixtieth minute). An electric field to the right (1 V cm^{-1}) was applied for one hour and was reversed (to the left) for another hour. Migration of cells was recorded continuously for two hours at six frames per minute. The circles indicate the end of the tracks and the arrows indicate when the electric field was reversed.

T CELL ELECTROTAXIS *IN VIVO*

To confirm the lymphocyte electrotaxis results *in vivo*, T cell migration in response to an applied DC electric field was tested in the skin of CXCR6-GFP transgenic mice (Geissmann et al. 2005) using intravital microscopy. Skin T cells in these transgenic mice express green fluorescent protein (GFP) under the promoter of the chemokine receptor CXCR6 (Geissmann et al. 2005), allowing their movement to be followed by time-lapse confocal microscopy. In the absence of an electric field, the GFP+ skin T cells migrate randomly. When an electric field was applied (1 to 2.5 V across ~5 mm in the peripheral ear tissue), the GFP+ cells migrated toward the cathode of the electric field for the duration of the experiment. T cell electrotaxis toward the cathode of an exogenously applied electric field, therefore, can be observed both *in vitro* and *in vivo*. Phospho-flow-cytometry analysis further showed that several key signaling molecules for cell motility, including Akt and Erk1/2, are possibly involved in lymphocyte electrotaxis (Lin et al. 2008).

MURINE NEUTROPHIL ELECTROTAXIS

Mouse neutrophils, isolated by peritoneal lavage, can also respond robustly to applied electric fields of 200 mV/mm and migrated directionally toward the cathode of an applied field with a directedness (cosine θ) of approximately 0.8. Loss of PI3K p110γ reduced directional migration by approximately 45% to 0.45, suggesting that PI3K plays an important role in the electrical compass. In addition, differentiated HL60 cells, which are somewhat neutrophil-like, also migrated toward the cathode of an applied electric field. As the electric field increased from 200 to 800 mV/mm, the cells steadily increased their directedness on fibronectin-coated coverslips until they were migrating with a directedness of 0.9 at 800 mV/mm. A fluorescent probe to detect PIP3 production, PHAkt-GFP, was transfected into HL60 cells and polarized within two minutes of field application toward the cathode-facing pole of the cell. Cells treated with latrunculin still polarized PHAkt-GFP toward the cathode, suggesting that the distribution of PIP3 toward the leading edge was independent of actin polymerization (Zhao et al. 2006).

MACROPHAGE AND POLYMORPHONUCLEAR CELL ELECTROTAXIS

Interestingly, studies on macrophages (Orida and Feldman 1982) and polymorphonuclear cells (PMNs) (Franke and Gruler 1990), suggest that some inflammatory cells can migrate toward the anode of an applied physiological strength electric field. The likelihood of murine peritoneal macrophages extending their pseudopod and migrating toward the anode increased slightly in fields between 150 and 390 mV/mm, while larger fields (780 mV/mm to 1.17 V/mm) significantly increased the number of macrophages in the field with their pseudopods extended toward the anode (Orida and Feldman 1982). While PMNs also exhibited anodal galvanotaxis, the concentration of calcium in the bathing medium had a significant effect on the direction of migration. The PMNs moved toward the anode at 2.5 mM Ca^{2+} and toward the cathode at 0.1 mM Ca^{2+} (Franke and Gruler 1990).

CONCLUSIONS

In summary, inflammatory cell electrotaxis studies demonstrate that the majority of inflammatory cells migrate directionally toward the cathode of an applied electric field *in vitro*, and lymphocytes can be directed by an externally applied electric field *in vivo*. Such electric-field-guided cell migration points to a new mechanism for controlling leukocyte recruitment and positioning in tissues. In addition, lymphocyte electrotaxis may contribute to physiological processes such as wound healing and immune responses.

REFERENCES

Chang, P. C., G. I. Sulik, H. K. Soong, and W. C. Parkinson. 1996. Galvanotropic and galvanotaxic responses of corneal endothelial cells. *J Formos Med Assoc* 95(8):623–27.

Cooper, M. S., and R. E. Keller. 1984. Perpendicular orientation and directional migration of amphibian neural crest cells in dc electrical fields. *Proc Natl Acad Sci USA* 81(1):160–14.

Djamgoz, M. B. A., M. Mycielska, Z. Madeja, S. P. Fraser, and W. Korohoda. 2001. Directional movement of rat prostate cancer cells in direct-current electric field: Involvement of voltage-gated Na+ channel activity. *J Cell Sci* 114(Pt 14):2697–705.

Franke, K., and H. Gruler. 1990. Galvanotaxis of human granulocytes: Electric field jump studies. *Eur Biophys J* 18(6):335–46.

Geissmann, F., T. O. Cameron, S. Sidobre, N. Manlongat, M. Kronenberg, M. J. Briskin, M. L. Dustin, and D. R. Littman. 2005. Intravascular immune surveillance by CXCR6+ NKT cells patrolling liver sinusoids. *PLoS Biol* 3(4):e113.

Li, X., and J. Kolega. 2002. Effects of direct current electric fields on cell migration and actin filament distribution in bovine vascular endothelial cells. *J Vasc Res* 39(5):391–404.

Lin, F., F. Baldessari, C. C. Gyenge, T. Sato, R. D. Chambers, J. G. Santiago, and E. C. Butcher. 2008. Lymphocyte electrotaxis *in vitro* and *in vivo*. *J Immunol* 181(4):2465–71.

McCaig, C. D., A. M. Rajnicek, B. Song, and M. Zhao. 2005. Controlling cell behavior electrically: Current views and future potential. *Physiol Rev* 85 (3):943–78.

Mycielska, M. E., and M. B. Djamgoz. 2004. Cellular mechanisms of direct-current electric field effects: Galvanotaxis and metastatic disease. *J Cell Sci* 117(Pt 9):1631–39.

Onuma, E. K., and S. W. Hui. 1985. A calcium requirement for electric field-induced cell shape changes and preferential orientation. *Cell Calcium* 6(3):281–92.

Orida, N., and J. D. Feldman. 1982. Directional protrusive pseudopodial activity and motility in macrophages induced by extracellular electric fields. *Cell Motil* 2(3):243–55.

Rapp, B., A. de Boisfleury-Chevance, and H. Gruler. 1988. Galvanotaxis of human granulocytes. Dose-response curve. *Eur Biophys J* 16(5):313–19.

Robinson, K. R., and M. A. Messerli. 2003. Left/right, up/down: The role of endogenous electrical fields as directional signals in development, repair and invasion. *Bioessays* 25(8):759–66.

Sheridan, D. M., R. R. Isseroff, and R. Nuccitelli. 1996. Imposition of a physiologic DC electric field alters the migratory response of human keratinocytes on extracellular matrix molecules. *J Invest Dermatol* 106(4):642–46.

Zhao, M., A. Agius-Fernandez, J. V. Forrester, and C. D. McCaig. 1996a. Directed migration of corneal epithelial sheets in physiological electric fields. *Invest Ophthalmol Vis Sci* 37(13):2548–58.

Zhao, M., A. Agius-Fernandez, J. V. Forrester, and C. D. McCaig. 1996b. Orientation and directed migration of cultured corneal epithelial cells in small electric fields are serum dependent. *J Cell Sci* 109(Pt 6):1405–14.

Zhao, M., H. Bai, E. Wang, J. V. Forrester, and C. D. McCaig. 2004. Electrical stimulation directly induces pre-angiogenic responses in vascular endothelial cells by signaling through VEGF receptors. *J Cell Sci* 117(Pt 3):397–405.

Zhao, M., A. Dick, J. V. Forrester, and C. D. McCaig. 1999. Electric field-directed cell motility involves up-regulated expression and asymmetric redistribution of the epidermal growth factor receptors and is enhanced by fibronectin and laminin. *Mol Biol Cell* 10(4):1259–76.

Zhao, M., C. D. McCaig, A. Agius-Fernandez, J. V. Forrester, and K. Araki-Sasaki. 1997. Human corneal epithelial cells reorient and migrate cathodally in a small applied electric field. *Curr Eye Res* 16(10):973–84.

Zhao, M., B. Song, J. Pu, T. Wada, B. Reid, G. Tai, F. Wang, A. Guo, P. Walczysko, Y. Gu, T. Sasaki, A. Suzuki, J. V. Forrester, H. R. Bourne, P. N. Devreotes, C. D. McCaig, and J. M. Penninger. 2006. Electrical signals control wound healing through phosphatidylinositol-3-OH kinase-gamma and PTEN. *Nature* 442(7101):457–60.

9 Effects of DC Electric Fields on Migration of Cells of the Musculoskeletal System

Najmuddin J. Gunja and Clark T. Hung
Department of Biomedical Engineering
Columbia University
New York, New York

J. Chlöe Bulinski
Department of Biological Sciences
Columbia University
New York, New York

CONTENTS

INTRODUCTION

Injuries to soft tissues such as ligaments, tendons, and the knee meniscus can lead to concomitant injury to surrounding joint tissues, eventually resulting in osteo-arthritis. Arthritis today is the leading cause of disability among adults in the United States, according to the U.S. Centers for Disease Control and Prevention (CDC Data and Statistics, 2009). Thus, the development of strategies to aid in the repair or regeneration of injured tissues is paramount to minimize and delay joint degeneration. A popular method of tissue regeneration is to use bioreactors that attempt to mimic *in vivo* conditions for generating new tissues *de novo* (Abousleiman and Sikavitsas, 2006). In certain instances, however, tissue repair may be a more feasible option, and this is usually performed through invasive surgery or by the body's own intrinsic defense mechanisms, such as migration of healthy cells into the damaged area. Cell migration may be an important component of soft tissue repair, as previous work has shown that meniscal allografts (Jackson and Simon, 1993) and devitalized meniscal plugs (Kambic et al., 2000) can be repopulated with cells when placed *in vivo*. One possible stimulus to enhance and direct cell migration is electric fields, which have been shown to induce galvanotaxis. *In vivo*, endogenously generated electric field gradients are known to guide cell migration during wound healing and embryonic development (Robinson, 1985); thus, they may provide a similar benefit for damaged soft tissue (see Chapters 3–8 and 10).

Damage to soft tissues such as articular cartilage, knee meniscus, ligaments, and tendons leads to a poor healing response due to a lack or limited supply of vasculature in these tissues. Thus, a vicious cycle ensues, in which the normal inflammatory response and clotting cascade responsible for tissue healing and repair does not occur, and reparative cells and chemical factors fail to be delivered effectively to the wound site. As such, techniques to recruit cells to the injury site may be particularly important for these tissues, and exogenous stimuli via electric fields may provide a mode to achieve this.

This chapter provides an overview of the effects of galvanotaxis on cells derived from the musculoskeletal system, specifically chondrocytes, meniscus cells, ligament fibroblasts, synovial mesenchymal stem cells, osteoblasts, and osteoclasts. Possible mechanisms involved in galvanotaxis are discussed briefly. Future directions are also discussed, such as the use of electric fields to aid in the differentiation of stem cells for tissue engineering (see Chapter 4).

STRUCTURE-FUNCTION RELATIONSHIP OF TISSUES AND CELLS

To better understand the variations in the effects of electric fields on cells from the musculoskeletal system, it is useful to first outline some key anatomical and functional characteristics of the tissues of interest and their component cells. For this purpose, the knee joint is highlighted as a diarthrodial joint to place these cells and tissues in context. The human knee consists of articular cartilage, knee menisci, tendons, ligaments, and bones (Figure 9.1). Absent from the figure is

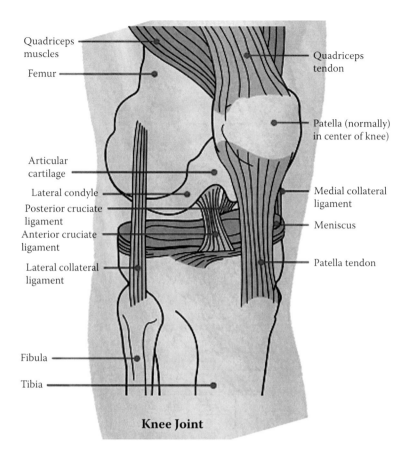

Quadriceps muscles
Femur
Articular cartilage
Lateral condyle
Posterior cruciate ligament
Anterior cruciate ligament
Lateral collateral ligament
Fibula
Tibia
Quadriceps tendon
Patella (normally) in center of knee)
Medial collateral ligament
Meniscus
Patella tendon

Knee Joint

FIGURE 9.1 **See color insert.** Anatomy of the human knee. (Modified from http://upload. wikimedia.org/wikipedia/commons/thumb/0/09/Knee_diagram.svg/2000px-Knee_diagram. svg.png.)

another tissue called the synovial membrane, which encloses the intra-articular portion of the knee joint.

ARTICULAR CARTILAGE: CHONDROCYTES

Articular cartilage is an avascular and aneural tissue located at the ends of diar-throdial joints; thus, it is immunologically privileged. It is composed of several layers of a dense extracellular matrix (ECM) of collagen II and proteoglycans that provide tensile strength and compressive resistance to the tissue, respectively (Almarza and Athanasiou, 2004). The tissue is composed of cells of one type, chondrocytes, which are responsible for upkeep of the ECM. Chondrocytes comprise about 10% of the tissue volume, by weight. In the knee joint, the morphology of the chondrocyte can vary depending on location within the tissue; that is, chondrocytes on the superficial layer are flattened and elongated, while those in the middle and bottom layers are more spherical or ellipsoid. The metabolic activity of chondrocytes is known to increase with depth and decrease with age. During development, it is believed that chondrocytes are formed during the interzone phase, by the differentiation of mesenchymal cells (Pacifici et al., 2000).

LIGAMENT: FIBROBLASTS

Ligaments are nonelastic soft connective tissues that function as load transmitters from bone to bone and act as stabilizers for the knee joint by resisting tensile forces. The ECM is composed of highly oriented dense fibrous tissue consisting of mainly type I collagen and proteoglycans. The main cellular components of ligaments are spindle-shaped (elongated) fibroblasts that synthesize and secrete ECM. Normal physiologic loading is essential for the balance between degenerated ECM and ECM undergoing repair by the fibroblast population (Loitz and Frank, 1993). The origin of ligament fibroblasts is still unclear, although they are thought to arise from tendon primordia during embryonic development (Tozer and Duprez, 2005). The cruciate ligaments of the knee, in particular, may form from a population of intermediate zone cells that have ceased Col2a1 expression (Hyde et al., 2008).

MENISCUS: INNER AND OUTER MENISCUS CELLS

The knee meniscus is a fibrocartilaginous semi-lunar-shaped disc located on the tibial plateau of the knee joint (Figure 9.1). Vascular elements are limited to the outer regions of the meniscus. The meniscus serves several functions, including load transmission, shock absorption, and joint lubrication. Similar to ligaments, the major ECM component of the meniscus, within the outer regions, is collagen I. Both collagens I and II are present in the inner regions of the meniscus, in a 2:3 ratio. Proteoglycans are also found in the tissue, with the majority present in the inner regions (Sweigart and Athanasiou, 2001). Although several subtypes of cells have been identified in the meniscus, the tissue can be divided into two regions: (1) an outer region containing mainly fibroblast-like cells that are elongated or spindle shaped, and (2) an inner region containing mainly chondrocyte-like cells that are rounded or spherical. Meniscus cells are responsible for maintaining the homeostasis of the meniscus, and their metabolic activity is dependent on the forces exerted on the tissue in the knee joint. The origin of the different cell types is unclear; however, it is thought that during embryonic development, the meniscus is formed by precursor cells expressing collagen type IIa as well as by cells that invade the joint that are not part of the anlagen (Hyde et al., 2008).

SYNOVIAL MEMBRANE: TYPE B SYNOVIOCYTES/SYNOVIAL MESENCHYMAL STEM CELLS (sMSCs)

The synovial membrane is a layer of soft tissue that lines noncartilaginous surfaces of joints. It is believed to arise by condensation of the mesenchymal layer during embryonic development and forms during the cavitation and morphogenesis of the joint (Pacifici et al., 2000). The functions of the synovial membrane include lubricating the joint, removing cellular waste products from the joint space, and providing nutrition for the surrounding avascular cartilage. The synovial membrane can be divided into two thin layers. The outer layer contains a mix of fibrous and fatty tissues, while the inner layer consists of a single cell sheet. Consequently, the outer layer consists of mainly fibroblast-like and adipose cells, whereas the inner layer consists of fibroblasts (type B fibroblasts or sMSCs) and macrophages (type A fibroblasts). The inner fibroblasts are responsible for producing hyaluronan and lubricin that together contribute to joint lubrication. The macrophages are responsible for removing undesirable substances from the synovial fluid. The type B fibroblasts in the inner layer are particularly relevant for tissue engineering, as they have been shown to migrate from the synovium to the knee meniscus under certain conditions of injury and to modulate a healing response there (Hunziker and Rosenberg, 1996). In addition, these cells are also sometimes referred to as sMSCs, as they demonstrated ability to produce a nonnative matrix *in vitro* when exogenously stimulated (Pei et al., 2008).

BONE: OSTEOBLASTS AND OSTEOCLASTS

Bone is a hard, brittle, and dense connective tissue that serves multiple functions, such as to protect vital body organs, to provide a frame for the body, and to generate the cellular components of blood, via the hematopoietic bone marrow. Bone is composed of an inorganic phase, consisting of hydroxyapatite, and an organic phase, consisting of type I collagen. The two major cell types present during bone formation and resorption are osteoblasts and osteoclasts. Osteoblasts are immature mononucleated bone cells that are responsible for the production of alkaline phosphatase, which has been implicated in bone mineralization (Rodan, 1992). They originate from osteo-progenitor cells. Osteoclasts are multinucleated cells that are equipped with phagocyte-like mechanisms that account for bone resorption. Osteoclasts are derived from the monocyte–stem cell lineage (Suda et al., 2001).

GALVANOTAXIS

Different cell types are known to exhibit different responses to endogenously and exogenously applied chemical, mechanical, and electrical stimuli. Of particular interest are direct current (DC) electric fields that have been implicated in directing cell migration during development, wound healing, or tissue regeneration. The phenomenon of directed cell movement under an applied electric field is known as electrotaxis or galvanotaxis (Mycielska and Djamgoz, 2004). In this section, we review the effects of electric fields of cells derived from the musculoskeletal system. Table 9.1 provides a synopsis of the effects of electric fields on musculoskeletal cells as well as on other commonly studied cell types, allowing a useful comparison.

TABLE 9.1
Musculoskeletal Cell Migration Induced by Applied DC Electric Fields

Musculoskeletal Cells

Articular chondrocytes (Chao et al. 2000)	Cathodal migration at E = 0–10 V/cm; alignment perpendicular to applied field.	Speed: 13.74 μm/hr (6 V/cm)
Anterior cruciate ligament (knee) fibroblasts (Chao et al. 2007)	Cathodal migration at E = 6 V/cm; modulated by collagen substrate; alignment perpendicular to applied field.; role of IP3 pathway assessed with U-71322 and neomycin	Speed: 15 μm/hr
Inner and outer meniscus cells (Gunja et al. 2010)	Cathodal migration at E = 6 V/cm	Speed: Outer cells: 14.2 μm/hr; Inner cells: 8 μm/hr
Osteoblasts and osteoclasts (Ferrier et al. 1986)	Cathodal migration of osteoblasts at E = 1–10 V/cm Anodal migration of osteoclasts at E = 1–10 V/cm	Osteoblast velocity: 3–32 μm/hr Osteoclasts velocity: 72–140 μm/hr
Bovine osteoblast-like cells (Curtze et al. 2004)	Cathodal migration at E = 10–20 V/cm Increased calcium levels, electric field changed traction forces of cells	No speed or velocity measured
Rat osteoblast-like cells (Wang et al. 1998)	Direct current = 100 μA/cm^2 Enhanced proliferation and calcification of cells; increased intracellular free calcium ion concentration	No speed or velocity measured

Other Selected Cell Types

Lens epithelial cells (primary bovine and human; transformed human cell line) (Wang et al. 2000)	Anodal migration of primary bovine and transformed human at E = 0–2.5 V/cm line Cathodal migration of primary human at E = 0–2.5 V/cm	Speed: 4.5 μm/hr at 2 V/cm
Human dermal fibroblasts and keratinocytes (Sillman et al. 2003)	No migration or alignment of dermal fibroblasts on collagen I-coated substratum at E = 0.1 V/mm Cathodal migration of fibroblasts +/– serum, high magnesium (the latter did not affect speed) at E = 0.1 V/mm Cathodal migration of keratinocytes at E = 0.1 V/mm	Fibroblast speed: 15–19 μm/hr in media lacking or containing sera Keratinocyte speed: 48 μm/hr
3T3 fibroblasts (Finkelstein et al. 2004)	Cathodal migration and perpendicular alignment at E = 0.6–6 V/cm Microtubules and adhesive components modulated migration of fibroblasts, as did signals in wounded monolayers	Speed: 10–22 μm/hr at 0.6–6 V/cm

TABLE 9.1 (continued)
Musculoskeletal Cell Migration Induced by Applied DC Electric Fields

Other Selected Cell Types

3T3 and HeLa fibroblasts (Finkelstein et al. 2007)	Cell surface charge altered using neuraminidase and biotin-avidin. At E = 6 V/cm, neuraminidase did not affect cell speed but affected directionality (no cathodal migration). Biotin-avidin did not affect cell speed or directionality.	3T3 speed: 14.4 µm/hr 3T3 + neuraminidase speed: 13 µm/hr HeLa speed: 7.4 µm//hr HeLa + neuraminidase speed: 6 µm/hr HeLa + biotin-avidin speed: 6 µm/hr
Prostate cancer cells (MAT-LyLu, AT-2 lines) (Djamgoz et al. 2001)	Cathodal migration at E = 3 V/cm	MAT-LyLu speed: 23 um/hr AT-2 speed: 7.5 µm/hr
Human adipose-derived stem cells (HADSC) Human epicardial fat-derived stem cells (Tandon et al. 2009)	Elongation and perpendicular alignment at E = 6 V/cm HADSC disassembled gap junctions and upregulated expression of bFGF, VEGF, thrombomodulin and connexin-43.	No speed or velocity measured
Neuronal stem/progenitor cells (Li et al. 2008)	Cathodal migration at E = 0–2.5 V/cm NMDAR/Rac1/actin transduction pathway needed for EF-induced migration	Speed: 9–16 µm/hr at 0.5–2.5 V/cm.
Gingival fibroblasts (human) (Ross, Ferrier, and Aubin 1989)	Perpendicular alignment but not migration at E = 1–15 V/cm. Role of Ca^{2+} explored with ionophore, lanthanum, channel blocker	No speed or velocity measured

CHONDROCYTES

Chondrocytes are known to migrate preferentially in response to chemokines such as nitric oxide (Frenkel et al., 1996) or ECM molecules such as fibronectin or collagen (Shimizu et al., 1997). More recently, the effects of galvanotaxis on chondrocyte migration were examined in order to test the effects of varying the electric field strengths, the presence of serum factors, the temperature, and the efficacy of phosphoinositide signaling pathways (Chao et al., 2000). In the first experiment, the authors applied six different electric field strengths, 0 to 6 V/cm, to bovine chondrocytes, and measured cell orientation, speed, and directed velocity of migration, using time-lapse microscopy. As expected, chondrocytes were found to move toward the cathode regardless of field strength, and increasing field strength significantly increased both cell speed and directed velocity. Interestingly, this behavior persisted at a higher field strength (10 V/cm), although statistics were not performed in this case due to cell death that was also induced by this high field strength. In the second experiment, the effects of serum on chondrocyte migration were investigated. Although cathodal migration was observed with or without serum, directed velocity was decreased threefold in its absence. In the third experiment, electric-field-induced migration of chondrocytes was compared at room and physiological temperatures. Physiological temperature (37°C) increased cell velocity approximately twofold, relative to room temperature (approximately 25°C). However, no significant differences in directed velocity were observed between groups in the absence of an electric field. In the fourth experiment, the authors utilized (1) a phospolipase C inhibitor, U-73122, and (2) an aminoglycoside antibiotic, neomycin, to investigate the importance of the inisitol phospholipid pathway during chondrocyte migration. Both the phospholipase C inhibitor and neomycin were found to decrease cathodal migration significantly. In addition, neomycin, which is a positively charged molecule that binds to the cell surface, was also shown to promote anodal migration in some cells.

LIGAMENT FIBROBLASTS

Electric fields have shown promise in the healing of injured ligaments. For instance, the application of DC electric fields to rabbit patellar ligaments, with full-thickness defects, restored tensile stiffness to the tissue, while it concomitantly decreased the percentage of type III collagen (Akai et al., 1988). A recent *in vitro* study compared the effects of static and pulsed DC electric fields on anterior cruciate ligament fibroblasts (Chao et al., 2007). In the first of four parts of the study, the effects of a static electric field on ligament fibroblasts were examined at several field strengths. Ligament cell migration toward the cathode and alignment of cells perpendicular to the field were observed. Ligament cell speed was significantly enhanced with increasing field strengths, and ligament fibroblasts migrated significantly more rapidly than chondrocytes (Figure 9.2). In the second part of the study, substratum coatings were explored, to optimize migration and better mimic *in vivo* migration of cells on collagen. Here, cells plated on a collagen I substrate showed enhanced cell migration with and without electric fields, compared to cells migrating on uncoated glass slides. The third phase of the study tested pulsing electric fields that are often used therapeutically. Pulsed fields affected both the migration speed and the expression of collagen I. Cells exposed to a pulsed field showed greatest speed at a frequency of 0.025 Hz, although this was significantly lower than the static field control. Interestingly, the authors found that collagen I gene expression levels were higher in the 0.025 Hz group when compared to the static field control. In the original series of experiments, a wound-healing experiment was performed by scraping a strip of cells from a dense monolayer, and exposing the wound model to static or pulsed electric fields. Pulsed electric fields of 0.008 Hz significantly enhanced the speed of

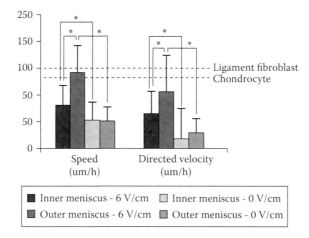

FIGURE 9.2 See color insert. Cell speed and directed velocity of inner and outer meniscus cells. Asterisks represent conditions in which mean values show honestly significant differences (HSD) between groups. Data for each group were analyzed by one-way ANOVA, followed by Tukey's post hoc test of the mean for each group to determine whether differences between them exceeded the computed HSD. Dashed lines represent average speed of ligament fibroblasts and chondrocytes at an applied DC electric field strength of 6 V/cm.

migration into the denuded area during the three-hour experiments, compared to static electric fields of the same voltage (6 V/cm).

INNER AND OUTER MENISCUS CELLS

Electric fields may be amenable to the induction of cell migration toward the site of meniscal injury. For example, radio frequency energy treatments of injured rabbit menisci were found to significantly enhance autocrine motility factor (AMF) in meniscal fibrochondrocytes. AMF has been shown to stimulate cell motility in fibroblasts and cancer cells and may also aid in meniscus cell migration (Hatayama et al., 2007). A recent *in vitro* study examined the effects of a static electric field (6 V/cm) on the migration, directionality, and adhesion strength of inner and outer meniscus cells plated on glass slides (Gunja et al., 2010). Outer, fibroblast-like meniscus cells were found to migrate more quickly than inner chondrocyte-like meniscus cells; both were enhanced by the electric field, compared to the no-field controls (see Figure 9.2). Inner meniscus cells exhibited higher adhesion strength than the outer meniscus cells. These results are consistent with the observed cell morphology and phenotype, and they confirm previous findings that in general fibroblasts migrate more quickly than chondrocytes (Chao et al., 2007). Interestingly, both inner and outer meniscus cells migrated toward the cathode and aligned perpendicular to the electric field, as previously observed with ligament fibroblasts and chondrocytes (Chao et al., 2000, 2007).

Synovial Mesenchymal Stem Cells

Cells from the synovial membrane have been implicated in modulating a healing response in injured cartilage and menisci. TGF-β1 and bFGF have been shown to induce synovial cells to migrate into defects through the partial thickness of articular cartilage (Hunziker and Rosenberg, 1996). Recently, radio frequency energy treatment was shown to enhance the capacity of synovial fibroblasts to repopulate a meniscus defect in rabbits (Hatayama et al., 2007). Our laboratory is currently investigating the effects of direct current electric fields on sMSCs following "priming" with and without growth factors. Preliminary studies suggest that growth factors (TGF-β1, bFGF, and PDGF-B) significantly enhance cell migration, in the presence or absence of an electric field; however, the electric field significantly enhances cell speed and directionality toward the cathode. This suggests that under appropriate conditions, sMSCs may be induced to migrate toward a site of injury using a combination of DC electric fields and growth factors.

Osteoblasts and Osteoclasts

Electric fields may be effective in bone induction and regeneration. They have been used to treat fractures (Brighton et al., 1985) and nonunions (Brighton and Pollack, 1985), and to promote osseo-integration (Shigino et al., 2001). Several *in vitro* studies have been performed to study bone cell galvanotaxis. In a study comparing the effects of electric fields on osteoblasts and osteoclasts (using low and high field strengths, 1 and 10 V/cm, respectively), it was observed that osteoclasts migrated toward the anode while osteoblasts migrated toward the cathode (Ferrier et al., 1986). The migration pattern of osteoclasts was consistent with other monocyte-derived cell types (Orida and Feldman, 1982), while osteoblasts mimicked the directionality observed in other cell types derived from the primitive mesenchyme (chondrocytes, ligaments, inner and outer meniscus cells, and sMSCs (Chao et al., 2000, 2007; Gunja et al., 2010)). In another study examining effects of stronger, nonphysiological electric fields (10 to 20 V/cm) on bovine osteoblast-like cells, increased calcium levels were observed in addition to changes in traction forces of the cells (Curtze et al., 2004). However, cell speeds were not measured in this study. Interestingly, rat osteoblast-like cells exposed to an electric field of 100 μA/cm^2 showed increased intracellular calcium levels but no preferential direction of migration (Wang et al., 1998). This suggests that the effect of electric fields may be species dependent, although one cannot rule out other differences in the measurement systems used in these two studies.

MECHANISMS INVOLVED IN GALVANOTAXIS

To better understand the mechanisms involved in galvanotaxis, it is useful to review the steps that occur in cell migration in the absence of an electric field. In response to an extracellular stimulus, cells develop a polarized morphology (i.e., distinct front and rear ends), and this is mediated by Rho family GTPases, microtubules, and integrins (Vicente-Manzanares et al., 2005). At the front end of the cell, actin polymerization controls the protrusion of flat membrane particles called lamellipodia and filopodia (Pollard and Borisy, 2003). This process is regulated by the Arp2/3 complex (Weaver et al., 2003) and formins (Watanabe and Higashida, 2004). The next step involves assembly and disassembly of adhesions at the leading and trailing edges of the cell, respectively. This process is mediated by structural molecules such as talin (Nayal et al., 2004) that link integrins to actin or signal transduction molecules like zyxin (Zaidel-Bar et al., 2004) that bind directly to α-actinin. At the molecular level, Rho GTPases such as Rho (Nobes and Hall, 1999) and dynamin (Ezratty et al., 2005) are thought to control the decision of a focal adhesion to undergo maturation or turnover. The contractile force required to move the remainder of the cell forward is controlled by myosin-actin interactions (Jay et al., 1995). This may occur through contraction of filaments that connect to adhesion complexes within the cell or by propelling adhesion complexes along cortical actin filament tracks. In the final step, focal contacts are disassembled in a microtubule-dependent step, and the rear of the cell lifts off, leaving major fractions of integrins and focal adhesions on the substratum (Ezratty et al., 2005; Franco et al., 2004).

The cell migration mechanisms mentioned above likely occur during galvanotaxis as well. However, other cellular factors and mechanisms may be used by electric fields to change cell speed and directionality. For instance, the introduction of a DC electric field has been shown to influence the influx of ions into the cell passively (Cooper and Keller, 1984) or through voltage-gated (Cho et al., 1999; Djamgoz et al., 2001) and stretch-activated (Lee et al., 1999) channels. In addition, electrical fields have been shown to redistribute and activate growth factor receptors on the cell surface (Fang et al., 1999; Zhao et al., 2002); the resulting polarization of the cell and activation of its protein kinase signaling pathways may contribute to galvanotaxis. All cells also contain charged molecules on the surface of their plasma membrane. Exposure to electric fields may shift these charged molecules, altering and perhaps polarizing the plasma membrane (Patel and Poo, 1982). However, at least one study has shown that shifting the surface charge on some cell types does not alter membrane polarization and directional migration (Finkelstein et al., 2007).

These mechanisms have been investigated more closely in fibroblast cell lines. All types of fibroblasts have been found to migrate toward the cathode when exposed to DC electric fields (Brown and Loew, 1994; Erickson and Nuccitelli, 1984; Finkelstein et al., 2004, 2007; Harris et al., 1990; Onuma and Hui, 1988), with the exception of stromal fibroblasts (Soong et al., 1990) and dermal fibroblasts (Zhao et al., 2006), which both migrate to the anode. Cells of a mouse fibroblast cell line, C3H/10T1/2, increased their levels of intracellular calcium in response to electric fields; the high level of calcium was maintained throughout duration of exposure to the field (Onuma and Hui, 1988). Additionally, compounds that block calcium channels inhibited galvanotaxis without influencing spontaneous cell movement (Onuma and Hui, 1988). An influx of calcium causes cell contraction and membrane polarization; molecular mechanisms for these intracellular changes are detailed elsewhere (Funk and Monsees, 2006; Mycielska and Djamgoz, 2004). Interestingly, cathodal migration of two fibroblast cell lines, NIH 3T3 and SV101, was found to be independent of extracellular calcium or intracellular calcium changes (Brown and Loew, 1994; Hahn et al., 1992). This disparity most likely results from differing components of the culture medium that may affect action or strength of the applied electric fields. For example, serum is known to change kinetics of ion channels (Ding and Djamgoz, 2004) or induce hyperpolarization of membrane potential (Dixon and Aubin, 1987). Other ions, such as sodium, have also been implicated in directing migration in electric fields. In fibroblasts lacking the sodium-hydrogen exchanger (NHE-null PS120 fibroblasts), migration is dependent on the sodium-hydrogen exchanger isoform 1 (NHE1) (Denker and Barber, 2002), as NHE-null PS120 fibroblasts containing mutations in NHE1 exhibit loss of directionality, polarity, and anchoring (Denker and Barber, 2002).

Cytoskeletal components such as actin, myosin, and microtubules also play important roles in cell alignment and migration. Mouse embryo fibroblasts elongate their cell bodies and align their cytoskeletons (i.e., stable microtubules and actin stress fibers) perpendicularly to the electric field (Harris et al., 1990). This interesting phenomenon, also observed in several other cell types, is known as galvanotropism. One plausible explanation is that perpendicular alignment reduces the electric field gradient across the cell (Chao et al., 2000) (see Chapter 2). In 3T3 fibroblasts, inhibition of actin microfilaments disrupted cell migration, while inhibition of microtubules reduced cell speed but did not halt motility (Finkelstein et al., 2004). Cytoskeletal rearrangement is most likely influenced by intracellular ions such as calcium, which controls actin polymerization and depolymerization dynamics, as well as myosin contractility (Mycielska and Djamgoz, 2004).

TISSUE REGENERATION AND FUTURE DIRECTIONS

The majority of studies published to date have focused on uses of chemical and mechanical stimuli to regenerate tissue; however, the potential for bioelectrical stimulation of the regeneration process has been largely unexplored. This is surprising since endogenous electric current has been implicated in tissue healing and regeneration (Borgens, 1982).

In the majority of cell types found in the musculoskeletal system, the application of DC electric fields results in galvanotropism and enhanced migration. These results have practical implications for musculoskeletal tissue engineering studies in which spatial orientation of cells within a scaffold system is thought to influence ECM alignment and the overall functional properties of the tissue. As proof of concept, a recent study examined migration of fibroblasts within a three-dimensional collagen gel at various electric field strengths (Sun and Cho, 2004). The authors were able to modulate and control fibroblast adhesion and migration in three dimensions. An ensuing experiment comparing fibroblasts and bone-marrow-derived mesenchymal stem cells showed that fibroblasts were able to reorient the collagen matrix to a greater degree than were the mesenchymal stem cells (Sun et al., 2006). This result is consistent with the higher adhesion strengths of the mesenchymal stem cells, which presumably allows them to resist reorientation in the field. Indeed, different cell types may require different levels of electrical stimulus to produce the same response. Alternatively, the surface properties of the scaffold may be modulated to achieve the desired adhesion strengths required for optimal cell migration or reorientation.

Electric fields may also be useful tools for sorting of cells in scaffolds that contain multiple cell types. For example, one could devise a meniscus tissue engineering scaffold that includes cells that are fibroblast-like and chondrocyte-like, corresponding to the outer and inner regions of the meniscus, respectively. Since fibroblasts migrate more quickly than chondrocytes in DC electric fields, this difference in cell migration speed could affect the desired spatial sorting of cells within the scaffold after the electric field had been applied.

REFERENCES

Abousleiman, R. I., and Sikavitsas, V. I. 2006. Bioreactors for tissues of the musculoskeletal system. *Adv Exp Med Biol* 585:243–59.

Akai, M., Oda, H., Shirasaki, Y., and Tateishi, T. 1988. Electrical stimulation of ligament healing. An experimental study of the patellar ligament of rabbits. *Clin Orthop Relat Res* 235:296–301.

Almarza, A. J., and Athanasiou, K. A. 2004. Design characteristics for the tissue engineering of cartilaginous tissues. *Ann Biomed Eng* 32(1):2–17.

Borgens, R. B. 1982. What is the role of naturally produced electric-current in vertebrate regeneration and healing. *Int Rev Cytol Surv Cell Biol* 76:245–98.

Brighton, C. T., Hozack, W. J., Brager, M. D., Windsor, R. E., Pollack, S. R., Vreslovic, E. J., et al. 1985. Fracture-healing in the rabbit fibula when subjected to various capacitively coupled electrical fields. *J Orthopaedic Res* 3(3):331–40.

Brighton, C. T., and Pollack, S. R. 1985. Treatment of recalcitrant non-union with a capacitively coupled electrical-field—A preliminary-report. *J Bone Joint Surg-Am Vol* 67A(4):577–85.

Brown, M. J., and Loew, L. M. 1994. Electric field-directed fibroblast locomotion involves cell surface molecular reorganization and is calcium independent. *J Cell Biol* 127(1):117–28.

Centers for Disease Control Data and Statistics. 2009. National Health Survey. Published only on http://www.cdc.gov/nchs/data/hus/hus09.pdf.

Chao, P. H., Lu, H. H., Hung, C. T., Nicoll, S. B., and Bulinski, J. C. 2007. Effects of applied DC electric field on ligament fibroblast migration and wound healing. *Connect Tissue Res* 48(4):188–97.

Chao, P. H., Roy, R., Mauck, R. L., Liu, W., Valhmu, W. B., and Hung, C. T. 2000. Chondrocyte translocation response to direct current electric fields. *J Biomech Eng* 122(3):261–67.

Cho, M. R., Thatte, H. S., Silvia, M. T., and Golan, D. E. 1999. Transmembrane calcium influx induced by AC electric fields. *FASEB J* 13(6):677–83.

Cooper, M. S., and Keller, R. E. 1984. Perpendicular orientation and directional migration of amphibian neural crest cells in DC electrical fields. *Proc Natl Acad Sci USA* 81(1):160–64.

Curtze, S., Dembo, M., Miron, M., and Jones, D. B. 2004. Dynamic changes in traction forces with DC electric field in osteoblast-like cells. *J Cell Sci* 117(13):2721–29.

Denker, S. P., and Barber, D. L. 2002. Cell migration requires both ion translocation and cytoskeletal anchoring by the Na-H exchanger NHE1. *J Cell Biol* 159(6):1087–96.

Ding, Y., and Djamgoz, M. B. 2004. Serum concentration modifies amplitude and kinetics of voltage-gated Na+ current in the Mat-LyLu cell line of rat prostate cancer. *Int J Biochem Cell Biol* 36(7):1249–60.

Dixon, S. J., and Aubin, J. E. 1987. Serum and alpha 2-macroglobulin induce transient hyperpolarizations in the membrane potential of an osteoblastlike clone. *J Cell Physiol* 132(2):215–25.

Djamgoz, M. B. A., Mycielska, M., Madeja, Z., Fraser, S. P., and Korohoda, W. 2001. Directional movement of rat prostate cancer cells in direct-current electric field: Involvement of voltage-gated Na+ channel activity. *J Cell Sci* 114(Pt 14):2697–705.

Erickson, C. A., and Nuccitelli, R. 1984. Embryonic fibroblast motility and orientation can be influenced by physiological electric fields. *J Cell Biol* 98(1):296–307.

Ezratty, E. J., Partridge, M. A., and Gundersen, G. G. 2005. Microtubule-induced focal adhesion disassembly is mediated by dynamin and focal adhesion kinase. *Nat Cell Biol* 7(6):581–90.

Fang, K. S., Ionides, E., Oster, G., Nuccitelli, R., and Isseroff, R. R. 1999. Epidermal growth factor receptor relocalization and kinase activity are necessary for directional migration of keratinocytes in DC electric fields. *J Cell Sci* 112(Pt 12):1967–78.

Ferrier, J., Ross, S. M., Kanehisa, J., and Aubin, J. E. 1986. Osteoclasts and osteoblasts migrate in opposite directions in response to a constant electrical-field. *J Cell Physiol* 129(3):283–88.

Finkelstein, E., Chang, W., Chao, P. H., Gruber, D., Minden, A., Hung, C. T., et al. 2004. Roles of microtubules, cell polarity and adhesion in electric-field-mediated motility of 3T3 fibroblasts. *J Cell Sci* 117(Pt 8):1533–45.

Finkelstein, E. I., Chao, P. H., Hung, C. T., and Bulinski, J. C. 2007. Electric field-induced polarization of charged cell surface proteins does not determine the direction of galvanotaxis. *Cell Motil Cytoskel* 64(11):833–46.

Franco, S. J., Rodgers, M. A., Perrin, B. J., Han, J., Bennin, D. A., Critchley, D. R., et al. 2004. Calpain-mediated proteolysis of talin regulates adhesion dynamics. *Nat Cell Biol* 6(10):977–83.

Frenkel, S. R., Clancy, R. M., Ricci, J. L., Di Cesare, P. E., Rediske, J. J., and Abramson, S. B. 1996. Effects of nitric oxide on chondrocyte migration, adhesion, and cytoskeletal assembly. *Arthritis Rheum* 39(11):1905–12.

Funk, R. H., and Monsees, T. K. 2006. Effects of electromagnetic fields on cells: Physiological and therapeutical approaches and molecular mechanisms of interaction. A review. *Cells Tissues Organs* 182(2):59–78.

Gunja, N. G., Fong, J. V., Vunjak-Novakovic, G., and Hung, C. T. 2010. Migration responses of meniscus cells to applied direct current electric fields. Paper presented at the Orthopedic Research Society, New Orleans, LA.

Hahn, K., DeBiasio, R., and Taylor, D. L. 1992. Patterns of elevated free calcium and calmodulin activation in living cells. *Nature* 359(6397):736–38.

Harris, A. K., Pryer, N. K., and Paydarfar, D. 1990. Effects of electric fields on fibroblast contractility and cytoskeleton. *J Exp Zool* 253(2):163–76.

Hatayama, K., Higuchi, H., Kimura, M., Takeda, M., Ono, H., Watanabe, H., et al. 2007. Histologic changes after meniscal repair using radiofrequency energy in rabbits. *Arthroscopy* 23(3):299–304.

Hunziker, E. B., and Rosenberg, L. C. 1996. Repair of partial-thickness defects in articular cartilage: Cell recruitment from the synovial membrane. *J Bone Joint Surg Am* 78(5):721–33.

Hyde, G., Boot-Handford, R. P., and Wallis, G. A. 2008. Col2a1 lineage tracing reveals that the meniscus of the knee joint has a complex cellular origin. *J Anat* 213(5):531–38.

Jackson, D. W., and Simon, T. 1993. Assessment of donor cell survival in fresh allografts (ligament, tendon, and meniscus) using DNA probe analysis in a goat model. *Iowa Orthop J* 13:107–14.

Jay, P. Y., Pham, P. A., Wong, S. A., and Elson, E. L. 1995. A mechanical function of myosin II in cell motility. *J Cell Sci* 108(Pt 1):387–93.

Kambic, H. E., Futani, H., and McDevitt, C. A. 2000. Cell, matrix changes and alpha-smooth muscle actin expression in repair of the canine meniscus. *Wound Repair Regen* 8(6):554–61.

Lee, J., Ishihara, A., Oxford, G., Johnson, B., and Jacobson, K. 1999. Regulation of cell movement is mediated by stretch-activated calcium channels. *Nature* 400(6742):382–86.

Li, L., El-Hayek, Y. H., Liu, B., Chen, Y., Gomez, E., Wu, X., et al. 2008. Direct-current electrical field guides neuronal stem/progenitor cell migration. *Stem Cells* 26(8):2193–200.

Loitz, B. J., and Frank, C. B. 1993. Biology and mechanics of ligament and ligament healing. *Exerc Sport Sci Rev* 21:33–64.

Mycielska, M. E., and Djamgoz, M. B. 2004. Cellular mechanisms of direct-current electric field effects: Galvanotaxis and metastatic disease. *J Cell Sci* 117(Pt 9):1631–39.

Nayal, A., Webb, D. J., and Horwitz, A. F. 2004. Talin: An emerging focal point of adhesion dynamics. *Curr Opin Cell Biol* 16(1):94–98.

Nobes, C. D., and Hall, A. 1999. Rho GTPases control polarity, protrusion, and adhesion during cell movement. *J Cell Biol* 144(6):1235–44.

Onuma, E. K., and Hui, S. W. 1988. Electric field-directed cell shape changes, displacement, and cytoskeletal reorganization are calcium dependent. *J Cell Biol* 106(6):2067–75.

Orida, N., and Feldman, J. D. 1982. Directional protrusive pseudopodial activity and motility in macrophages induced by extracellular electric fields. *Cell Motil* 2(3):243–55.

Pacifici, M., Koyama, E., Iwamoto, M., and Gentili, C. 2000. Development of articular cartilage: What do we know about it and how may it occur? *Connect Tissue Res* 41(3):175–84.

Patel, N., and Poo, M. M. 1982. Orientation of neurite growth by extracellular electric fields. *J Neurosci* 2(4):483–96.

Pei, M., He, F., and Vunjak-Novakovic, G. 2008. Synovium-derived stem cell-based chondrogenesis. *Differentiation* 76(10):1044–56.

Pollard, T. D., and Borisy, G. G. 2003. Cellular motility driven by assembly and disassembly of actin filaments. *Cell* 112(4):453–65.

Robinson, K. R. 1985. The responses of cells to electrical fields: A review. *J Cell Biol* 101(6):2023–27.

Rodan, G. A. 1992. Introduction to bone biology. *Bone* 13(Suppl 1):S3–6.

Ross, S. M., Ferrier, J. M., and Aubin, J. E. 1989. Studies on the alignment of fibroblasts in uniform applied electrical fields. *Bioelectromagnetics* 10(4):371–84.

Shigino, T., Ochi, M., Hirose, Y., Hirayama, H., and Sakaguchi, K. 2001. Enhancing osseointegration by capacitively coupled electric field: A pilot study on early occlusal loading in the dog mandible. *Int J Oral Maxillofacial Implants* 16(6):841–50.

Shimizu, M., Minakuchi, K., Kaji, S., and Koga, J. 1997. Chondrocyte migration to fibronectin, type I collagen, and type II collagen. *Cell Struct Funct* 22(3):309–15.

Sillman, A. L., Quang, D. M., Farboud, B., Fang, K. S., Nuccitelli, R., and Isseroff, R. R. 2003. Human dermal fibroblasts do not exhibit directional migration on collagen I in direct-current electric fields of physiological strength. *Exp Dermatol* 12(4):396–402.

Soong, H. K., Parkinson, W. C., Bafna, S., Sulik, G. L., and Huang, S. C. 1990. Movements of cultured corneal epithelial cells and stromal fibroblasts in electric fields. *Invest Ophthalmol Vis Sci* 31(11):2278–82.

Suda, T., Kobayashi, K., Jimi, E., Udagawa, N., and Takahashi, N. 2001. The molecular basis of osteoclast differentiation and activation. *Novartis Found Symp* 232:235–47; discussion, 247–50.

Sun, S., and Cho, M. 2004. Human fibroblast migration in three-dimensional collagen gel in response to noninvasive electrical stimulus. II. Identification of electrocoupling molecular mechanisms. *Tissue Eng* 10(9–10):1558–65.

Sun, S., Titushkin, I., and Cho, M. 2006. Regulation of mesenchymal stem cell adhesion and orientation in 3D collagen scaffold by electrical stimulus. *Bioelectrochemistry* 69(2):133–41.

Sweigart, M. A., and Athanasiou, K. A. 2001. Toward tissue engineering of the knee meniscus. *Tissue Eng* 7(2):111–29.

Tandon, N., Goh, B., Marsano, A., Chao, P. H., Montouri-Sorrentino, C., Gimble, J., et al. 2009. Alignment and elongation of human adipose-derived stem cells in response to direct-current electrical stimulation. *Conf Proc IEEE Eng Med Biol Soc* 1:6517–21.

Tozer, S., and Duprez, D. 2005. Tendon and ligament: Development, repair and disease. *Birth Defects Res C Embryo Today* 75(3):226–36.

Vicente-Manzanares, M., Webb, D. J., and Horwitz, A. R. 2005. Cell migration at a glance. *J Cell Sci* 118(Pt 21):4917–19.

Wang, E., Zhao, M., Forrester, J. V., and McCaig, C. D. 2000. Re-orientation and faster, directed migration of lens epithelial cells in a physiological electric field. *Exp Eye Res* 71(1):91–98.

Wang, Q., Zhong, S. Z., Jun, O. Y., Jiang, L. X., Zhang, Z. K., Xie, Y., et al. 1998. Osteogenesis of electrically stimulated bone cells mediated in part by calcium ions. *Clin Orthopaedics Related Res* 348:259–68.

Watanabe, N., and Higashida, C. 2004. Formins: Processive cappers of growing actin filaments. *Exp Cell Res* 301(1):16–22.

Weaver, A. M., Young, M. E., Lee, W. L., and Cooper, J. A. 2003. Integration of signals to the Arp2/3 complex. *Curr Opin Cell Biol* 15(1):23–30.

Zaidel-Bar, R., Cohen, M., Addadi, L., and Geiger, B. 2004. Hierarchical assembly of cell-matrix adhesion complexes. *Biochem Soc Trans* 32(Pt 3):416–20.

Zhao, M., Pu, J., Forrester, J. V., and McCaig, C. D. 2002. Membrane lipids, EGF receptors, and intracellular signals colocalize and are polarized in epithelial cells moving directionally in a physiological electric field. *FASEB J* 16(8):857–59.

Zhao, M., Song, B., Pu, J., Wada, T., Reid, B., Tai, G., Wang, F., Guo, A., Walczysko, P., Gu, Y., Sasaki, T., Suzuki, A., Forrester, J. V., Bourne, H. R., Devreotes, P. N., McCaig, C. D., and Penninger, J. M. 2006. Electrical signals control wound healing through phosphatidylinositol-3-OH kinase-gamma and PTEN. *Nature* 442(7101):457–60.

10 Neuronal Growth Cone Guidance by Physiological DC Electric Fields

Ann M. Rajnicek
School of Medical Sciences, Institute of Medical Sciences
University of Aberdeen
Aberdeen, Scotland

CONTENTS

INTRODUCTION

Neurons exist within a complex milieu, in which they are exposed to many factors that affect their growth, survival, and function. These factors can be grouped into three general categories: chemical (growth factors, neurotransmitters, or adhesion molecules, which can be either soluble or substratum bound), physical (contours of the three-dimensional space surrounding the cell), and electrical (standing extracellular voltage gradients). During development and repair, neurons use these environmental factors to guide the tips of growing axons, called growth cones, to appropriate targets (e.g., another neuron or a muscle), where they form synaptic connections, effectively hardwiring the nervous system. Guiding growth cones to correct positions is therefore fundamental to proper nervous system function. Although much attention has been devoted to the roles of chemical cues in growth cone guidance (reviewed in Hong and Nishiyama, 2010; Mortimer et al., 2008) and the study of nerve guidance by physical cues is emerging, especially in the context of tissue engineering (Katz and Burdick, 2009; Park et al., 2007), the responses of neurons to extracellular electrical signals are less well studied and sometimes controversial (McCaig et al., 2002, 2005, 2009; Robinson and Cormie, 2007). This chapter summarizes evidence that electrical cues are present in the developing and damaged nervous systems, that electrical signals direct nerve growth, and outlines the cellular basis for directed growth. Additionally, it considers the potential for electrical stimulation in tissue engineering and clinical therapies to treat nervous system damage.

ENDOGENOUS DC ELECTRIC FIELDS IN THE DEVELOPING AMPHIBIAN NERVOUS SYSTEM

Fundamental to the hypothesis that electrical signals control normal nervous system development is evidence that an electric field (EF) (a standing gradient of voltage in a defined space) exists naturally within the developing central nervous system (CNS). How would such an EF arise? The basis for any endogenous EF is the separation of charge between the inside and outside of a tightly sealed epithelium, coupled with spatially localized ionic "leaks" (described elsewhere in this volume). In amphibian skin the transepithelial potential (TEP) measured directly using glass microelectrodes immediately beneath the ventral or lateral epithelium is ~60 mV, with the embryonic tissues positive relative to the extra-embryonic medium (artificial pond water) (Rajnicek et al., 1988). Spatial changes in the TEP accompany neurulation, the developmental period when the dorsal ectoderm thickens and folds on itself medially to form the neural tube, which is the precursor of the entire CNS (Hotary and Robinson, 1991, 1994; Shi and Borgens, 1995). Glass microelectrodes inserted beneath the epithelium in the region of the developing brain revealed a rostrocaudal gradient (EF) of 75 to 100 mV/mm, in the center of the neural groove, and a mediolateral gradient of 24 mV/mm across the neural ridges. The electrical profile of the early amphibian nervous system was mapped further using a noninvasive, highly sensitive electrode (the vibrating probe; Jaffe and Nuccitelli, 1974). Consistent with conventional microelectrode recordings, current loops were located at the neural ridges of the early neural tube (Figure 10.1A). Current, taken to be the flow of positively charged ions, exited the tip of each neural ridge, entering the center of the neural groove medially and entering the neighboring intact epithelium laterally (Hotary and Robinson, 1994; Shi and Borgens, 1995).

Later, once the neural tube has sealed completely (Figure 10.1B), it maintains a TEP such that its central lumen is negative relative to the extracellular spaces

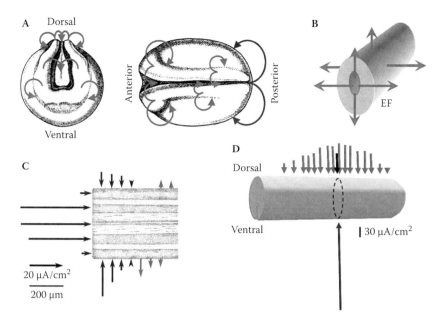

FIGURE 10.1   See color insert. Ionic currents in the developing and regenerating nervous systems. (A,B) A stage 17 *Xenopus* embryo (Nieuwkoop and Faber, 1994). The neural ridges have not yet fused completely, particularly at the anterior region where the brain will develop. The red arrows show ionic current loops around the developing neural tube. By convention biological ionic currents are drawn to show the direction of flow of positive ions. (A) The drawing on the left is a face-on view of the anterior end of the embryo, and the drawing on the right is a dorsal view. The neural ridges and the central neural groove can be seen dorsally. The blue arrow represents the outward blastopore current. (B) At a later developmental stage (> stage 21), after the neural tube has sealed and has moved away from the overlying skin. Na$^+$ ions are transported away from the lumen. (C) Ionic current pattern near the severed lamprey spinal cord (Borgens et al., 1980). Blue arrows are inward current and red arrows are outward current. (D) Current pattern measured in an *ex vivo* preparation of guinea pig spinal cord (Zuberi et al., 2008). Red arrows are inward currents measured on the dorsal side of the cord near a crush injury (black oval). The black arrow indicates the median size of the maximum dorsal current. The maximum current near the injury is higher at the ventral side of the cord (blue arrow).

outside it (Hotary and Robinson, 1991; Shi and Borgens, 1994). The physiological basis for both the skin and transneural TEPs is the directional movement of sodium ions—a consequence of the polarized distribution of ion pumps and channels in restricted membrane domains of individual epithelial cells (reviewed in McCaig et al., 2005, 2009) (see Chapter 1). Na$^+$ enters the epithelial cell from the pond water (or neural tube lumen) via apically located Na$^+$ channels, and Na$^+$ leaves the epithelial cell (or neural tube cell) via basolateral Na$^+$/K$^+$ ATPase pumps. A direct link between Na$^+$ transport and TEP in amphibian neurula stage embryonic skin was demonstrated by experiments in which the Na$^+$ channel blockers amiloride and benzamil were added to the medium bathing the embryo: the skin TEP and the current leaving a large skin wound (or the blastopore) were depleted by ~80%, recovering after amiloride was washed out (Rajnicek et al., 1988; Hotary and Robinson, 1994). If the neural tube TEP is crucial for normal development, then disrupting it should perturb embryogenesis, and it does. Injecting amiloride (or its analogue benzamil) into the neural tube lumen collapsed the neural tube TEP substantially, and the developing embryos exhibited profound cranial and nervous system defects (Borgens and Shi, 1995). Therefore, in amphibians, the early nervous system develops within an endogenous EF with its main axis aligned parallel to the embryonic rostrocaudal axis and a second EF axis orthogonal to it (Figure 10.1). Later, neurogenesis within the neural tube occurs in an EF ranging from 28 to 1,600 mV/mm (reviewed in McCaig et al., 2002, 2005).

Endogenous embryonic EFs are not limited to amphibian embryos. Vibrating probe measurements in four-day-old chick embryos revealed a large outward current (maximum 112 mA/cm^2) at the posterior intestinal portal with an intraembryonic EF of 33 mV/mm measured with glass microelectrodes in the tail region (Hotary and Robinson, 1990) (see Chapter 1). This endogenous EF is crucial for morphogenesis. When the posterior current was shunted anteriorly by implanting a hollow glass tube in the mid-trunk region (laterally, at the level of the wing bud), 92% of the embryos developed abnormally (Hotary and Robinson, 1992). Although minor defects were observed in limb or head formation, severe defects were observed posteriorly. In 81% of cases the tail, including the spinal cord, failed to develop, with the body often terminating at the level of the hind limb buds. Intriguingly, vibrating probe measurements of *rumpless* mutant chick embryos, which also lack a tail and posterior spinal cord, had a similarly reduced posterior current, which may contribute to their phenotype (Hotary and Robinson, 1992).

It has been proposed that growth cones in embryos are not exposed to physiological EFs, defined as ≤100 mV/mm (Robinson and Cormie, 2007). While this may apply to some embryonic tissues, in the case of *Xenopus* and chick embryos, EFs less than 100 mV/mm are present in the vicinity of developing neurons. It is reasonable, therefore, to propose that normal amphibian and chick neurogenesis is influenced by steady EFs, since EF-induced responses of *Xenopus* neurons (directional growth to cathode, enhanced neuronal differentiation, and neurite sprouting—see below) occur at EFs < 10 mV/mm (Hinkle et al., 1981). Even more remarkably, transient, ten-minute exposure of embryonic chick dorsal root ganglion (DRG) neurons to an EF of 24 or 44 mV/mm (similar to the 33 mV/mm measured in chick embryos) stimulates the rate of growth cone advance *in vitro* even twenty-four or forty-eight hours after EF termination (Wood and Willits, 2006, 2009). Therefore, even short-term exposure to a weak EF *in situ* may have long-lasting consequences on subsequent neuronal development. It is interesting that outward current at the posterior (blastopore) of *Xenopus* embryos increases throughout neurulation, peaking at the time when the first motor neuron growth cones reach their targets (Hotary and Robinson, 1994), further suggesting that growth cones navigate in an endogenous EF.

ENDOGENOUS DC ELECTRIC FIELDS
IN THE DAMAGED NERVOUS SYSTEM

As outlined above, endogenous EFs are present in the developing nervous system, and these are critical for normal neuronal development, but EFs are also present in tissues where axons regenerate, suggesting the possibility that they also participate in nervous system repair and that they can be manipulated to enhance it. The first such evidence came from studies using the vibrating probe to map the severed ends of the spinal cord of the larval lamprey *Petromyzon marinus*. Currents (carried principally by Na^+ and Ca^{2+}) as high as 500 $\mu A/cm^2$ were recorded entering the cord (Figure 10.1C) immediately after injury, falling to 100 $\mu A/cm^2$ by two days and subsequently to a stable 4 $\mu A/cm^2$ for the next four days (Borgens et al, 1980). The consequence of axonal transection was an EF of 10 mV/mm near the cut ends of axons (Strautman et al., 1990). Wick electrodes from an external battery were used to apply 10 μA of current to the completely severed lamprey spinal cord *in vivo* (Borgens et al., 1981) or 10 mV/mm to the severed cord *ex vivo* (Strautman et al., 1990) to nullify the injury current. Lamprey have some spontaneous CNS regenerative capacity, but about two months postinjury electrophysiological recordings in the regenerating animals demonstrated significant improvement of EF-treated over sham-treated controls. Action potential conduction was evident in both directions across the lesion site, and tracer dyes revealed nerve processes and abundant branching at or beyond the plane of transection. In *ex vivo* preparations the externally applied EF prevented the Ca^{2+} component of the injury current at the transection site (Strautman et al., 1990). Since axonal dieback is triggered by elevated cytoplasmic Ca^{2+}, the injury current may contribute to axonal growth and survival.

The role of an endogenous injury current in spinal cord regeneration has been explored in amphibians. At most stages of development *Xenopus* tadpoles regenerate their tails spontaneously following amputation. Regenerated tissue is cosmetically and functionally complete, including the spinal cord (Beck et al., 2003). Work from the Levin lab pioneered exploration of the bioelectric basis for tail regeneration by studying H^+ ion flow in the early tail stump (Adams et al., 2007) (see Chapter 3). Following amputation, the H^+ V-ATPase pumps H^+ ions out of the regenerating stump. Its inactivation, either pharmacologically or genetically, prevented tail regeneration (but not wound healing), and nerves growing into the stump tissue were tangled and misrouted. Since *Xenopus* neurons are sensitive to very small EFs (Hinkle et al., 1981), one interpretation is that the H^+ efflux is sufficient to establish an EF within the regenerating tissues such that the EF (stump tip is the cathode) directs regrowing neurons into the tip (McCaig et al., 2009). Indeed, immediately after amputation an outward current of ≤ 5 $\mu A/cm^2$ leaves the tail stump (Reid et al., 2009). However, ion substitution experiments (Reid et al., 2009) showed that current was carried by Na^+ ions, as demonstrated previously for younger *Xenopus* embryos (Rajnicek et al., 1988; Hotary and Robinson, 1994), so the direct contribution of the H^+ component of the tail current to axonal regeneration remains to be explored.

Injury currents are also present in the damaged mammalian spinal cord. Borgens' group have used the vibrating probe to map currents at guinea pig spinal cord crush and transection injuries (Zuberi et al., 2008) (see Chapter 11). Negligible inward current was measured orthogonal to the long axis of the intact cord, but it increased to 200 to 300 μA/cm^2 immediately after crush injury, with one sample driving more than 1,000 μA/cm^2 into the injury site (Figure 10.1D; Zuberi et al., 2008). The current density started to decline within a few minutes, but even four hours after injury it was stable at about 40 μA/cm^2. When the cord was severed, rather than crushed, a similar injury current entered the cut ends of the cord, but the level at which it stabilized by four hours was slightly higher than for crush injuries. In the context of this steady injury current, anisotropy in white matter (myelinated axons) tissue resistivity in the longitudinal axis of the spinal cord (300 to 400 Ωcm) compared to that in the transverse direction (1,500 to 2,500 Ωcm) favors establishment of a longitudinal standing voltage gradient (EF) in damaged cord tissues. Therefore, in the damaged cord an endogenous EF would be aligned with the long axis such that the injury site would represent the anode. These data are important because they represent the first measurements of endogenous injury currents in the mammalian spinal cord. Since the measurements were, of necessity, performed on spinal cords *ex vivo*, it remains possible that the vascularized cord *in vivo* would sustain even higher injury currents than those reported by Zuberi et al. (2008).

Until recently it was believed that the adult CNS could not generate new neurons, but neural progenitor cell populations have been identified in the spinal cord, and the mature mammalian brain and progenitor cells are recruited to sites of ischemic stroke damage (Hawryluk and Fehlings, 2008; Zhao, 2008; Jin et al., 2006). These observations, coupled with evidence that EFs are associated spatially and temporally with damage in the mammalian CNS (Zuberi et al., 2008), suggest that progenitor cell migration or development may be influenced by an injury-induced EF. Even in the intact brain, firing of hippocampal granule cells induces an EF of 50 mV/mm that can persist for hours or days (Lomo, 1971). This EF is sufficient to direct hippocampal axonal projections (Figure 10.2H; Rajnicek et al., 1992), suggesting the possibility that synaptic plasticity (activity-dependent reorganization of synaptic contacts) might also be influenced by an endogenous activity-related EF. This area is ripe for further study.

The damaged mammalian cornea is another place where adult mammalian nerves encounter a naturally occurring EF. Electrical control of corneal epithelial cell migration and wound healing has been covered in Chapter 5, so here discussion is limited to nerve responses. The cornea is ideal for these studies because it is one of the most densely innervated tissues in the body, and in common with the epithelia discussed above, it uses directed ion transport through the polarized epithelium to maintain a TEP of ~40 mV, with the stromal side positive relative to the tear film (Song et al., 2004). Upon wounding the epithelial seal is broken and the TEP collapses instantaneously and catastrophically to zero. However, because the TEP is maintained at +40 mV distally in the unwounded epithelium, a voltage gradient (EF) is established near the wound in the subepithelial tissues

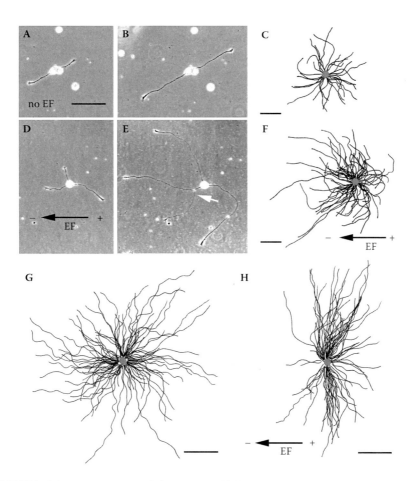

FIGURE 10.2   Neuronal growth in an extracellular EF. (A–F) *Xenopus* neurons in culture. (A–C) No EF applied. (A) Zero hours, scale is 100 μm and applies to A, B, D, and E. (B) The same cell as A but three hours later. (C) Composite drawing made by superimposing images of many cells after three hours of observation. The spot in the center marks where the somas were superimposed and each neurite was traced separately. Neurites grow randomly. Scale bar is 100 μm. (D–F) Cultured neurons in an EF of 150 mV/mm with the cathode at the left. (D) At zero hours. (E) The same cell after three hours of EF exposure. White arrow indicates new branch emerging toward the cathode. Scale as in A. (F) Composite drawing of EF-treated cells. 150 mV/mm for three hours. Scale is 100 mm. Neurites often curve dramatically toward the cathode. (G,H) Composite drawings of axon growth for embryonic rat hippocampal neurons in culture. (G) No EF applied for twenty-four hours. Scale is 50 μm. (H) After twenty-four hours in an EF of 180 mV/mm. Scale bar is 50 μm. Axons orient predominantly orthogonal to the EF vector and fewer face the cathode.

parallel to the epithelial surface, with the wound as the cathode. In *ex vivo* bovine cornea EFs of 42 and 30 mV/mm have been measured at distances of 125 and 500 μm from the wound edge, respectively (Chiang et al., 1992; Sta Iglesia and Vanable, 1998). Observation of whole mounted rat corneas revealed that by twenty-four hours regrowing nerve processes and new branches were directed toward the wound edge (Song et al., 2004). The directional growth was related directly to the wound-induced EF; it was enhanced by pharmacological treatments that increased the TEP (hence the endogenous EF), and it was inhibited by treatments that collapsed the TEP. Pharmacological manipulation of the EF was presumed to leave other coexisting directional cues intact (e.g., putative growth factor gradients or contact-dependent cues), so the interpretation was that the EF was the dominant cue directing axon regeneration (Song et al., 2004). However, it is possible that the EF and any standing chemical gradients (perhaps growth factors) interact to reinforce each other (see Figure 10.5). Interestingly, nerve growth toward the wound edge was prevented by d-tubocurare and neomycin, which also prevented cathodal electrotropism of *Xenopus* growth cones *in vitro*, suggesting that similar cellular cathodal guidance mechanisms operate in these diverse neuronal systems (Erskine et al., 1995; Erskine and McCaig, 1995).

NEURONAL GROWTH CONE RESPONSES TO PHYSIOLOGICAL DC EFS APPLIED *IN VITRO*

Having established that EFs exist within the damaged and regenerating nervous systems, it was important to determine whether neuron growth was affected by appropriately sized EFs. Sven Ingvar (1920), using chick neural explants, was the first to report orientation of nerve processes in the presence of an EF *in vitro*. Indeed, these were among the first experiments performed using the newly developed tissue culture technique. In a brief publication he reported that neurite outgrowth occurred "along the lines of force" with "morphological differences" in cathode- vs. anode-facing fibers. Ariens Kappers embraced the idea that electrical cues determined neuronal polarity, making the idea central to his theory of "neurobiotaxis," stating "thus it seems that the developmental character of the whole neurone may be explained by bioelectric forces" (Ariens Kappers, 1921, p. 139). This view was opposed by Weiss, a proponent of the theory that neurites are directed by contact guidance, because Weiss was unable to repeat Ingvar's result (Weiss, 1934), but subsequently Marsh and Beams (1946) showed that avian neurites were influenced by electric fields. Fibers deflected toward the cathode and growth on the cathode-facing sides of chick medullary explants were faster and more abundant than on the anode-facing sides of the same explants at current densities $> 120 \mu A/mm^2$. Growth toward the anode was suppressed at current densities $\geq 100 \mu A/mm^2$. Neurites even deflected toward the new cathode when the EF orientation was rotated by 90°.

This issue of whether chick neurons respond directly to an applied EF was resolved through a series of well-controlled experiments using chick dorsal root ganglia (DRG) explants (Jaffe and Poo, 1979). DC EFs of 70 to 140 mV/mm caused the region facing the cathode to grow faster than the anode-facing side of the same explant, but in contrast to the observation of Marsh and Beams (1946), neurites did not curve toward the cathode. Increasing the medium's viscosity (to hold the central mass of the explant stationary) eliminated the possibility that the differential growth was an artefact of anodal migration of the explant rather than the selective cathodal extension of the neurites.

Application of EFs to disaggregated *Xenopus laevis* neural tube cells allowed the responses of single cells to be studied, thus eliminating movement of the tissue explant itself, which confounded analysis in the earlier reports. The *Xenopus* preparation continues to be popular for studies of EF-directed neurite growth because it is simple to generate large numbers of embryos in the laboratory, the developing nervous system is readily accessible, neurons from the CNS are easy to grow in tissue culture (at room temperature, without the need for a CO_2 atmosphere, and on a variety of substrates, including bare tissue culture plastic), the neurons are robust and relatively large (cell body and growth cones are about twice as large as mammalian or zebrafish CNS neurons), they grow rapidly in culture, and importantly, they exist normally in an EF in intact and wounded embryos (see above). Furthermore, the *Xenopus* neural tube culture system has been used widely for studies of growth cone guidance by chemical gradients (chemotropism), thus allowing direct comparisons to be made between guidance by chemical and electrical gradients (Jin et al., 2005; Wang and Poo, 2005).

The first experiments using *Xenopus* neurons yielded dramatic results (Hinkle et al., 1981). Dissociated cells were exposed to a DC electric field under conditions in which the cells were isolated from the electrodes by agar/salt bridges, therefore eliminating the possibility of exposure to toxic electrode products or pH changes at the electrodes (McCaig et al., 2005). Neurites grew preferentially toward the cathode at a threshold of 7 mV/mm, with some neurites changing direction by as much as 180° to face the cathode (Figure 10.2). Crucially, cells still responded cathodally when fresh culture medium was perfused orthogonal to the EF vector during EF exposure, thus eliminating the possibility that the EF established gradients of charged tropic molecules (e.g., growth factors or neurotransmitters) in the medium, which biased the direction of growth by chemotropism rather than the EF itself (electrotropism).

Focal application of current from a glass pipette demonstrated that the EFs affected the growth cone (the dynamic tip of the growing neurite) specifically; the general conclusion was that growth cones are attracted toward a current sink (cathode) and are deflected from a current source (anode) (Patel, 1986; Patel et al., 1985; Freeman et al., 1985). The threshold EF that elicits the response in *Xenopus* growth cones is low: 0.3 to 3.0 mV/mm at the growth cone center, though this might be an underestimate of the EF (Patel and Poo, 1984). Considered together with evidence that an EF of 450 to 1,600 mV/mm exists in the amphibian neural tube, it is likely that growth cones use this natural voltage gradient as a directional cue during development to guide axon outgrowth, and possibly even the axis of cell division (McCaig et al., 2005; Yao et al., 2009).

The EF also had a profound effect on the fate of neurite outgrowths, with cathode-facing growth cones advancing significantly faster than those facing the anode, which tended to either advance slowly or retract, sometimes completely (McCaig, 1987; Patel and Poo, 1982). Reversing the EF polarity stimulated previously reabsorbed anode-facing neurites to regrow toward the new cathode (McCaig, 1987). This observation could have important consequences for clinical efforts to stimulate CNS repair following injury, and was an inspiration for using an alternating DC electric field to treat adult mammalian spinal cord lesions (see below and Chapter 11). Neurite branching was also controlled by the EF. More branches emerged from the cathode-facing sides of neurites in a uniform EF or when current was applied from a micropipette positioned orthogonal to the parent neurite (McCaig, 1990).

Directional growth of neurons is not restricted to the *Xenopus* spinal neuron preparation. Table 10.1 lists examples of a wide variety of neuronal and nervous system–related cell types, both primary cultures and cell lines, whose behavior has been explored in DC EFs, the range of EFs tested, and the direction of cell growth. The EF response varies with cell type, and in some cases, with the developmental age of the neurons. This is not surprising since the cells express different receptor proteins in their plasma membranes, and they produce different neurotransmitters, whose activity may affect the potential for EF-directed growth (see below). They would also be expected to reside normally within environments comprising different endogenous EF sizes and geometries.

The evidence for endogenous EFs in damaged mammalian CNS tissues described above, together with the clinical potential for DC EF application in CNS repair, necessitates study of how (or if) mammalian CNS neurons are affected by EFs (Zuberi et al., 2008; Shapiro et al., 2005; Chapter 11). The first evidence that DC EFs influence mammalian CNS neuron growth was from neurons isolated from the embryonic rat hippocampus (Rajnicek et al., 1992). The longest process in these cultures, which becomes the axon in the mature neuron (Dotti et al., 1988), aligned orthogonal to an EF as low as 28 mV/mm (but not 9 mV/mm) when examined after twenty-four hours of EF exposure (Figure 10.2G). This was accompanied by a decrease in the number of axons emerging from the cathode-facing sides of somas and a decrease in the length of cathode-facing axons (Rajnicek et al., 1992). Orthogonal axon alignment was also observed for cells in which axons had already extended for twenty-four hours prior to the EF exposure, implying that the EF can reshape neurons that have already established their polarity (Rajnicek, unpublished). Orthogonal alignment also occurs in neurons isolated from the embryonic mouse cortex, which share an embryonic origin with hippocampus (Stoney, 2010). Rat hippocampal axons and dendrites respond differently to EFs (Davenport and McCaig, 1993); growth cones of dendrites were

TABLE 10.1
Examples of Cells Exposed to DC Electric Fields *In Vitro*

Species	Neuronal Type	Growth Direction	EF Range (mV/mm)[a]	References
Xenopus	Spinal cord	Cathode	6–150	Hinkle et al. (1981), McCaig et al. (2000), Rajnicek et al. (2006a)
	Spinal cord	Cathode	0.3–3 (focal EF)	Patel and Poo (1984)
Zebrafish	Spinal cord	No directed response	100	Cormie and Robinson (2007)
Chicken	Medulla, explant	Cathode	100–120 µA/mm²	Marsh and Beams (1946)
	DRG, explant	Faster to cathode	70–140	Jaffe and Poo (1979)
	DRG, dissociated	Faster growth[b]	24 or 44 (transient EF[b])	Wood and Willits (2009)
	Sympathetic	Cathode	1,000	Gonzalez-Agosti and Solomon (1996)
	Postganglionic sympathetic	Orthogonal	300–500	Pan and Borgens (2010)
	Schwann cells	Anode	3–100	McKasson et al. (2008)
Rat	Hippocampus (embryonic)	Axons orthogonal, fewer face cathode	28–219	Rajnicek et al. (1992)
	Hippocampus (embryonic)	Dendrites to cathode, axons not directed	580–4,730 (focal EF)	Davenport and McCaig (1993)
	Hippocampus (postnatal)	Soma migrates to cathode	120–300	Yao et al. (2008, 2009)
	Cortical astrocyte (neonate)	Orthogonal	50–500	Borgens et al. (1994)
	Radial glia	Orthogonal alignment, soma to anode	100–300	Rajnicek and McCaig (unpublished)
Mouse	Cortex (embryonic)	Orthogonal	200	Stoney (2010)
Mouse cell line	N1E-115 neuroblastoma	Cathode	100–1,000	Bedlack et al. (1992)
Rat cell line	PC12 (sympathetic-like)	Anode	5–200	Cork et al. (1994)

[a] Range of EFs tested, not necessarily the threshold in which EF induced a response.

[b] The EF was applied for only ten minutes. Growth was examined twenty-four hours later, but the direction of growth was not assessed.

attracted toward a glass micropipette when it was a cathode, but axonal growth cones on the same cells were not affected, even though the EF magnitude was at least twenty-fold higher than is sufficient to induce orthogonal axonal alignment (Rajnicek et al., 1992). Collectively, this suggests that the EF has the potential to control both the structural and functional polarity of developing (and regenerating) mammalian CNS neurons.

TISSUE CULTURE CAVEATS

Tissue culture studies, in which conditions can be manipulated, and where cell responses can be monitored quantitatively, are imperative to understanding the cellular basis for EF responses, and studies using *Xenopus* neurons in culture have proven instructive for identifying the molecular pathways that underpin chemotropic growth cone steering (Lohof et al., 1992; Hong and Nishiyama, 2010). Indeed, since neurons reside in an endogenous EF *in vivo*, it could be argued that for tissue culture to be considered truly physiological, a small DC EF (whose magnitude would depend on tissue type) should always be present. The observation that simply changing the substratum (Rajnicek et al., 1998) alters EF responses, coupled with the varied responses of different neuronal types to similar EFs, raises an important point. Does the culture system itself affect the ability of growth cones to respond to an EF *in vitro* (or even the direction of the response)?

Cell culture conditions are optimized for each cell type, and this may affect cell behavior in an EF. Changing the particular batch of foetal bovine serum used to supplement the *Xenopus* neuron culture medium sometimes affects the intensity of the response, even though it comprises only 1% of the final medium composition (McCaig and Erskine, 1996; McCaig et al., 2000). Consequently, experiments are performed only after batch testing the serum and sufficient serum for several years is frozen. This practice, which ensures optimum cell survival and reduces biological variation in cultures, is used in laboratories around the world. Importantly, the serum batch that is chosen for *Xenopus* EF experiments is *not* the one that yields the strongest cathodal response; rather, it is selected to yield a response consistent quantitatively (neurite growth rate and extent of cathodal turning) with data published previously, therefore allowing direct comparison with previous work. The ability to reproduce experimental data over many years using varied batches of serum justifies this protocol.

Additionally, the surface on which the neurons are grown can impact the response. All of the responses for *Xenopus* neurons described above were for cells grown on bare tissue culture plastic (Falcon brand), the standard method for this culture, but if the dishes were coated with poly-L-lysine before plating the neurons, *Xenopus* neuronal morphology changed. Neurons had fewer neurites per cell, neurites were shorter and thicker, and their growth cones bore few filopodia, advancing slowly because of strong adhesion to the substratum. But in the context of EFs, the most dramatic effect was that their migration direction was reversed. Growth cones migrated toward the anode, rather than the cathode of the EF, and this was related to both the positive substratum charge density and the substratum "stickiness" (Rajnicek et al., 1998). Conversely, if the substratum was coated with Matrigel, which mimics the extracellular matrix, *Xenopus* growth cones and neurites adhered less well to it (even compared to bare Falcon plastic) and growth cones migrated toward the cathode.

This ability of growth cones to change the direction of electrotropism, depending on the charge or adhesiveness of the surfaces they encounter, has a potentially important biological implication for pathfinding. In its simplest interpretation, an endogenous EF provides a binary (cathode vs. anode) long-range guidance cue, with little scope for generation of stereotyped pathway complexity. However, modifying the surface chemistry of the extracellular terrain locally could affect how growth cones "read" the endogenous EF in different regions, providing a degree of specificity absent in a homogeneous EF. For example, neural cell adhesion molecule (NCAM), whose expression defines particular developing axon tracts, is modified by addition of highly negatively charged polysialic acid (PSA) residues in developing and regenerating axon tracts (Rutishauser, 2008). Therefore, even in a uniform endogenous EF, spatial and temporal variation in PSA-NCAM expression patterns, which alter surface charge and stickiness, has the potential to confer different growth cone electrotropic responses (anodal or cathodal) in different regions of an embryo. It is interesting in this context that perturbation of the normal firing frequency of motoneurons innervating the chick limb affects PSA-NCAM expression and leads to pathfinding defects (Hanson and Landmesser, 2004).

The biggest challenge continues to be describing EF responses of mammalian CNS neurons to EFs more comprehensively *in vitro*. These experiments are important because they will inform stimulator design and optimum EF application protocols for clinical studies. The adult CNS is notorious for its inability to repair itself, and so it is perhaps not surprising that adult CNS neurons do not survive in long-term culture. The culture system of choice for mammalian CNS is dissociated embryonic (usually rat) brain cultures. Under the conditions required to test for EF responses, however, these cells are delicate and their metabolic requirements make EF experiments difficult. In particular, the requirements for a 37°C, 5% CO_2 environment tend to cause dehydration and the need to maintain sterility for several days while agar bridges are in place (connecting electrode baths to the culture medium bathing the cells). It is possible that CNS tissue explants or tissue slice cultures may improve cell survival. But even in the absence of a robust mammalian CNS tissue culture system, there is substantial evidence that EFs applied *in situ* to the surgically or accidentally damaged spinal cord enhanced axonal growth in guinea pigs (anatomical and functional improvement), dogs (functional improvement), and sensory improvement in humans (Borgens et al., 1986, 1987, 1990, 1993, 1999; Shapiro et al., 2005). Therefore, it seems that the onus is on researchers to develop a useful tissue culture model with which to explore and extend the cellular basis for these responses.

In some cases the failure to detect EF responses can be attributed to a lack of experimental rigor and lack of understanding of the pitfalls of the technique. Mistakes evident in the literature (sometimes even in reports describing EF responses) include failure to isolate the cells from the stimulating electrodes (which prevent altered pH or accumulation of potentially cytotoxic electrode products), nonuniform EFs (making it difficult to relate directional response to the EF vector), large EFs (nonphysiological), inappropriate chamber design (chamber needs to be thin to reduce resistive heating and permit rapid heat dissipation; see McCaig et al., 1994), and inappropriate analysis (must allow for time-dependent changes in responses).

THE CELLULAR BASIS FOR DIRECTED GROWTH CONE GUIDANCE

A 20 μm wide growth cone on a *Xenopus* neuron migrates cathodally in an EF as small as 7 mV/mm, a voltage drop of <0.2 mV across the 20 μm wide central region of the growth cone, or ~0.5 mV across the entire span of the growth cone if filopodia parallel to the EF vector are considered (but the highly dynamic nature of filopodia means this is likely to be only transient). So how does a growth cone sense this very small extracellular voltage gradient and translate it into directed movement? Directed cell migration in an EF is sometimes mistaken for passive electrophoresis of the whole cell under the influence of the EF. This explains the directed movement of negatively charged cells in suspension (e.g., erythrocytes), and why freshly plated cells move anodally in an EF until they make effective substratum contacts and begin to crawl cathodally (Patel and Poo, 1982; Finkelstein et al., 2007). However, this does not explain cathodal migration of growth cones (or cells) because the growth cone surface charge is negative, so passive electrophoresis would drag the growth cone toward the anode. Additionally, growth cones maintain dynamic, focal contacts that would anchor them to the substratum and prevent passive electrophoresis. So the electrotropic mechanism must result from active migration across surfaces.

Growth cone steering is controlled by the cytoskeleton, so any proposal to explain directed movement must include a way to generate a cytoplasmic signaling gradient across the central growth cone cytoplasm with downstream consequences on asymmetric cytoskeletal dynamics. The plasma membrane is a good insulator, so rather than penetrating into the cytoplasm, most of the transcellular voltage gradient would be extracellular (Poo, 1981). Therefore, the most likely EF targets are the plasma membrane itself or its component proteins. Three proposals are considered here: (1) direct EF-induced perturbation of membrane potential, (2) redistribution of charged membrane components, and (3) interplay between signaling mechanisms that underpin chemotropism and membrane redistribution of receptors by the EF.

ASYMMETRY OF MEMBRANE POTENTIAL INDUCED BY THE EF

Theoretically, for a cell in a uniform DC EF, the membrane on the cathode-facing side of the cell will be depolarized and the membrane on the anode-facing side will be hyperpolarized (Robinson, 1985), and this has been demonstrated for mouse neuroblastoma cells in large (100 to 1,000 mV/mm) EFs (Bedlack et al., 1992). One consequence could be activation of voltage-gated K^+ channels, whose activation stimulates integrin signals locally (Brown and Dransfield, 2008), or opening of Ca^{2+} channels on the cathode-facing membrane and their inactivation anodally, leading to a cytoplasmic gradient of Ca^{2+}. A model incorporating Ca^{2+} is attractive because transient Ca^{2+} spikes control movement of *Xenopus* spinal neuron growth cones arriving at directional "decision regions" in the neural tube *in vivo* (Gomez and Spitzer, 1999). Furthermore, asymmetric Ca^{2+} influx could amplify the EF-induced spatial cytoplasmic Ca^{2+}gradient by stimulating localized Ca^{2+}-induced Ca^{2+} release from intracellular stores, elevating it even further in the cathode-facing cytoplasm. *Xenopus* growth cones fail to turn cathodally in the presence of pharmacological inhibitors that block N-, P-, or T-type voltage-gated Ca^{2+} channels, or Ca^{2+} release from intracellular stores (Stewart et al., 1995), suggesting that this two-tiered mechanism might play a role in cathodal steering. But the role for voltage-gated channels is unclear because it is unlikely that the small EFs that stimulate cathodal steering would be sufficient to have a direct effect on voltage-gated channel opening (Robinson, 1985). Furthermore, there is conflicting data about the absolute requirement for Ca^{2+} in growth cone turning. Stimulation by a physiological EF increases cytoplasmic growth cone Ca^{2+} rapidly (Rajnicek, unpublished; McCaig et al., 2002), and Co^{2+} and La^{3+} blockade of plasma membrane Ca^{2+} channels inhibited cathodal turning (McCaig, 1989b). However, another study found that after twelve hours of EF exposure, growth cones were oriented cathodally, even in the absence of extracellular Ca^{2+} or when cytoplasmic Ca^{2+} gradients were damped with the calcium buffer 1,2-bis(o-aminophenoxy)ethane-N,N,N*,N*-tetraacetic acid (BAPTA) (Palmer et al., 2000). It is possible, however, that during prolonged EF exposure the cells adapted to low Ca^{2+} conditions, invoking other signals to restore cathodal electrotropism, analogous to the response after inhibition of Rho GTPase signaling (Rajnicek et al., 2006a). Alternatively, a short-term asymmetric Ca^{2+}-dependent signal may trigger a sequence of subsequent Ca^{2+} independent signals that retain a "memory" of the initial asymmetric stimulus. Time-lapse imaging of growth cone Ca^{2+} and correlating it with directed behavior would resolve this issue.

Asymmetry of Charged Membrane Receptors and Intracellular Cascades

In the absence of a role for direct membrane potential perturbation causing cytoplasmic gradients, how else could electrotropism arise? More than thirty years ago Lionel Jaffe (1977) proposed the idea of "lateral electrophoresis" in this context. By this model the extracellular EF moves proteins within the plane of the membrane, causing asymmetric cytoplasmic signaling downstream of their, now asymmetric, local activation (Figure 10.3A). Most proteins have a net negative charge at physiological pH, so for a cell in a steady extracellular DC, EF electrophoresis would tend to pull proteins toward the anode within the fluid plasma

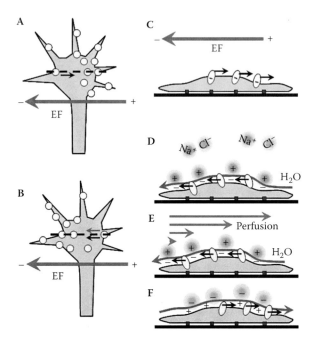

FIGURE 10.3 See color insert. Mechanisms for EF-induced redistribution of membrane proteins in growth cones. (A,B) Membrane receptors (yellow) are moved within the plane of the plasma membrane by lateral electrophoresis (A) or electro-osmosis (B). The black arrow indicates the direction of receptor movement, and the blue arrow in B indicates the direction of electro-osmotic fluid flow over the cell surface. (C–F) View through a cross section of the growth cones in A and B at the level of the dashed lines. Growth cone to substratum contacts are indicated by squares. (C) Lateral electrophoresis. The receptors, which have a net negative charge on the extracellular domain, are drawn toward the anode. (D) Electro-osmosis. Na^+ ions accumulate near the membrane to counter its negative charge. The Na^+ ions are drawn toward the cathode of the EF, dragging their shells of hydration with them passively. The result is fluid flow (blue arrow) very near the cell surface in the direction of the counter-ion flow (in this case, toward the cathode). This is sufficient to move membrane proteins in the direction of fluid flow. (E) Perfusion of the bulk medium in a direction opposite to electro-osmotic fluid flow does not affect electro-osmotic membrane protein redistribution because the bulk fluid flow does not extend sufficiently near the cell surface. (F) Changing the charge of the cell surface reverses the direction of electro-osmotic fluid flow and the direction of receptor accumulation in an EF.

membrane. How far and how fast they move depends on factors that include EF strength, membrane fluidity, and the net charge of the extracellular region of the protein. Rapid (within fifteen minutes) anodal accumulation of integrins occurs in zebrafish keratocytes in a physiological EF (Huang et al., 2008), but several other membrane proteins, including receptors for acetylcholine (AChR) and concanavalin A (Con A), accumulate at the cathodal side of *Xenopus* muscle cells in culture, and these have been implicated in cathodal growth cone guidance (Poo and Robinson, 1977; Orida and Poo, 1978; Poo, 1981; McCaig, 1989b; Erskine and McCaig, 1995). Lateral electrophoresis is a feasible explanation for cathodal migration of receptor proteins only if these proteins are positively charged, but most are negative. The phenomenon of electro-osmosis explains this apparent paradox (Figure 10.3B) (Poo, 1981).

Lateral electrophoresis in the membrane is analogous to the movement of proteins in an electrophoresis gel, so it is familiar to most laboratory biologists, but electro-osmosis, the flow of water that occurs very near the cell membrane of a cell in a steady DC extracellular EF, is less well known. When a cell is bathed in physiological medium the major ions in the bulk medium are Na^+ and Cl^-, but very near the membrane Na^+ ions accumulate to counterbalance the relatively fixed net negative charge of the plasma membrane lipids. When an EF is applied the Na^+ counter-ions and their associated shells of hydration are drawn toward the cathode by electrophoresis (Figure 10.3B). The consequence is a flow of water very near the cell surface, and proteins that extend into this layer can be pulled in the direction of fluid flow like a sailboat moving on a lipid sea (Figure 10.3D). Therefore, the proteins would move in the direction of the counter-ion (Na^+) flow, possibly even against their electrophoretic potential. Whether electrophoresis or electro-osmosis dominates will depend on the charge of the membrane surface, the charge and size of the membrane protein, and its mobility in the membrane. Indeed, it is likely that both phenomena occur in the same membrane.

The observation that constant perfusion of medium during EF exposure (parallel to the EF vector, opposite to the vector, or orthogonal to the vector) failed to affect cathodal accumulation of AChRs or cathodal growth cone turning might be interpreted as evidence against an electro-osmotic mechanism (Hinkle et al., 1981; Patel and Poo, 1982). However, electro-osmosis occurs within a few hundred angstroms of the membrane surface, and laminar bulk medium flow would not occur this close to the membrane (McLaughlin and Poo, 1981), so electrophoretic redistribution of membrane receptors remains a viable hypothesis (Figure 10.3E).

The electro-osmotic hypothesis was tested using *Xenopus* muscle cells; Con A receptors accumulated in the cathode-facing membrane, but the receptors accumulated at the anode-facing side if the cell membrane was made relatively positive by treatment with neuraminidase to remove negatively charged sialic acids (Figure 10.3F), or by incubation with the positively charged lipid 3,3′-dioctadecylindocarbocyanine iodide (diI) (McLaughlin and Poo, 1981; Stollberg and Fraser, 1990). Conversely, addition of the negatively charged ganglioside GM1 to membranes increased cathodal accumulation of Con A receptors. In mammalian cell lines (HeLa cells and CHO fibroblasts), electro-osmosis has been implicated in cathodal accumulation of wheat germ agglutinin (WGA)–labelled sials and integrin receptors (Finkelstein et al., 2007).

What molecules are influenced by lateral electrophoretic and electro-osmotic forces? Studies of the mechanism for EF-induced growth cone steering have focused on the *Xenopus* growth cones, which have the best studied EF responses. Pharmacological inhibitors were used to test the roles of a variety of extracellular membrane receptors and intracellular signaling pathways. A predominant

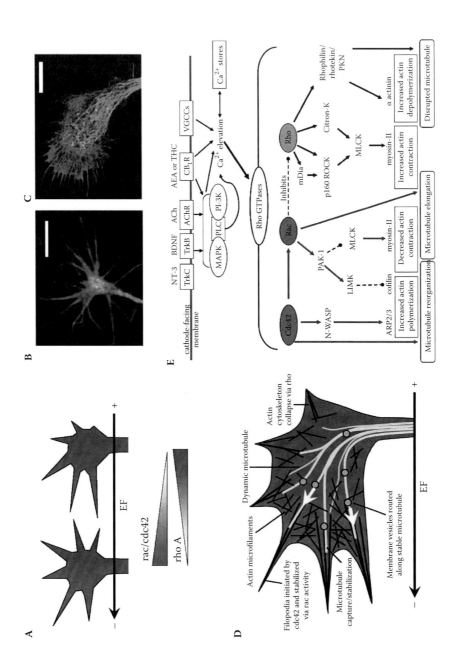

electric field receptor has not been identified, and is unlikely to exist, but pharmacological studies point to roles for several transmembrane proteins in cathodal steering (Figure 10.4E). These include AChR (McCaig 1989b; Erskine and McCaig, 1995), nerve growth factor receptor (McCaig et al., 2000), TrkB (the brain-derived neurotrophic factor receptor) (McCaig et al., 2000), TrkC (the neurotrophin-3 receptor) (McCaig et al., 2000), CB_1R (cannabinoid receptor 1) (Berghuis et al., 2007), and voltage-gated Ca^{2+} channels (Stewart et al., 1995). Though the absolute requirement for Ca^{2+} is uncertain (above and Palmer et al., 2000), most of the signaling cascades converge at Ca^{2+}, suggesting that it contributes, even if only transiently. Intracellularly, key signaling molecules include phospholipase C (PLC; Erskine et al., 1995), intracellular Ca^{2+} release (Stewart et al., 1995; but see Palmer et al., 2000), cAMP signaling (Rajnicek and McCaig, 2001; McCaig et al., 2002), and the Rho GTPases rho A, cdc42, and rac1 (Rajnicek et al., 2006a), but not MAPK or PI-3 kinase (Rajnicek et al. 2006a).

An EF-induced AChR gradient (high cathodally) can be induced in *Xenopus* muscle cells (Orida and Poo, 1978), but the relatively small size and dynamic nature of growth cones makes it especially challenging to visualize a gradient of any molecule in the growth cone, especially in an EF of physiological size. Recently, however, a gradient of Rho A (highest anodally) was detected in *Xenopus* spinal neuron growth cones exposed to an EF at least threefold smaller than the EF measured in the *Xenopus* neural tube (Rajnicek et al. 2006a). Rho A is a member of the Rho GTPase family of molecules, which also includes rac 1 and cdc42. In signaling terms these molecules are downstream of the candidate membrane receptors listed above and upstream of effectors of cytoskeletal dynamics described below (Figure 10.4).

Furthermore, they have been implicated as "molecular switches" in guidance of growth cones by extracellular chemical gradients (Jin et al., 2005). Their role in EF guidance was tested by inhibition of rac 1, cdc42, and rho A signaling collectively using toxin B from *Clostridium difficile*. Toxin B prevented cathodal growth cone guidance, so the three GTPases were then inhibited singly to determine their relative contribution to the mechanism (Rajnicek et al., 2006a). Rho A was inhibited selectively using C3 transferase (also called C3 exoenzyme),

---

FIGURE 10.4   *(see facing page)* **See color insert.** Hypothetical mechanism for cathodal turning of *Xenopus* growth cones. (A) The EF establishes a gradient of rac/cdc42, activities relative to rhoA, such that rac/cdc42 is higher cathodally and rho A is higher anodally. This gradient is amplified by their mutual antagonism. (B,C) Rho GTPase signaling affects local cytoskeletal organization in response to the EF. Red staining shows actin microfilaments and green shows microtubules. Scale bars are 10 µm. (B) Control growth cone with no EF applied. (C) Growth cone exposed to 150 mV/mm for five hours, cathode is at left. This cell was treated with a peptide that inhibits cdc42 signaling, reducing the extent of filopodia at the periphery. The growth cone has turned toward the cathode and the anode-facing side has collapsed. (D) Cytoskeletal events underpinning cathodal turning. (E) Molecular signals at the cathode-facing membrane. (Modified from Rajnicek et al., 2006a, 2006b.)

and cdc42 and rac 1 were inhibited selectively using custom-made peptides that blocked their effector binding sites. This strategy revealed temporally distinct roles for rho A, cdc42, and rac 1. Although all three are essential for turning during the first two hours, only rac 1 signaling is essential throughout the entire five hours (Rajnicek et al., 2006a).

Regardless of the molecules involved, if the growth cone is to change direction, the downstream consequence must be asymmetric assembly and disassembly of the actin- and tubulin-based cytoskeleton. The cytoskeletal anatomy of the growth cone is striking (Figure 10.4). Because the neurite terminates in the growth cone, microtubules from the neurite extend into the center of the growth cone, splaying outward. The center of the growth cone also contains a network of actin microfilaments and a pool of membrane-bound vesicles and mitochondria that shuttle along the microtubules. The most prominent features of the growth cone periphery are the highly dynamic finger-like extensions called filopodia. These comprise long bundles of actin filaments and are analogous to antennae, constantly sampling the extracellular environment to detect attractive or repulsive (or permissive or nonpermissive) cues, informing the decision about which direction to advance. Flattened, veil-like, actin-rich structures, called lamellipodia, extend and retract dynamically between the filopodia.

The role of actin filaments and microtubules in cathodal steering has been tested. Pharmacological inhibition of actin microfilament dynamics with cytochalasin failed to prevent cathodal turning of *Xenopus* growth cones, but in that study the central region of the growth cone retained actin filaments (McCaig, 1989a). A subsequent study, using the more potent inhibitor latrunculin A, showed that cathodal turning was attenuated significantly when actin labelling was restricted to a few small punctae (Rajnicek et al., 2006b). Attenuated turning was not attributable to paralysis of the growth cone because latrunculin-treated growth cones advanced at the same rate as untreated controls, possibly via microtubule-based motility. A direct link was found between microfilaments and Rho A activity. Rho A stimulates cytoskeletal collapse in growth cones (Kozma et al., 1997; Kim et al., 2003), and in EF-treated growth cones anodal elevation of Rho A correlated spatially with selective loss of filopodia and lamellipodia (Rajnicek et al., 2006b). The role of microtubules was explored using the microtubule destabilizing drugs nocodazole or vinblastine and the microtubule stabilizing drug taxol. Damping microtubule dynamics with any of these drugs attenuated cathodal turning without any obvious effect on microfilaments in the lamellipodia or filopodia (Rajnicek et al., 2006b). The data suggest that microfilaments and microtubules act coordinately because growth cones failed to turn cathodally either when microfilaments were depleted while sparing microtubules, or when microtubule dynamics were abolished but peripheral microfilaments were spared.

Temporal analysis of growth cone filopodial and lamellipodial orientation during EF exposure revealed that filopodial reorientation to the cathode preceded cathodal growth cone turning by several minutes, but lamellipodial reorientation accompanied it (Rajnicek et al., 2006b). Filopodial reorientation and subsequent growth cone turning toward the cathode were eliminated by a peptide that prevented effector binding to the rac 1 CRIB domain (hence downstream signaling) and a peptide that prevented signaling via cdc42 suppressed filopodial formation, increased lamellipodial area, and permitted cathodal lamellipodial orientation (Rajnicek et al., 2006b). However, lamellipodial reorientation in the absence of filopodial reorientation was not sufficient to induce turning of the whole growth cone.

The working model for *Xenopus* growth cone guidance by an EF is shown in Figure 10.4. The basis of the model is graded activity of rho GTPases. Rho A is elevated on the anode-facing side of the growth cone compared to the cathode-facing side, antagonizing activity of rac and cdc42 anodally. This signaling gradient is amplified by the mutual antagonism of cdc42/rac 1 and rho A. This gradient of activities means that lamellipodia (stimulated by rac 1) and filopodia (stimulated by cdc42) extend predominantly toward the cathode and the cytoskeleton collapses anodally (stimulated by rho A). Cathode-facing filopodia are stabilized by an unresolved mechanism that might include Pak1, which mediates growth cone filopodia (Kim et al., 2003). Stabilized filopodia could then guide the directed assembly of microtubules toward the cathode or capture and stabilize the ends of dynamic microtubules that actively probe the peripheral region, perhaps via the IQGAP-CLIP170 complex whose stability is enhanced by activated rac/cdc42 (Schaefer et al., 2002; Gordon-Weeks, 2003; Fukata et al., 2002). Membrane vesicles would then be routed preferentially toward the cathode along the stabilized microtubules, and membrane expansion would be biased in that direction (Zakharenko and Popov, 1998), perhaps underpinning the observed increase in the cathode-facing lamellipodial area (Rajnicek et al., 2006b). Filopodial, not lamellipodial, reorientation is crucial in this scheme because stabilized microtubules colocalize with stable microfilaments in filopodia, not lamellipodia (Schaefer et al., 2002; Zhou et al., 2002). While this model explains cathodal responses in the *Xenopus* system for which most data are available, it does not explain the less well-studied perpendicular orientation observed in rodent hippocampal neurons or chick sympathetic neurons (Rajnicek et al., 1992; Stoney, 2010; Pan and Borgens, 2010). It seems, however, that at least for chick sympathetic neurons perpendicular alignment arises by selective retraction of anode- and cathode-facing processes (Pan and Borgens, 2010).

ELECTRIC FIELDS IN COMBINATION WITH OTHER DIRECTIONAL CUES

Together, the evidence for an electrical gradient in embryos (varying with space and time), the presence of chemical gradients in the developing nervous system (Farrar and Spencer, 2008), and the similarity in signaling cascades proposed to explain growth cone guidance by electrical and chemical gradients (O'Donnell

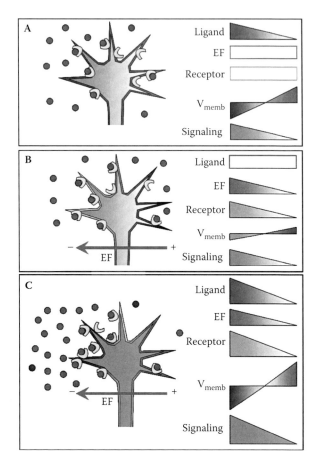

FIGURE 10.5   See color insert. Hypothetical mechanism for interaction of chemical and electrical gradients in growth cone guidance. (A) A growth cone in a gradient of a chemoattractive ligand (high at left). Ligand (blue dots) binds to appropriate receptors (green) and the membrane potential (V_{memb}) is affected; the membrane becomes depolarized by ~15 mV on the side of the growth cone facing the ligand source (left) (Hong and Nishiyama, 2010). Since there is more ligand at the left side of the growth cone, "attractive" signaling is stimulated in the cytoplasm and the growth cone turns toward the left, up the ligand gradient. (B) A growth cone bathed in both a uniform ligand distribution and an extracellular voltage gradient (EF) (cathode at left). The EF on its own does not affect V_{memb} significantly at physiological EFs (Patel and Poo, 1982), but receptors for the ligand have been redistributed toward the cathode by the EF (see Figure 10.3). Even though ligand is uniformly distributed, attractive cytoplasmic signaling is biased cathodally due to locally enhanced ligand-receptor interactions. Therefore, the cathode is analogous to a chemoattractant. (C) Interaction of electrical and chemical signals could amplify the cytoplasmic signaling gradient. The EF could establish, stabilize, or amplify a gradient of a charged chemoattractant while also redistributing its membrane receptors cathodally. Receptors may cluster (e.g., AChRs) once they reach a critical density, forming stable aggregates resistant to back-diffusion. *(continued on facing page)*

et al., 2009) raise an important question: Do these gradients interact to sculpt the embryonic nervous system? One possibility is that the chemical gradients are established (or maintained) as a consequence of the EF in the extracellular space. The endogenous EF in the flank of *Xenopus* embryos is ~40 mV/mm, and it is at least tenfold larger across the developing *Xenopus* neural tube (Hotary and Robinson, 1991, 1994). Fluorescently labelled bovine serum albumin (BSA), a negatively charged protein, injected into the flank region of *Xenopus* embryos, distributed itself along the EF lines, and on theoretical grounds BSA can produce a twofold gradient over a distance of 200 μm in an EF as small as 20 mV/mm (Robinson and Messerli, 1996, 2003). Therefore, the EFs present naturally in embryos are sufficient to modulate chemical gradients, which are considered fundamental to nervous system morphogenesis.

There may also be direct interplay between electrical and chemical signaling gradients resulting from a membrane potential gradient in the growth cone induced by asymmetry in ligand-receptor interactions (Figure 10.5).

For example, upon binding to their receptors the chemorepulsive molecules Sema3A, Slit2, and netrin-1 stimulate membrane hyperpolarization of ~15 mV, and the chemoattractive molecules BDNF and netrin-1 cause a similar size depolarisation in *Xenopus* spinal neurons (Nishiyama et al., 2008; Hong and Nishiyama, 2010). Since a cathode-facing membrane would be depolarized and the anode-facing one would be hyperpolarized in an external voltage gradient (Robinson et al., 1985; Bedlack et al., 1992), this polarity is consistent with the cathode acting like a chemoattractant and the anode as a chemorepellent. An EF of physiological magnitude would not perturb the membrane potential sufficiently to open voltage-gated channels, so this mechanism has been dismissed as the basis for directed migration in an EF (Robinson, 1985). But an EF of 100 mV/mm would depolarize the cathode-facing membrane of a 10 μm diameter cell or growth cone by ~0.5 mV with a corresponding hyperpolarization anodally (Pater and Poo, 1982). For a neuron surrounded by both a steady voltage gradient and a stable chemical gradient, there could be additive effects on membrane potential with consequent asymmetric cytoplasmic signaling. This may be sufficient to amplify or stabilize the signaling gradients that underpin directional growth cone migration, perhaps even sufficiently to open voltage-gated Ca^{2+} channels, whose activity has been implicated in growth cone turning (Stewart et al., 1995). Further

---

FIGURE 10.5 *(continued)*   This would increase receptor-ligand binding frequency with consequent depolarization of the cathode-facing membrane. This polarized V_{memb} change would boost the modest effects on V_{memb} by the EF itself (depolarized cathodally). This additive effect would amplify the cathode-facing attractive cytoplasmic signaling, perhaps sufficiently to open voltage-gated channels asymmetrically. The growth cone would therefore turn to the left (up the concentration gradient and toward the cathode). Although this model describes an attractive cue, the anode is analogous to a repulsive cue (including membrane hyperpolarization), so both sets of events could happen simultaneously in spatially restricted sides of the growth cone, further amplifying the signaling gradient.

amplification could be achieved as a consequence of EF-induced asymmetry of receptors (Figure 10.3), or the extracellular ligand gradient itself could be stabilized by the EF (Figure 10.5). In either case, increased local ligand availability or increased density of membrane receptors for the ligand would perturb the local membrane potential. This electrochemical gradient may be particularly relevant in clinical situations where relatively large EFs could be applied.

Another example of how chemical and electrical gradients may work cooperatively is endocannabinoid signaling. Endocannabinoids are endogenous ligands for the cannabinoid receptors that are present in neuronal membranes. The first to be identified, anandamide (AEA), is produced spontaneously by neurons, and an endogenous AEA gradient has been proposed to control cortical neuron growth cone pathfinding in the mammalian brain (Kano et al., 2009; Berghuis et al., 2007). AEA binds to and activates the cannabinoid receptor 1 (CB_1R), which is enriched in growth cones of *Xenopus* neurons and in neurons of the developing mammalian cortex. Growth cones *in vitro* are repelled by a gradient of the CB_1R antagonist WIN55,212-2 because asymmetric activation of Rho A causes collapse of the side of the growth facing the high end of the gradient (Berghuis et al., 2007). This is analogous to Rho A–mediated collapse of the anode-facing side of *Xenopus* growth cones and subsequent cathodal steering (Rajnicek et al., 2006b) (Figure 10.4). Interestingly, AEA stimulates Rho A activation via CB_1R signaling, and when AEA was present in the growth medium *Xenopus* neurons were "blinded" to the EF (Berghuis et al., 2007). The working hypothesis is that AEA activates CB_1Rs and stimulates Rho A over the entire growth cone, thereby swamping out the cytoplasmic Rho A signaling gradient induced normally by the EF (Figure 10.4D).

This observation has potentially important consequences for the developing nervous system. AEA is structurally similar to Δ^9 tetrahydrocannabinol (THC), the psychoactive component of *Cannabis sativa* (marijuana), which also exerts its effects via the CB_1R. If an electrical gradient or an AEA gradient (or both) is important for embryonic neural development, then foetal exposure to THC, which crosses the placenta (Harbison and Mantilla-Plata, 1972), has the potential to perturb normal axonal pathways and subsequent synaptic activity. Indeed, cognitive effects that extend into adolescence have been observed in children exposed to THC *in utero* (Spano et al., 2007), and 100 nM THC prevents cathodal orientation of *Xenopus* growth cones in an EF *in vitro* (Tait et al., 1995; McCaig and Erskine, 1996).

Collectively, these data also have important consequences for attempts to improve human spinal cord repair using electrical stimulation (Shapiro et al., 2005; Borgens, 2010, this volume). Spinal-injured patients often self-medicate with marijuana (Tate et al., 2004), so it would be important to notify patients involved in clinical trials that marijuana use may decrease the efficacy of the electrical treatment. Furthermore, it is essential to eliminate marijuana users from studies because those data may attenuate the clinical outcome of the studies.

RELEVANCE FOR TISSUE ENGINEERING AND CLINICAL THERAPIES

Despite global research efforts spanning several decades, the prospect for full recovery following serious spinal cord trauma or brain injury remains bleak. Since no single therapy has yet proven effective, significant progress is likely only to emerge from novel combination therapies. An oscillating electric field (DC EF with its polarity reversed every twenty to thirty minutes to encourage growth of both the ascending and descending spinal tracts) enhances functional recovery after surgical or accidental spinal cord injury (Chapter 11; Shapiro et al., 2005; Borgens et al., 1999). This therapy may prove even more effective when combined with drug treatments. For example, the nucleotide inosine improved collateral growth in a rat model of spinal injury (Benowitz et al., 1999), and inosine in combination with an oscillating EF improved functional recovery more than inosine alone in a guinea pig model of spinal injury (Bohnert et al., 2007). The mechanism for the improvement is unknown because the responses of inosine-treated growth cones to EFs *in vitro* have not been tested. Another strategy could target molecules such as rho A, which have been implicated in EF-induced growth cone steering (Rajnicek et al., 2006a). Rho A inhibition enhances repair in a mammalian model of spinal cord injury (Dergham et al., 2002), and human clinical trials are under way using the Rho A antagonist Cethrin® (licensed by Alseres Pharmaceuticals) on spinal injury (Kwon et al., 2010). Perhaps a combined therapy would be more effective.

In the context of spinal cord repair, an attractive idea is to fabricate a bridge that spans the lesion, offering structural support and guidance to growing axons in combination with an EF. A decade ago Borgens used a strategy similar to this in a guinea pig spinal cord (Borgens, 1999). A silicone "guidance channel" (a tube 6 mm long × 1 mm wide) was implanted into the severed cord with a cathode in the center of the tube and the anode positioned outside the vertebral column. The EF (~2.5 mV/mm in the tube) was applied for about three weeks. After one to two months tissue plugs with significant axon growth were observed at both ends of the guidance channels in EF-treated animals, but the tissue plugs did not contain significant nerve growth in sham-treated controls.

This approach could be developed and refined to exploit the ability to synthesize textured materials with particular electrical and chemical properties (Park et al., 2007; Straley, 2010). For instance, it would be useful to dope the scaffold material with molecules that nullify repulsive or nonpermissive molecules present in the glial scar or with growth factors that enhance EF responses (McCaig et al., 2000).

CONCLUSIONS

Electric fields exist within the developing and damaged nervous systems. Since the rate and direction of neuronal growth in culture can be controlled by EFs similar to those measured in the vicinity of the developing nervous system, it is reasonable to expect that a DC electric field contributes to neurogenesis and connectivity. Several key signaling molecules have been identified that underpin directed growth effects. An exciting prospect is that tissue engineering in combination with an applied EF may lead to an effective therapy for CNS repair.

ACKNOWLEDGMENTS

Supported by a European Commission Framework 6 New and Emerging Science and Technology grant, Contract 028437. Zhiqiang Zhao performed the rat radial glial experiments cited in Table 10.1 while employed on this grant.

REFERENCES

Adams DS, Masi A, and Levin M. 2007. H⁺ pump-dependent changes in membrane voltage are an early mechanism necessary and sufficient to induce tail regeneration. *Development* 134:1325–35.

Ariens Kappers CU. 1921. On structural laws in the nervous system: The principles of neurobiotaxis. *Brain* 44:125–49.

Beck CW, Christen B, and Slack JM. 2003. Molecular pathways needed for regeneration of spinal cord and muscle in a vertebrate. *Dev. Cell* 5:429–39.

Bedlack RS, Wei MD, and Loew LM. 1992. Localized membrane depolarizations and localized calcium influx during electric field-guided neurite growth. *Neuron* 9:393–403.

Benowitz LI, Goldberg DE, Madsen JR, Soni D, and Irwin N. 1999. Inosine stimulates extensive axon collateral growth in the rat corticospinal tract after injury. *Proc. Natl. Acad. Sci. USA* 96:13486–90.

Berghuis P, Rajnicek AM, Morozov YM, Ross RA, Mulder J, Urban GM, Monory K, Marsicano G, Matteoli M, Canty A, Irving AJ, Katona I, Yanagawa Y, Rakic P, Lutz B, Mackie K, and Harkany T. 2007. Hardwiring the brain: Endogenous cannabinoids shape neuronal connectivity. *Science* 316:1212–16.

Bohnert DM, Purvines S, Shapiro S, and Borgens RB. 2007. Simultaneous application of two neurotrophic factors after spinal cord injury. *J. Neurotrauma* 24:846–63.

Borgens RB. 1999. Electrically-mediated regeneration and guidance of adult mammalian spinal axons into polymeric channels. *Neuroscience* 91:251–64.

Borgens RB. 2010. Can applied voltages be used to produce spinal cord regeneration and recovery in humans? This volume.

Borgens RB, Blight AR, and McGinnis ME. 1987. Behavioral recovery induced by applied electric fields after spinal cord hemisection in guinea pig. *Science* 238:366–69.

Borgens RB, Blight AR, and McGinnis ME. 1990. Functional recovery after spinal cord hemisection in guinea pigs: The effects of applied electric fields. *J. Comp. Neurol.* 296:634–53.

Borgens RB, Blight AR, and Murphy DJ. 1986. Transected dorsal column axons within the guinea pig spinal cord regenerate in the presence of an applied electric field. *J. Comp. Neurol.* 250:168–80.

Borgens RB, Jaffe LF, and Cohen MJ. 1980. Large and persistent electrical currents enter the transected lamprey spinal cord. *Proc. Natl. Acad. Sci. USA* 77:1209–13.

Borgens RB, Roederer E, and Cohen MJ. 1981. Enhanced spinal cord regeneration in lamprey by applied electric fields. *Science* 213:611–17.

Borgens RB and Shi R. 1995. Uncoupling histogenesis from morphogenesis in the vertebrate embryo by collapse of the transneural tube potential. *Dev. Dyn.* 203:456–67.

Borgens RB, Shi R, Mohr TJ, and Jaeger CB. 1994. Mammalian cortical astrocytes align themselves in a physiological voltage gradient. *Exp. Neurol.* 128:41–49.

Borgens RB, Toombs JP, Blight AR, McGinnis ME, Bauer MS, Widmer W, and Cook JR Jr. 1993. Effects of applied electrical fields on clinical cases of complete paraplegia in dogs. *Restor. Neurol. Neurosci.* 5:305–22.

Borgens RB, Toombs JP, Breur G, Widmer WR, Water D, Harbath AM, March P, and Adams LG. 1999. An imposed oscillating electrical field improves the recovery of function in neurologically complete paraplegic dogs. *J. Neurotrauma* 16:639–57.

Brown SB and Dransfield I. 2008 Electric fields and inflammation: May the force be with you. *Sci. World J.* 8:1280–94.

Chiang M, Robinson KR, and Vanable JW Jr. 1992. Electrical fields in the vicinity of epithelial wounds in the isolated bovine eye. *Exp. Eye Res.* 54:999–1003.

Cork RJ, McGinnis ME, Tsai J, and Robinson KR. 1994. The growth of PC12 neurites is biased toward the anode. *J. Neurobiol.* 25:1609–16.

Cormie P and Robinson KR. 2007. Embryonic zebrafish neuronal growth is not affected by an applied electric field *in vitro*. *Neurosci. Lett.* 411:128–32.

Davenport RW and McCaig CD. 1993. Hippocampal growth cone responses to focally applied electric fields. *J. Neurobiol.* 24:89–100.

Dergham P, Ellezam B, Essagian C, Avedissian H, Lubell WD, and McKerracher L. 2002. Rho signaling pathway targeted to promote spinal cord repair. *J. Neurosci.* 22:6570–77.

Dotti CG, Sullivan CA, and Banker GA. 1988. The establishment of polarity by hippocampal neurons in culture. *J. Neurosci.* 8:1454–68.

Erskine L and McCaig CD. 1995. Growth cone neurotransmitter receptor activation modulates electric field-guided nerve growth. *Dev. Biol.* 171:330–39.

Erskine L, Stewart R, and McCaig CD. 1995. Electric field-directed growth and branching of cultured frog nerves; effects of aminoglycosides and polycations. *J. Neurobiol.* 26:523–36.

Farrar NR and Spencer G. 2008. Pursuing a 'turning point' in growth cone research. *Dev. Biol.* 318:102–11.

Finkelstein EI, Chao PH, Hung CT, and Bulinski JC. 2007. Electric field-induced polarization of charged cell surface proteins does not determine the direction of galvanotaxis. *Cell Motil. Cytoskel.* 64:833–46.

Freeman JA, Manis PB, Snipes GJ, Mayes BN, Samson PC, Wikswo JP, and Freeman DB. 1985. Steady growth cone currents revealed by a novel circularly vibrating probe: A possible mechanism underlying neurite growth. *J. Neurosci. Res.* 13:257–83.

Fukata M, Nakagawa M, and Kaibuchi K. 2003. Roles of rho-family GTPases in cell polarisation and directional migration. *Curr. Opin. Cell Biol.* 15:590–97.

Gomez TM and Spitzer NC. 1999. *In vivo* regulation of axon extension and pathfinding by growth-cone calcium transients. *Nature* 397:350–55.

Gonzalez-Agosti C and Solomon F. 1996. Response of radixin to perturbations of growth cone morphology and motility in chick sympathetic neurons *in vitro*. *Cell Motil. Cytoskel.* 34:122–36.

Gordon-Weeks PR. 2003. Microtubules and growth cone function. *J. Neurobiol.* 58:70–83.

Hanson MG and Landmesser LT. 2004. Normal patterns of spontaneous activity are required for correct motor axon guidance and the expression of specific guidance molecules. *Neuron* 46:687–701.

Harbison RD and Mantilla-Plata B. 1972. Prenatal toxicity, maternal distribution and placental transfer of tetrahydrocannabinol. *J. Pharmacol. Exp. Ther.* 180:446–53.

Hawryluk GWJ and Fehlings MG. 2008. The center of the spinal cord may be central to its repair. *Cell Stem Cell* 3:230–32.

Hinkle L, McCaig CD, and Robinson KR. 1981. The direction of growth of differentiating neurones and myoblasts from frog embryos in an applied electric field. *J. Physiol.* 314:121–35.

Hong K and Nishiyama M. 2010. From guidance signals to movement: Signaling molecules governing growth cone turning. *The Neuroscientist* 16:65–78.

Hotary KB and Robinson KR. 1990. Endogenous electrical currents and the resultant voltage gradients in the chick embryo. *Dev. Biol.* 140:149–60.

Hotary KB and Robinson KR. 1991. The neural tube of the *Xenopus* embryo maintains a potential difference across itself. *Dev. Brain Res.* 59:65–73.

Hotary KB and Robinson KR. 1992. Evidence for a role for endogenous electrical fields in chick embryo development. *Development* 114:985–96.

Hotary KB and Robinson KR. 1994. Endogenous electrical currents and voltage gradients in *Xenopus* embryos and the consequences of their disruption. *Dev. Biol.* 166:789–800.

Huang L, Cormie P, Messerli MA, and Robinson KR. 2008. The involvement of Ca^{2+} and integrins in directional responses of zebrafish keratocytes to electric fields. *J. Cell. Physiol.* 219:162–72.

Ingvar S. 1920. Reaction of cells to galvanic current in tissue culture. *Proc. Soc. Exp. Biol. Med.* 17:198–99.

Jaffe LF. 1977. Electrophoresis along cell membranes. *Nature* 265:600–2.

Jaffe LF and Nuccitelli R. 1974. An ultrasensitive vibrating probe for measuring steady extracellular currents. *J. Cell Biol.* 63:614–28.

Jaffe LF and Poo MM. 1979. Neurites grow faster towards the cathode than the anode in a steady field. *J. Exp. Zool.* 209:115–28.

Jin K, Wang X, Xie L, Mao XO, Zhu W, Wang Y, Shen J, Mao Y, Banwait S, and Greenberg DA. 2006. Evidence for stroke-induced neurogenesis in the human brain. *Proc. Natl. Acad. Sci.* 103:13198–2021

Jin M, Guan C, Jiang Y, Chen G, Zhao C, Cui K, Song Y, Wu C, Poo M, and Yuan X. 2005. Ca^{2+}-dependent regulation of Rho GTPases triggers turning of nerve growth cones. *J. Neurosci.* 25:2338–47.

Kano M, Ohno-Shosaku T, Hashimotodani Y, Uchigashima M, and Watanabe M. 2009. Endocannabinoid-mediated control of synaptic transmission. *Physiol. Rev.* 89:309–80.

Katz JS and Burdick JA. 2009. Hydrogel mediated delivery of trophic factors for neural repair. *Wiley Interdiscip. Rev. Nanomed. Nanobiotechnol.* 1:128–39.

Kim MD, Kamiyama D, Kolodziej P, Hing H, and Chiba A. 2003. Isolation of rho GTPase effector pathways during axon development. *Dev. Biol.* 262:282–93.

Kozma R, Sarner S, Ahmed S, and Lim L. 1997. Rho family GTPases and neuronal growth cone remodelling, relationship between increased complexity induced by Cdc42Hs, Rac1, and acetylcholine and collapse induced by RhoA and lysophosphatidic acid. *Mol. Cell. Biol.* 17:1201–11.

Kwon BK, Okon E, Plunet W, Baptiste D, Fouad K, Hillyer J, Weaver LC, Fehlings MC, and Tetzlaff W. 2010. A systematic review of directly applied biologic therapies for acute spinal cord injury. *J. Neurotrauma*, in press.

Lohof AM, Quillan M, Dan Y, and Poo MM. 1992. Asymmetric modulation of cytosolic cAMP activity induces growth cone turning. *J. Neurosci.* 12:1253–61.

Lomo T. 1971. Patterns of activation in a monosynaptic cortical pathway: The perforant path input to the dentate area of the hippocampal formation. *Exp. Brain Res.* 12:18–45.

Marsh G and Beams HW. 1946. *In vitro* control of growing chick nerve fibers by applied electric currents. *J. Comp. Physiol.* 27:139–57.

McCaig CD. 1987. Spinal neurite reabsorption and regrowth *in vitro* depend on the polarity of an applied electric field. *Development* 100:31–41.

McCaig CD. 1989a. Nerve growth in the absence of growth cone filopodia and the effects of a small applied electric field *J. Cell Sci.* 93:715–21.

McCaig CD. 1989b. Studies on the mechanism of embryonic frog nerve orientation in a small applied electric field. *J. Cell Sci.* 93:723–30.

McCaig CD. 1990. Nerve branching is induced and oriented by a small applied electric field. *J. Cell Sci.* 95:605–15.

McCaig CD. 2010. Electrical signals control epithelial cell physiology and wound repair. This volume.

McCaig CD, Allan DW, Erskine LE, Rajnicek AM, and Stewart R. 1994. Growing nerves in an electric field. *Neuroprotocols* 4:134–41.

McCaig CD and Erskine LE. 1996. Nerve growth and nerve guidance in a physiological electric field. In *Nerve growth and guidance*, ed. CD McCaig, 151–70. London: Portland Press Ltd.

McCaig CD and Rajnicek AM. 1991. Electrical fields, nerve growth and nerve regeneration. *Exp. Physiol.* 76:473–94.

McCaig CD, Rajnicek AM, Song B, and Zhao M. 2002. Has electrical growth cone guidance found its potential? *Trends Neurosci.* 25:354–59.

McCaig CD, Rajnicek AM, Song B, and Zhao M. 2005. Controlling cell behaviour electrically: Current views and future potential. *Physiol. Rev.* 85:943–78.

McCaig CD, Sangster L, and Stewart R. 2000. Neurotrophins enhance electric field-directed growth cone guidance and directed nerve branching. *Dev. Dyn.* 217:299–308.

McCaig CD, Song B, and Rajnicek AM. 2009. Electrical dimensions in cell science. *J. Cell Sci.* 122:4267–76.

McKasson MJ, Huang L, and Robinson KR. 2008. Chick embryonic Schwann cells migrate anodally in small electrical fields. *Exp. Neurol.* 211:585–87.

McLaughlin S and Poo MM. 1981. The role of electro-osmosis in the electric field-induced movement of charged macromolecules on the surfaces of cells. *Biophys. J.* 34:85–93.

Mortimer D, Fothergill T, Pujic Z, Richards LJ, and Goodhill GJ. 2008. Growth cone chemotaxis. *Trends Neurosci.* 31:90–98.

Nieuwkoop PD and Faber J. 1994. *Normal table of Xenopus laevis (Daudin)*. New York: Garland Publishing.

Nishiyama M, von Schimmelmann MJ, Togashi K, Findley WM, and Hong K. 2008. Membrane potential shifts caused by diffusible guidance signals direct growth-cone turning. *Nature Neurosci.* 11:762–71.

O'Donnell M, Chance RK, and Bashaw GJ. 2009. Axon growth and guidance: Receptor regulation and signal transduction. *Ann. Rev. Neurosci.* 32:383–412.

Orida N and Poo MM. 1978. Electrophoretic movement and localisation of acetylcholine receptors in the embryonic muscle cell membrane. *Nature* 275:31–35.

Palmer AM, Messerli MA, and Robinson KR. 2000. Neuronal galvanotropism is independent of external Ca^{2+} entry or internal Ca^{2+} gradients. *J. Neurobiol.* 45:30–38.

Pan L and Borgens RB. 2010. Perpendicular organization of sympathetic neurons within a required physiological voltage. *Exp. Neurol.* 222:161–64.

Park H, Cannizzaro C, Vunjak-Novakovic G, Langer R, Vacanti CA, and Farokhzad OC. 2007. Nanofabrication and microfabrication of functional materials for tissue engineering. *Tissue Eng.* 13:1867–77.

Patel NB. 1986. Reversible inhibition of neurite growth by focal electric currents. In *Ionic currents in development*, ed. R Nuccitelli, 271–78. New York: Alan R. Liss.

Patel N and Poo MM. 1982. Orientation of neurite growth by extracellular electric fields. *J. Neurosci.* 2:483–96.

Patel N and Poo MM. 1984. Perturbation of the direction of neurite growth by pulsed and focal electric fields. *J. Neurosci.* 4:2939–47.

Patel N, Xie ZP, Young SH, and Poo MM. 1985. Response of nerve growth cones to focal electric currents. *J. Neurosci. Res.* 13:245–56.

Poo MM. 1981. *In situ* electrophoresis of membrane components. *Ann. Rev. Biophys. Bioeng.* 10:245–76.

Poo MM and Robinson KR. 1977. Electrophoresis of concanavalin A receptors in the embryonic muscle cell membrane. *Nature* 265:602–5.

Rajnicek AM, Foubister LE, and McCaig CD. 2006a. Temporally and spatially coordinated roles for rho, rac, cdc42 and their effectors in growth cone guidance by a physiological electric field. *J. Cell Sci.* 119:1723–35.

Rajnicek AM, Foubister LE, and McCaig CD. 2006b. Growth cone steering by a physiological electric field requires dynamic microtubules, microfilaments and rac-mediated filopodial asymmetry. *J. Cell Sci.* 119:1736–45.

Rajnicek AM, Gow NAR, and McCaig CD. 1992. Electric field-induced orientation of rat hippocampal neurones *in vitro*. *Exp. Physiol.* 77:229–32.

Rajnicek AM and McCaig CD. 2001. cAMP and protein kinase A signalling underlie growth cone turning in a physiological electric field. *Soc. Neurosci. Abstr.* 27:795.19.

Rajnicek AM, Robinson KR, and McCaig CD. 1998. The direction of neurite growth in a weak electric field depends on the growth surface: Relationship to net surface charge and surface adhesivity. *Dev. Biol.* 203:412–23.

Rajnicek AM, Stump RF, and Robinson KR. 1988. An endongenous sodium current may mediate wound healing in *Xenopus* neurulae. *Dev. Biol.* 128:290–99.

Reid B, Song B, and Zhao M. 2009. Electric currents in tadpole regeneration. *Dev. Biol.* 335:198–207.

Robinson KR. 1985. The responses of cells to electrical fields: A review. *J. Cell Biol.* 101:2023–27.

Robinson KR and Cormie P. 2007. Electric field effects on spinal cord injury: Is there a basis in the *in vitro* studies? *Dev. Neurobiol.* 68:274–80.

Robinson KR and Messerli MA. 1996. Electric embryos: The embryonic epithelium as a generator of developmental information. In *Nerve growth and guidance*, ed. CD McCaig, 131–50. London: Portland Press Ltd.

Robinson KR and Messerli MA. 2003. Left/right, up/down: The role of endogenous electrical fields as directional signals in development, repair and invasion. *BioEssays* 25:759–66.

Rutishauser U. 2008. Polysialic acid in the plasticity of the developing and adult vertebrate nervous system. *Nature Rev. Neurosci.* 9:26–35.

Schaefer AW, Kabir N, and Forscher P. 2002. Filopodia and actin arcs guide the assembly and transport of two populations of microtubules with unique dynamic parameters in neuronal growth cones. *J. Cell Biol.* 158:139–52.

Shapiro S, Borgens R, Pascuzzi R, Roos K, Groff M, Purvines S, Rodgers RB, Hagy S, and Nelson P. 2005. Oscillating field stimulation for complete spinal cord injury in humans: A phase 1 trial. *J. Neurosurg. Spine* 2:3–10.

Shi R and Borgens RB. 1994. Embryonic neuroepithelium sodium transport, the resulting physiological potential, and cranial development. *Dev. Biol.* 165:105–16.

Shi R and Borgens RB. 1995. Three dimensional gradients of voltage during development of the nervous system as invisible coordinates for the establishment of embryonic pattern. *Dev. Dyn.* 202:101–14.

Song B, Zhao M, Forrester JV, and McCaig CD. 2004. Nerves are guided and nerve sprouting is stimulated by a naturally occurring electrical field *in vivo*. *J. Cell Sci.* 117:4681–90.

Spano MS, Ellgren M, Wang X, and Hurd YL. 2007. Prenatal cannabis exposure increases heroin seeking with allostatic changes in limbic enkephalin systems in adulthood. *Biol. Psychiatry* 61:554–63.

Sta Iglesia DD and Vanable JW Jr. 1998. Endogenous lateral electric fields around bovine corneal lesions are necessary for and can enhance normal rates of wound healing. *Wound Repair Regen.* 6:531–42.

Stewart R, Erskine L, and McCaig CD. 1995. Calcium channel subtypes and intracellular calcium stores modulate electric field-stimulated and -oriented nerve growth. *Dev. Biol.* 171:340–51.

Stollberg J and Fraser SE. 1990. Local accumulation of acetylcholine receptors is neither necessary nor sufficient to induce cluster formation. *J. Neurosci.* 10:247–55.

Stoney PN. 2010. The roles of Pax6 in neural precursor migration and axon guidance. PhD thesis, University of Aberdeen.

Straley KS, Foo CWP, and Heilshorn SC. 2010. Biomaterial design strategies for the treatment of spinal cord injuries. *J. Neurotrauma* 27:1–19.

Strautman AF, Cork RJ, and Robinson KR. 1990. The distribution of free calcium in transected spinal axons and its modulation by applied electrical fields. *J. Neurosci.* 10:3564–75.

Tait F, Kosterlitz HW, Pertwee RG, and McCaig CD. 1995. Cannabinoids inhibit growth cane turning in an electric field. *Soc. Neurosci. Abstr.* 21:322.2.

Tate DG, Forchheimer MB, Krause JS, Meade MA, and Bombardier CH. 2004. Patterns of alcohol and substance use and abuse in persons with spinal cord injury: Risk factors and correlates. *Arch. Phys. Med. Rehabil.* 85:1837–47.

Wang GX and Poo MM. 2005. Requirement of TRPC channels in netrin-1-induced chemotropic turning of nerve growth cones. *Nature* 434:898–904.

Weiss P. 1934. *In vitro* experiments on the factors determining the course of the outgrowing nerve fiber. *J. Exp. Zool.* 68:393–448.

Wood MD and Willits RK. 2006. Short-duration, DC electrical stimulation increases chick embryo DGR neurite outgrowth. *Bioelectromagnetics* 27:328–31.

Wood MD and Willits RK. 2009. Applied electric field enhances DRG neurite growth: Influence of stimulation media, surface coating and growth supplements. *J. Neural Eng.* 6:046003, doi:10.1088/1741-2560/6/4/046003.

Yao L, McCaig CD, and Zhao M. 2009. Electrical signals polarize neuronal organelles, direct neuron migration, and orient cell division. *Hippocampus* 19:855–68.

Yao L, Shanley L, McCaig CD, and Zhao M. 2008. Small applied electric fields guide migration of hippocampal neurons. *J. Cell. Physiol.* 216:527–35.

Zakharenko S and Popov S. 1998. Dynamics of axonal microtubules regulate the topology of new membrane insertion into the growing neurites. *J. Cell Biol.* 143:1077–86.

Zhao C, Deng W, and Gage FH. 2008. Mechanisms and functional implications of adult neurogenesis. *Cell* 132:645–60.

Zhou FQ, Waterman-Storer CM, and Cohan CS. 2002. Focal loss of actin bundles causes microtubule redistribution and growth cone turning. *J. Cell Biol.* 157:839–49.

Zuberi M, Liu-Snyder P, ul Haque A, Porterfield DM, and Borgens RB. 2008. Large naturally-produced electric currents and voltage traverse damaged mammalian spinal cord. *J. Biol. Eng.* 2:17, doi:10.1186/1754-1611-2-17.

11 Can Applied Voltages Be Used to Produce Spinal Cord Regeneration and Recovery in Humans?

Richard Ben Borgens
Center for Paralysis Research
Department of Basic Medical Sciences
School of Veterinary Medicine
Purdue University
West Lafayette, Indiana

CONTENTS

INTRODUCTION

It is now clear that the use of an applied voltage gradient can significantly, and beneficially, alter the course of human spinal cord injury (SCI). This opinion is based on the results of the first published Food and Drug Administration (FDA)–sanctioned phase 1 human trials (Shapiro et al. 2005) and a new analysis of efficacy provided by comparison to expectant levels of recovery and a published control group. These data come from recent guidelines establishing a framework for both planning human clinical trials in SCI and *post hoc* analysis (Fawcett et al. 2007; Lammertse et al. 2007). These issues will be discussed below. I will first discuss the basis for understanding how the mechanism of this therapy works. It is likely due to white matter regeneration in compromised spinal cords, even though one cannot test this directly in humans. Next I will only briefly review the supportive *in vitro* data—covered extensively in Chapter 10. I will end on a new evaluation of the clinical data and what the future holds for this form of therapy.

TO REGENERATE OR NOT TO REGENERATE?

In a recent review of SCI appearing in *Nature*, authors Horner and Gauge made this statement in the abstract: "Clear and indisputable evidence for adult functional regeneration remains to be shown" (Horner and Gage 2000). This statement is patently false, as will be shown below—not only for the case of experimentally applied voltages, but also for other, unrelated techniques. More often than not, "opinion leaders" do a great disservice to students of central nervous system (CNS) regeneration by ignoring entire literatures and holding ideological positions favoring the "buzz of the day"—or their own contributions.

To unequivocally prove that regeneration of white matter of the adult mammal leads to synaptogenesis and functional recovery in the same individual is likely an impossible task. First, to actually demonstrate in an adult guinea pig or rat that a behavioral recovery has indeed occurred is significantly problematic, particularly given the poor resolution of the perennial "walking tests," such as the BBB—an evaluation of walking named after its inventors, Basso, Beattie, and Bresnahan, all colleagues at Ohio State University (Basso et al. 1995). SCI produces a hyper-reflexia of the lower limbs that easily leads to a false positive if one is focused on determining any movement of the hind limb after injury/experimental treatment. The BBB scale of 0–21 is helpful by providing a breakpoint (between grades 8 and 10) above apparent coordination of the fore and hind limbs leads to a reasonable expectation of supraspinal control.

Unfortunately, much, if not most, use of the BBB is by persons who collapse the scale to 0 to 6, or 9, or 12, which depresses a proper analysis of movement, and inflates the grades of low-performing animals. These animals are typical of every SCI trial. It is also true that one cannot determine "regeneration" in compressed white matter due to the robust sprouting of surviving axons. Anterograde or retrograde uptake of intracellular dyes does not easily permit a distinction between these. Therefore, a complete or partial transection of the cord is required in a true regeneration study for its conclusions to be unequivocal. This procedure is not representative of the clinic; it negatively impacts animal survival, and increases the generally heightened reflexia of the lower limbs. We provide an alternative approach, the cutaneous trunci muscle (CTM; below) reflex, and a means to determine the actual plane of transection (also not usually revealed in the SCI literature).

"Proof of concept" of axonal regeneration in the spinal cord best comes from clearly determining a functional recovery, and then exhaustively evaluating the anatomy and physiology of treated animals. *Sometimes this can be achieved in the same animal, but more often not.* For example, anterograde fills of white matter must be several vertebral segments from the lesion so that dorsal root entry of peripheral nervous system (PNS) axons (which grow nicely into a local area of cord at the injury site) does not lead to a false interpretation of regeneration. These "drop back" fills then permit only relatively few labeled axons of hundreds of thousands available for labeling to make it to, or through, the plane of transection. This would indeed be proof of white matter regeneration. However, obtaining this result for an identified tract of white matter together with unequivocal proof of behavioral recovery (usually in a small subset of treated animals) in one individual is extremely unlikely. Retrograde uptake studies also share this possible problem—unless the fill is so complete as to identify the parent axons or, better, the cell bodies giving rise to these sprouts and branches. Though there are interesting new techniques available to study axonal regeneration in the mammalian cord (such as green fluorescent protein (GFP) technology), these also have their drawbacks, such as the latter requiring xenografting to be employed (Davies et al. 1997, 1999). Any attempt at anatomical identification is also hampered by the "sprouting" nature of regenerating axons. I say "sprouts" because there is little to no documentation since the time of Ramon y Cajal that large-caliber myelinated axons (~8 to 10 microns in diameter) actually regenerate from their terminal ends within the white matter tissue after retrograde dieback by extending equally large-caliber fibers into now vacant territory. The new outgrowth created by experimental techniques is usually small, sometime submicron, branching sprouts.

It is instructive to consider an example of a significant effort at producing such a clear linkage of anatomy to functional recovery before moving to voltage-mediated regeneration. This evaluation stems from an older study of CNS regeneration through peripheral nerve bridges.

A most relevant study from the Aguayo group involved the implantation of a segment of peripheral nerve, replacing the optic nerve in the adult hamster. The new segment bridged the superior colliculus with the eye (retinal ganglion cells). The latter projected axons through the peripheral nerve bridge, making contact with the appropriate layers of the superior colliculus. These new connections were physiologically functional and the investigators were able to record inhibitory postsynaptic potentials and excitatory postsynaptic potentials in response to bursts of light at the eye (Bray et al. 1987; Vidal-Sanz et al. 1987; Keirstead et al. 1989).

ANATOMICAL EVIDENCE FOR VOLTAGE-MEDIATED REGENERATION

IDENTIFIED TRACTS, AXONS, OR PATHWAYS

We have come to believe that through the use of invertebrates one can precisely identify regeneration of an axon to its appropriate target. For example, in the leech, the entire central circuits, projections, and targets are identified. Thus, regeneration of a single axon can be traced and the appropriateness of its target synapse determined (Muller et al. 1981). To a more limited extent, this is also true of the ammocoete larvae of the sea lamprey (Rovainen 1967, 1974; Wood and Cohen 1979). The Mauthner and Mueller cells projecting from the third ventricle of the brain are identified as having many of the target cells within the lower spinal cord. Thus, in these animals, the investigator can perform a "tour de force," measuring the appropriateness of the regeneration of spinal axons, recording from them both intracellularly and extracellularly (Borgens et al. 1981), and to a more limited extent, even perform rudimentary behavioral tests such as prodding the rostrum of the ammocete to elicit a rearward undulating escape behavior. For the medical researcher this biology may not be so useful, since spinal cord axons of the larval lamprey regenerate naturally, often to their targets.

In the mammal, our choices are restricted. Though axons of the spinal cord may begin to regenerate—particularly in the immature mammal—they cease to elongate. This was called abortive regeneration by Santiago Ramon y Cajal (1928). Therefore, it is not a black-and-white issue of axonal regeneration or the lack of it when studying the comparative zoology of CNS regeneration. One must also be mindful of the fact that robust sprouting from undamaged or untransected fibers in adult mammalian white matter occurs. We will address this issue below. One can create a special pathway for mammalian axons to regenerate into or through—proving the ability of central fibers to regenerate in response to injury. A good example is the use of a peripheral nerve "bridge" implanted into a spinal cord at both ends, or the brainstem on the rostral end of the segment. As shown since the time of Ramon y Cajal, axons that gain entrance to the tube or bridge grow through it, often exiting at the other end. That regeneration of CNS elements had occurred was proven since retrograde labeling was used to demonstrate the origin of the labeled fibers in the bridge. This notion found favor for a short while with clinical researchers—even supporting pilot trials of a similar technique in the spinal cords of humans by Carl Kao (Kao 1974; Kao and Chang 1977; Kao et al. 1977). These notions eventually played out and are no longer considered feasible in the human due to the extensive surgery involved and the very few fibers that project through the bridge. We have used this theme to provide formal proof that CNS axon regeneration can be initiated, and guided by electrical fields (Borgens 1999; see below).

Though uncommon in the adult mammal, a particular spinal cord long tract can be identified to a specific behavioral mode, such as the corticospinal tract's control of forearm and digit manipulation. The behaviors sampled can be reaching, purchase of food pellets, marble burying behavior in hamsters and mice, and other esoteric functions. Often, however, spontaneous recovery in control lesions eventually undoes the attempt at discerning a difference between the experimental and control therapies. Most significant behaviors, motor and even sensory appreciation, show a remarkable ability at restoration (sometimes by unknown mechanisms) over the long-term observation of laboratory rodents. Below we shall discuss a clear alternative, the CTM pathway, where damage to an identified tract in the ventral spinal cord leads to lifelong deficits in behavior.

MARKING THE PLANE OF TRANSECTION

Obtaining anatomical evidence for regeneration of white matter axons in the mammal is not as seemingly facile as one might guess. The first issue before any other is to unambiguously demonstrate where the plane of transection occurred—sometimes many months later. Contusion or crush injuries are not acceptable in this regard, and indirect measures, such as using the formation of a fibroglial scar to mark the plane of regeneration, are also fraught with difficulty. If the scar that forms is greater than even a millimeter in extent (in the longitudinal plane) and the actual elongation of axons or their branches extend a fraction of this, one would not be able to label these as "regenerating," as they may not have passed the plane of transection.

These issues call for the use of a marker to actually, and exactly, mark the plane of transection of the spinal cord. Without this, branching patterns, degenerating end bulbs, or what appear to be growth cones have no meaning. It is commonplace to see reports where several labels are used in association with an undefined transection plane, or worse yet, a contusion site. End bulbs may be marked with β-amyloid precursor protein, yet this is evidence of a transection of the fiber—and dissolution of the distal segment and proximal end bulb. Such labeling does not speak to the regenerative process.

Fudicial markers have a long history of use in embryology/developmental biology, but curiously not in the modern literature of axonal regeneration in mammals. For example, cactus spines were once used to pierce the block of an embedded specimen, leaving a hole in the support material after histological sectioning. In modern times it appears that researchers in CNS regeneration have for too long

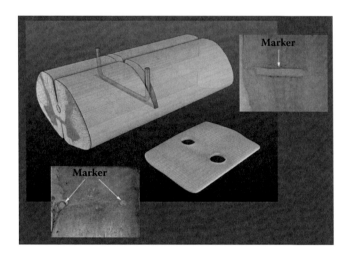

FIGURE 11.1   See color insert. The placement of a U-shaped spinal cord marking device. The device is removed just prior to histology. The illustration and actual histological images show the fudicial holes left in the tissue. In the top right, the elongated hole reveals the base of the device.

relied on the presence of scar tissue at the site of a transection or compression injury, or attempts at three-dimensional reconstruction of the whole specimen that was embedded. In the latter, most three-dimensional views of spinal cord samples rely upon digital images gathered every tenth serial section. Thus, the final view is 90% computer derived. Furthermore, the fibroglial scar may extend rostral and caudal of the injury site by many millimeters, even up to one complete vertebral segment. The inability to define the exact plane of the lesion would make it difficult to draw meaningful conclusions if CNS fibers were to regenerate less than this distance in response to an experimental therapy.

In her studies of regeneration in the adult rat brain, Ann Foerester (formerly of McMaster University) made a creative and significant advancement. She understood that to be able to clearly define, and quantitate, axonal regeneration in the mammalian brain, a marker was required. She employed a staple-shaped device that was used to cut a local region of the brain. This was left in place until it was removed at the stage of histological processing. Small holes were left in the brain tissues where the marker had been (Figure 11.1). By using the location of the holes, she could precisely determine the boundaries of her original transection. This allowed her to reveal that brain axons do regenerate past the plane of a cut, and that there are conditions wherein this may not occur and those when it does occur (Foerster 1982). We have borrowed this interesting concept to mark the vertical level, and horizontal boundaries, of transection lesions to the guinea pig spinal cord (Borgens et al. 1986a).

One of the criticisms of the Foerester device was that axons projecting downward, under the device, and then arising on the other side might be an artifact of the device itself. When the parenchyma was cut with the device, perhaps it bent the axons into these abnormal positions, giving the appearance of fibers project-ing around the lesion, when in fact they might have been pulled into these posi-tions. This idea is unlikely. This critique attempts to treat delicate axons as if they were strands of flexible spaghetti. There is no literature that suggests that axons *in situ* can be forced into such tortuous pathways. Nevertheless, to escape this criticism, we did not use the U-shaped marker device to cut the spinal cord tissue, but rather placed it into the hemisection only as a marker device. The results of this technique relative to the regeneration of axons within the cord in response to Efs are described below.

ANATOMICAL EVIDENCE FOR VOLTAGE-MEDIATED REGENERATION

I will abandon the chronology of these investigations to make an important first point: an imposed weak gradient of voltage will induce *anatomically defined* axonal regeneration in the adult mammalian spinal cord. The experimental evi-dence for this statement is not necessarily of clinical significance, nor will it bear on the functional significance of this regeneration. The following experiment does, however, constitute formal scientific proof that applied Efs can induce, and guide, significant axonal elongation in electrically treated animals, but not in controls.

THE BRIDGE EXPERIMENT

Borrowing heavily from the classical use of peripheral nerve bridges described above, this technique can reveal the capability of adult mammalian axons to regenerate within a permissive environment. Thus, we designed our own version of this test. In the classical studies, the peripheral nerve segment was chosen because PNS axons regenerate naturally within the trunks of peripheral nerves. Perhaps this environment is permissive while the CNS was not? The results of all tests of this notion were positive apart from those of le Gros Clark (1942), Barnard and Carpenter (1950), and Feigin et al. (1951). An incomplete list of the clear evi-dence of regeneration through peripheral nerve bridges can be found in Ramon y Cajal (1928) quoting Tello, Heinicke (Heinicke and Kiernan 1978; Heinicke 1980), Chi et al. (1980), and Sugar and Gerard (1940). Formal scientific proof of regeneration would await the use of intracellular labels by David and Aguayo (1981) to unmistakably demonstrate the axon's type and origin.

   In our case, we did not wish to test this notion, and chose a hollow silicon tube (6 × 1 mm) that would not itself provide any substantive permissive environment for axonal regeneration (Figure 11.2). The tube was placed surgically into the guinea pig's back, and the ends of the tube embedded in the spinal cord at both ends. The surgery to accomplish this was extremely damaging to the cord and the axons within it. Early histological observations confirmed massive transection and damage to spinal cord white matter. Uncountable end bulbs were observed using intracellular tracers applied at both ends (i.e., rostral and caudal) of the silicon tube shortly after surgery. Over the next month the tubes were left in place. In a

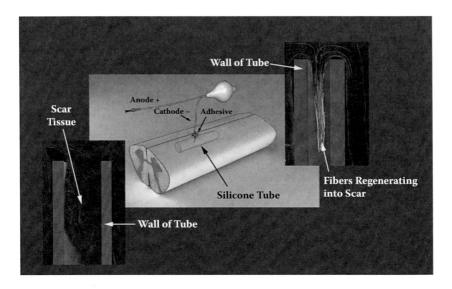

FIGURE 11.2   See color insert. A cathode inserted within the hollow tube implanted within the dorsal spinal cord. The two insets show the plug of scar tissue in control tubes and the robust growth of axons into electrically active tubes. (From Borgens, *Neuroscience*, 91, 251–64, 1999.)

few animals, one end of the tube lifted out of the spinal cord parenchyma, but in most both ends remained in place. Prior to implantation, a minute Teflon-insulated electrode was placed in the middle of the tube—its uninsulated tip within the hollow center. This electrode was in continuity with a small current-regulated power supply surgically located under the skin of the neck. An Ef of ~2.5mV/mm was produced inside the tube by this circuit; however, the field within nearby tissues would have been significantly reduced—estimated to be on the order of hundreds of μV/mm. We were only able to place a cathode within the confined environment of the tube since this electrode was not incompatible with living cells and tissues, while anodes produce toxic electrode products, gasses, and a fall in pH, which would make the center of the tube a cell-free zone. Control animals received the same stimulator, wire, and tubes, except that current was not passed through the circuit.

After ~1 month the lesion was exposed and the tiny electrode withdrawn from the tube. Using a small probe, a few crystals of rhodamine dextran were inserted into the hole where the electrode had been inserted and removed—and the hole once again sealed with a fast-set dental adhesive. This effort would retrogradely label any axons within the tube. The wounds were closed and twenty-four hours passed until the sacrifice of the animal and the observation of the tubes and the spinal cord. In fourteen of sixteen control animals, nerve fibers were not found within the scar tissue filling the tube. Only a total of four small axons were found coursing with blood vessels in the balance, but could not be identified. In every control but these two exceptions, the ends of the tube filled with a dense astroglial scar tissue—sometimes meeting at the center of the tube, protruding into it from each end. The scar penetrated to a depth of several millimeters. In electrically treated animals, brightly filled fibers penetrated the scar in large numbers. These coursed through the scar tissue in twisting/tightly turning patterns, sometimes arching back on themselves. They were too numerous to be counted. The caliber of these labeled axons was mixed, though many axons were on the order of 5 μm (likely myelinated). The use of the retrograde label demonstrated decisively where most of these fibers originated. Most labeled axons in the electrically active tubes deviated from nearby white matter tracts of the cord. In other cases the fibers originated from labeled nerve cell bodies of intact gray matter (Borgens 1999). The four fibers in control tubes were believed to be autonomic, as their origin could not be traced and they were only found in association with blood vessels formed within the scar.

Beginning in 1986 we published several reports tracing the course of regenerating axons *in situ* in response to applied DC voltages. The method of applying the electric field was by means of a small implantable stimulator, which delivered ~1.0 to 50 μA of current for up to four weeks. The seminal means of assay was anterograde filling of ascending axons with horse-radish peroxidase (HRP) to trace them around the level of the transection. This application was useful in that it provided a glimpse into the character of regeneration of long-tract white matter. At higher levels of current (35 μA) fibers were traced around the margins of the hemisections, or descended in a ventral direction to pass beneath the lowest boundary of the section. These fibers grew through undamaged parenchyma to terminate in the rostral segment. At lower levels of current/Ef the response was less robust or absent (Borgens et al. 1986b). The case for regeneration was made, but the numbers of fiber traced were very small—given the distance that was required for the intracellular label to travel. The label was injected into the spinal cord ~2 cm from the transection plane. Moreover, HRP is a dependable but large (~41,000 Daltons) intracellular label, and its use further reduces the potential to label many axons. We abandoned its use in the early 1990s in favor of fluorescent dextrans (mw ~ 8,000). This labeling was more useful and, in association with the marker device, greatly enlarged our understanding of the developmental biology of Ef-induced axonal regeneration. Summarizing this long series of reports:

- Axons arising from nearby severed white matter indeed underwent retrograde degeneration to an end point ≤ 250 microns from the transection plane. This could be shown by averaging the termination of filled endings at this level, less than 200 microns, and at the plane of transection. The terminals (end bulbs) of fibers were statistically significantly further from the plane of transection in control animals (carrying dummy voltage sources) than in electrically treated animals.
- This observation was in complete agreement with the results of experiments where distally negative electrical fields limited the acute dieback of labeled lamprey reticulospinal axons relative to the transection compared to controls (Roederer et al. 1983; Strautman et al. 1990). Moreover, distally positive Efs produced a greater distance of dieback

than occurred naturally in controls (Roederer et al. 1983; Strautman et al. 1990; Borgens and Bohnert 1997). The mechanisms of action underlying this polarized response are understood. The Ca^{2+} component of an ionic current of injury entering the end of these giant axons after damage is responsible for the collapse of the cytoskeleton and retrograde degeneration of the fiber (Schlaepfer 1974). Distally negative extracellular Efs limit this Ca^{2+} entry into the terminals of damaged axons, while distally positive Efs enhance penetrance of Ca^{2+} into the damaged axon as it pours down its electrochemical gradient—leading to a greater distance of dieback (Borgens et al. 1980; Roederer et al. 1983; Strautman et al. 1990). The phenomenon of an inwardly directed Ca^{2+} injury current is not an esoteric facet of lamprey giant axons, but is characteristic of all cells—and recently demonstrated to occur in adult guinea pig spinal cords in organ culture (Zuberi et al. 2008).

- Regenerating axons that projected to the plane of transection often curved and grew in a horizontal direction along the boundaries of the scar tissue in adult guinea pig spinal cord, or were deviated in other unnatural projections (Borgens and Bohnert 1997). In most cases, but not all, the regenerating axons were fine branches originating from larger bore terminals of axons within the white matter.

- The fine branches of regenerating axons also penetrated the astroglial scar to project into the opposite segment of cord. This was shown by three-dimensional reconstruction of confocal images, and following the course of regenerating axons in optical slices (Borgens and Bohnert 1997). Thus, axons can indeed penetrate a mature fibroglial scar in the mammalian spinal cord, given the right conditions.

- The frequency, amount, and distance of regeneration initiated by Efs could be enhanced when in association with the administration of a neurotropic factor. We tested the dual use of Efs and the growth factor inosine, and the same techniques were used to reveal the extent and trajectories of fibers double-labeled as ascending and descending projections (Bohnert et al. 2007).

BEHAVIORAL EVIDENCE FOR VOLTAGE-MEDIATED REGENERATION

The task of behavioral analysis after spinal cord lesions is both difficult and controversial. At first glance, particularly to the neophyte, this statement would seem ridiculous based on the catastrophic loss of function in both man and animal after spinal cord damage. Surely anything, even a modest improvement, would be apparent if an experimental therapy was useful. Over the years there have been several reviews and critiques of this subject and significant work by students, such as the late Michael Goldberger (Goldberger et al. 1990), on the origins of motor recovery after spinal lesion. While there is understanding of cord-mediated reflex and supraspinal control of motor functioning, the anatomical circuits are still largely unknown. Said another way, knowledge of the anatomical basis for supraspinal control of motor behavior is very incomplete. This presents a difficulty in understanding the efficacy of a treatment by selective relesioning or other techniques. What one is left with is a retransection experiment designed to obliterate the recovery presumably mediated by the tested "therapy." These caveats apply not only to tests of walking, but also to tests such as toe spreading in free fall (Borgens et al. 1990). Here a test animal, suspended in a sling, is dropped a short distance. Typically there is a reflex-generated "spreading of the digits." This is apparently lost after spinal cord damage in the mid-thoracic region. The observers record toe spreading as a means to imply functional recovery in hind limbs. We have found this test to be wholly unsatisfactory, as the frequency of spontaneous recovery of the toe spreading behavior is too high (Borgens et al. 1990). As will be discussed, the level of local (to the segment of spinal cord) reflex activity also complicates an analysis of ambulation in quadruped animals.

Can We Effectively Study the Recovery of Walking in Experimental Animals?

There is a wide gulf in the balance between supraspinal control of motor function, and locally controlled and generated motor movements in limbs. The former is dominant in humans, as we are the world's only obligatory bipedal mammal. Quadruped locomotor activity is more dominated by reflexive movement of the limbs—aided in coordination by descending supraspinal influence on these local circuits in the cord. Therefore, we should expect some form of uncoordinated walking or rudimentary ambulation in normal animals after spinal cord damage in the absence of any treatment. Furthermore, the loss of descending inhibitory influences dampening local reflex circuits at the fore and hind limbs leads to a hyperreflexia in both animals and man. This complicates further a dissection of the effect of a therapy meant to improve motor recovery.

The most widely used behavioral analysis tool of the last decade is the BBB evaluation of walking, named after its inventors, Basso, Beattie, and Bresnahan (Basso et al. 1995) at that time, all colleagues at the Ohio State University. This testing involves placing a rat on a flat surface, with some degree of minor roughness to allow the animals "purchase." "Blinded" observers then grade the animal on a 0 to 21 scale relative to its ability to ambulate or not. Below a scale of 8 to 9, individual movement of the hind limbs cannot be separated from reflex movements that *do not* reflect nerve impulse transmission through the spinal cord lesion (indicative of true "recovery"). Rather, any movement here could be ascribed to reflex movement of the leg(s). Even a difference between control and experimental in scores whose means fall below 10 could mean only that the experimental application heightened—or initiated—reflex activity: it is not "proof" that the brain has directed/modified motor behavior below the lesion. In most reports, the original BBB rules are bent by reducing the scale to a 0–6 or 9 scale, where usually the first four or five grades are close to those descriptions in the BBB criterion. The higher grades more clearly reflect volitional movement. This "compression of scores" serves to inflate the appearance of what was the capability of the animals. These data look good on a graph, but fail in providing meaningful data about motor recovery. We have recently used the BBB scale with strict attention to blinding of two observers, and a tight adherence to what each of the twenty-one grades meant to signify in function (Shi et al. 2009). Thus, we believe that with considerable work the BBB is useful, if only it is followed as the inventors created it.

THE CUTANTOUS TRUNCI MUSCLE REFLEX

In the mid-1980s neuroscientist Jack Diamond (formerly of McGill University) tipped me off to the most interesting of behaviors in animals mediated by ascending tracts in the spinal cord (Theriault and Diamond 1988a, 1988b). My long-time colleague Andrew Blight and I looked at Diamond's group's data on the behavior (a rippling contraction of the skin of the back after a local stimulus), and his exploration of the neural circuit underlying it in the rat. We then dissected the same behavior in the adult guinea pig—with an emphasis on the spinal cord component of the behavior and its physiology (Blight et al. 1990).

The behavior is properly a "long tract" reflex with an intramedullary and extramedullary component. To elicit a local "twitching" or "contraction" of the back skin (produced by the cutaneous trunci muscle (CTM) subtending the dermis), the back skin is probed or pricked (Figure 11.3). This noxious cascade of compound action potentials (CAPs) enters each segment of the vertebral column (i.e., from its own dermatome) through the dorsal cutaneous nerves (DCNs) of the back—all lying parallel, and segmentally arranged roughly perpendicular to the long axis of the vertebral column. The DCNs locally enter the spinal cord at the dorsal root and immediately project into the ventral funiculus where they may synapse, but all first-order and second-order axons ascend the cord in large, long, unbranched axons via the CTM tract. This tract of axons lies just medial to the spinothalamic tract in rodents. There is a minimum of contralateral projections, but they do occur (see below). The bulk of axonal projections are ipsilateral to the entry point. This is the ascending leg, or afferent leg, of the CTM circuit. Higher, in the thoracocervical region, lie nuclei of CTM motor neurons on each side of the spinal cord. There are no detectable contralateral projections between these motor pools on the left and right sides. Thus, stimulation on one side of the skin has a small component of skin contraction at the midline or contralateral to it, but the bulk of the behavior is expressed ipsilateral to the local region of stimulation. The efferent (motor) component of the CTM circuit is the set of motor neuron projections that leave the spinal cord at the brachial plexus—and follows through to the cutaneous trunchi muscle via branching fibers of the lateral thoracic nerves (LTNs; Figure 11.3). A crush or a transection of the spinal cord can be used to interrupt the ascending CTM pathway. A partial lateral hemisection of the cord obliterates the CTM functioning for the life of the animal, but on one side only.

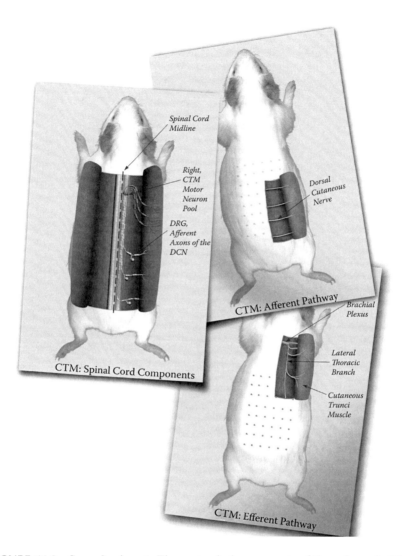

FIGURE 11.3 See color insert. The anatomical components of the guinea pig CTM reflex. The tattooed dots on the back skin provide quantitative information about movement of the skin after direct and local stimulation. The process of stimulation before and after injury is videotaped from above, and stop frame analysis of the movement of the dots when the skin contracts achieves the following data: (1) the boundaries of function and nonfunctional receptive fields, (2) the vector of skin movement relative to a local stimulation, (3) the speed of contraction, (4) the latency of contraction, and other such quantitative evaluation of this behavior. (From Blight et al., *Journal of Comparative Neurology*, 296(4), 614–33, 1990.)

When rare spontaneous recoveries have been reviewed, the histological sections revealed that the CTM tact was not completely severed. Therefore, in hemisections we adopted the use of a sharpened insect pin to further confirm the ventral completeness of the lesion after the original scalpel cut. Significant crushing of the cord also produces a significant loss of the CTM with approximately 17%—usually much less—spontaneous recovery in the population.

WHAT CONSTITUTES A CTM RECOVERY?

Review Figure 11.3 to understand that a receptive field of the CTM is dependent on the projections of DCNs into the cord at that vertebral level. Selective transection of segmentally arranged DCNs (with one trunk undisturbed in between) show an island of functioning skin surrounded rostrally and caudally by nonresponding skin in which there is no twitching or contraction after stimulation. In the parlance, this island of unresponsive dermatome is called a region of areflexia.

If we cut or crush the entire width of the spinal cord, then the region of areflexia is the full width of the back skin and below the level of the injury to the cord. If we lesion only one side of the cord (left or right of midline), then we produce a region of areflexia ipsilateral to the lesion and below its vertebral level. Any recovery of twitching within this region of areflexia is considered a recovery. But is it due to restored conduction through the spinal cord injury by regeneration—specifically in transection studies?

There are only two changes in the animal that can account for a recovery of CTM functioning in areflexive dermatomes. First is the aforementioned short-distance regeneration, synapse formation, and restoration of the condition pathway to the motor pool high in the neck. The second is a robust sprouting of cutaneous nerves of nearby undamaged skin into the areflexive region. These two possibilities can be separated by selective lesioning, as explained in Borgens and Bohnert, 1997.

APPLIED VOLTAGES INDUCE REGENERATION AND BEHAVIORAL RECOVERY IN LABORATORY ANIMALS

The first behavioral test of an applied DC voltage in adult guinea pigs used the CTM reflex as a behavioral assay. A right lateral hemisection was used to sever the ascending CTM tract in the cord; thus, a rostrally negative Ef was applied. Evaluation of the CTM receptive field was also coupled to electromyographic recordings of the functional—or nonfunctional—receptive fields.

This first experiment, published in *Science* (Borgens et al. 1987), clearly proved behavioral recovery could be restored by an applied Ef. In the current-treated group, 25% of the animals recovered some component of the CTM, but none of the control animals did. Behavioral evaluation was carried out after the removal of the electrical field stimulators. In a subsequent study (Borgens et al. 1990) analysis of the ventral spinal cord revealed regeneration of axons in the population of animals that also recovered the CTM reflex, but not in controls (that were completely refractory to behavioral recovery of the CTM, and followed for up to 139 days postsurgery). However, we must be cautious here: more often than not, regenerating axons were not observed in CTM positive animals. Furthermore, we had no evidence to suggest that the axons labeled by anterograde intracellular labeling were actually components of the ascending CTM tract. One would be quite lucky to exactly label the relevant tract in the mammalian spinal cord and have this positive result coincide with a behavioral recovery in this single animal. Still, it was a very positive result to see anatomical regeneration *and* behavioral recovery in the same population of animals and the complete absence of it in sham-operated animals. Moreover, distally positive fields were tested in another experimental group, and CTM recovery was not observed in this group—consistent with the hypothesis that the distally negative Efs promote axonal regeneration in this direction, where distally positive Efs may redirect axons in the opposite direction or lead to more extensive dieback (Borgens et al. 1990).

SUPPORTIVE EVIDENCE FROM NEURONS *IN VITRO*

The observations made in whole animal experiments described above find support from the *in vitro* observations of neurons. I direct interested readers to recent excellent reviews of this literature (McCaig et al. 2002, 2005) and Chapter 10. The two literatures have both been mutually beneficial. For example, the apparent difference in time course in degeneration and regrowth in neurites *in vitro* (McCaig 1987) provided the seed of an idea to produce *in vivo* stimulators that might influence axonal regeneration in both directions within the Ef. This notion was first successfully tested in cases of naturally injured canine paraplegics in two separate blinded and controlled clinical studies (Borgens et al. 1993, 1999).

There are some brief thoughts worth addressing on the relevance of *in vitro* studies to the use of Efs in animals—and especially humans.

- The beautiful asymmetries documented during the development of neurons and their neurites *in vitro* in the presence of an applied voltage are most convincing when a substrate for the cells is not used. This uncomplicates the interpretation that the voltage does not act indirectly on cells via an action on the substrate. At the same time, this method significantly complicates any inferences about the responses in animals where numerous endogenous cues for neurite guidance and growth exist (McCaig et al. 2002, 2005). The bulk of the *in vitro* literature supports the view that an applied voltage may induce and impose a direction on neurite growth, as well as impose a process of neurite degeneration at the antipode. When neurites of *Xenopus* CNS cells are directed toward the cathode, the fibers facing the anode either turn or are resorbed into the cell body. When fibers projecting from individual chick dorsal root ganglion neurons or postganglionic sympathetic neurons are parallel with the applied gradient of voltage, they are resorbed into the cell body, while new fibers exclusively project perpendicular to the voltage gradient (Pan and Borgens 2010).

- The magnitude of voltage tested to initiate turning or other interesting behaviors in cultured neurons is broad ranging, from a few mVs/mm to well over hundreds of mVs/mm. Often these levels of applied voltage depend on the practical consequences of following neurite growth over the relatively few hours of documentation. I have never been particularly impressed with the often reported claim of a low level of Ef where *in vitro* responses do not occur. One could reasonably retort that it may take many days in culture for this response to emerge. *In vivo*, it is clear that applied voltages as low as a few hundred μVs/mm suffice, and the responses may not be apparent for weeks to months (Borgens 1986; Borgens et al. 1990; Borgens and Bohnert 1997; Bohnert et al. 2007).

It is true that often over the last thirty years the literature on *in vitro* responses of neurons has been supportive of the simultaneously emerging literature on axonal regeneration and axonal turning in response to Efs in animals. The early studies of single neurons *in vivo*, cultured on plastic substrates by McCaig (Hinkle et al. 1981) and Patel and Poo (1982), nearly overlapped our first reporting of axonal responses in the lamprey spinal cord (Borgens et al. 1981). Thus, our studies seemed to form a unique basis one for the other, and were discussed in this context. Growth responses in both phenomena appeared to be mediated by distally negative Efs. In later years the *in vitro* literature became much more complicated (Rajnicek et al. 1992; Davenport and McCaig 1993; Pan and Borgens 2010), though there is still not a single convincing report of cultured neurons that are completely resistant to an applied Ef, to my knowledge. The recent negative report by Cormie and Robinson (2007) was based on the use of a single magnitude of Ef and a single substrate—and thus has no standing in this regard. The experiments did not test neurons attached to the "floor" of the culture dish (as all other reports, including Robinson's own laboratory). Rather, they used laminin as a substrate, which would have altered the response to the Ef (Britland and McCaig 1996) (see Chapter 10). Furthermore, a nearly simultaneous review by this duo also discusses the use of Efs to treat human spinal cord injury (Robinson and Cormie 2008).

CLINICAL UTILITY RELATIVE TO SCI

AN APPLIED DC VOLTAGE AS A THERAPY FOR HUMAN SPINAL CORD INJURY

An indwelling stimulator system, producing a slow duty cycle DC application (Figure 11.4), was submitted to the FDA device panel, an Investigational Device

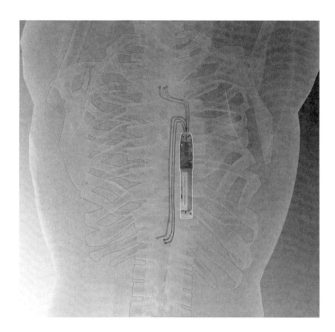

FIGURE 11.4 A long, cylindrical OFS unit is implanted into the back of the patient and three pairs of electrodes are routed rostral and caudal of the injury. These are sutured to bone, and do not enter the central canal and touch the spinal cord itself.

Exemption Number was obtained, human clinical trials in severe SCI were begun, and the results published in 2005 (Shapiro et al. 2005).

This part of the chapter guides the reader through the rationale for the development and use of the oscillating field stimulator (OFS) as a biomedical implant, the results of the first phase 1 clinical trial, and the status of the technology today.

Unidirectional, Steady DC Voltages Have Little Clinical Significance—What Might Be The Remedy?

Up to this point I have discussed the animal studies, from fish to rats and guinea pigs, revealing the possibilities for applied Efs to both initiate and guide CNS nerve regeneration and sprouting. This application was shown to be polarized; that is, growth (associated with a functional recovery) was initiated with the cathode placed in front of an injury zone to serve as a beacon at this local. The opposite polarity has been associated with greater destruction of axons facing the anode and the failure to produce a recovery of function from SCI. There are few regions of the human nervous system where such a polarized response based on a single polarity of Ef application would be useful. An example of one would be the PNS, where the large nerve trunks of the extremities contain a mixed population of axons with differing conduction pathways (i.e., motor and sensory), but the same projection to target cells (distal). These mixed nerves project from central locations (i.e., motor nerves of the gray matter and the DRG). Therefore, functional regeneration would be theoretically possible with only one polarity of an Ef application with the *cathode distal within the extremity*. Such an arrangement would coax the proximal segments of axons toward their downstream motor target cells—or the proximal segments of sensory fibers back to a region of their receptors. In the mammalian spinal cord this possibility does not exist.

In the cord, ascending first- and second-order fibers (such as pain and pressure afferents) climb the cord projecting to the hindbrain. Predominantly, motor tracts project from the brain downstream, to project onto their target's neurons. These "long tracts" of differing polarity of projection lie next to each other in the parenchyma of the white matter. A single polarity of an Ef imposed along the long axis of the spinal cord would likely produce effects for one group (facing the cathode) while enhancing an opposite response (facing the anode). Thus, the ability to initiate growth or degeneration of nerve fibers would remain a "laboratory parlor trick," unless a means to produce growth in both directions—and absent of degenerative effects—would be discovered. Only in this paradigm could functional recovery in the CNS be associated with a single polarity of application.

A clue to a possible remedy was reported by McCaig (1987). He revealed that the loss of neurites when facing the anode takes a relatively long time (~45 minutes) compared to the time in which growth cone activity was begun when terminals faced the cathode (~ in minutes). This window in time provided a possible means to reverse the polarity of the applied Ef where degenerative effects would not have begun—or at least would not be significant. After reversal of the polarity, the effects of nerve fiber terminals facing the cathode would be in the presence of a distally negative cathode for enough time to possibly initiate growth. We chose to develop this idea as a DC application with a very slow duty cycle—an Ef reversal every fifteen to twenty minutes. This should not be thought of as an AC field. We called the method oscillating field stimulation (OFS). Later, this acronym was extended to define the stimulators themselves.

A Jump to the Clinic: The Paraplegic Dog

The CPR laboratories are located behind the Purdue University School of Veterinary Medicine. This gave us a unique asset in the development of CNS therapies—the naturally injured paraplegic dog.

Dogs suffer severe clinical spinal cord injuries, as do people, and for the very worst of these injuries their prognosis may not be that much better than for humans. As discussed above, they may be able to ambulate awkwardly, without any sensory feedback, control of bladder and bowel, and other pathology associated with SCI. Most do not recover ambulation, however, as testified by the number of dog cart companies that produce a two-wheeled cart to suspend the animal's hindquarters over an axle and wheels. Actually, most dogs take to this contraption readily. Still, most paraplegic dogs in the late 1980s and early 1990s were euthanized when owners discovered they must express the dog's bladder daily and deal with persistent lower urinary tract infections, skin care, and rehospitilizations—likely for the life of their pet. I should mention this response by owners to these terrible accidents seems to be changing in today's culture, with more emphasis on animals in general and pets in particular.

In the late 1980s, Andrew Blight, Jim Toombs (a small animal surgeon par excellence), and I organized the first use of OFS in these clinical cases. Paradoxically, most spinal cord accidents in dogs are not due to motor vehicle impacts, but to spontaneous intervertebral disc herniation. This type of injury occurs in high frequency in specific dogs, such as the Dachshund. Other breeds that are susceptible to these problems are dogs with long backs and short legs, making the animal a biomechanical accident waiting to happen. Breaking forward motion with the front legs, while the long back is arching forward, puts significant pressure on intervetebral discs, which explode out of their position into the spinal cord. Moreover, discs in these breeds (chondrodystrophic animals) are more brittle with less water content, and prone to traumatic, explosive herniation. In humans, discs herniate outward, pressing on spinal roots and causing pain. Rarely do disc herniations cause spinal cord injury and paralysis in people. In dogs this can occur slowly over time (in animals described as having "disc disease") or traumatically. We chose to use the latter types of injuries in our studies, as the resultant pathology—even on an anatomical level—is similar to what is observed in people.

Dog Neurology: Choosing the Candidates— Conventional Treatment and Experimental Therapy

There is a useful literature on the neurology of disc diseases and traumatic spinal cord injury in the veterinary literature (Brown et al. 1977; Hoerlein 1978, 1979; Toombs and Bauer 1993). Briefly, the more serious the injury upon presentation in the clinic, the more serious the herniation appears on plain film, CAT, magnetic resonance imaging (MRI), and myelogram. Most of the time the offending mass of disc material can be seen pushing against the cord. The typical neurological examination upon presentation is divided into four parts: tests for proprioceptive placing of the hind limbs, deep pain, superficial pain, and ambulation. Of these, conventional clinical wisdom determined that the absence of deep pain was the most prophetic of a poor outcome. A "neurologically complete" dog failed all tests. This is not to be confused with "anatomically complete." Accidental transection of the human spinal cord is very rare—almost all spinal cord injuries are severe compression/contusion injuries. Complete transection of the cord in dogs is more common, particularly in motor vehicle impacts. We did not recruit transection dogs into our blinded trial testing compression injuries. Moreover, we had learned in pilot experiments that complete transections were refractory to OFS. Therefore, we built our patient exclusion criteria to sample these "worst of the worst" SCIs in the dogs—referred to in the literature as Hanson's type 1 herniations. The four outcome measures were assessed by blinded investigators, and totaled to produce a total neurological score. Somatosensory evoked potential (SSEP) conduction studies were evaluated separately. Neurological exams were at six weeks and six months postintervention. OFS-treated dogs were significantly and statistically more functional than controls (placebo implantations). At six weeks p = 0.031, and at six months, 0.036, comparing their total neurological scores between groups. Clinical measurements of nerve conduction through the lesion (SSEPs) occurred in four of twelve experimentally treated dogs, while none of the eleven controls recovered conduction—a trend toward significance—but at the low N, statistical significance was not met in this single examination. Though it seems paradoxical, functional animals after SCI may not register a positive SSEP given the variation in the placement of electrodes and the character of this extracellular recording technique. In the SSEP method, ascending potentials are stimulated at the sciatic nerve of the hind leg, and are measured arriving at the contralateral sensorimotor cortex of the brain by surface electrodes. Said another way, the recovery of a positive SSEP is significant, but a negative one is not of much inferential use.

The early 1993 test also examined two stimulator designs for surgical use. Over the next three years, a "human use" design was chosen incorporating fail-safe circuitry in the units, and other requirements for use in an FDA-sanctioned trial in humans. This human use OFS design was tested in another blinded clinical trial in paraplegic canines—using once again the same study plan as the 1993 publication. This second OFS study was completed and published in 1999 (Borgens et al. 1999). A larger group of animals was evaluated in this trial (twenty OFS, fourteen sham). The outcome was similar to the original 1993 study, if not a weaker outcome, in that only three of the five evaluation methods were statistically different when compared individually. The total neurological score was significant at the six-month recheck (p = 0.047). While the characteristics of the circuit and embodiment of the stimulator differed between each trial (1993 vs. 1999), the stimulation paradigm did not. This allowed a more refined comparison using a larger number of sample data by totaling all animals treated with OFS. (Refer to the discussion in Borgens et al. (1999).) The outcome measures were statistically significantly improved by OFS treatment. However, a surprise was the frequency and magnitude of recovered superficial pain appreciation. In the OFS-treated groups, more than half of all animals who received OFS therapy recovered superficial pain, while only two of fourteen placebo implant animals recovered this function in the 1999 trial alone (p = 0.0001). None of the 1993 sham-operated animals recovered ascending pain. As we will see below, this dominance of recovery in ascending pain function was discovered once again in human patients treated with OFS.

TREATING THE HUMAN SPINAL INJURY WITH APPLIED VOLTAGES

Treatment with OFS involves the surgical implantation of an OFS medical device in the subacute period of the injury (within approximately the first month postinjury), and its removal approximately fourteen to sixteen weeks later (Figures 11.4 and 11.5).

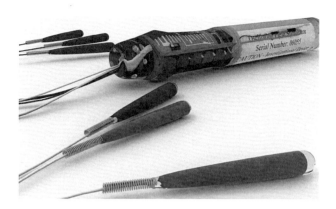

FIGURE 11.5 Close-up detail of an OFS implant as used in human spinal cord injury. The coils on each electrode are bare platinum iridium. The junctions to the pacemaker cables are first coated with an insulating compound and then covered with a medical grade Silastic. These insulated splices are useful for suturing to facets of the vertebral column.

Typical of an FDA phase 1 clinical study, the OFS trial was a single medical center study with a restricted number of patients (initially ten; one was lost to follow-up at the one-year checkup), and without a sham-treated comparator group. Thus, claims concerning efficacy were not made in that report. However, it is useful to point out that the individuals chosen for this first trial were all neurologically complete patients—most with cervical injuries (see Table 11.1).

These patients would not be expected to gain any function back whatsoever, while rarely a small subgroup might gain back one vertebral segment of function, or less, below the neurological level of the injury. We will explore this more thoroughly below when evaluating the historical clinical literature on recovery from SCI.

All patients received a thorough neurological examination using the American Spinal Cord Injury Association neurological scoring method (Figure 11.6). This is a standardized neurological test used to determine the prognosis for any one patient—and a tool to compare outcomes from different experimental therapies. The reader should acquaint himself or herself with this examination.

AMERICAN SPINAL INJURY ASSOCIATION (ASIA) SCORING

This analysis of the severity of a SCI was developed by the American Spinal Injury Association for all patients with SCI. It provides some baseline data on the minimal functionality of a patient as assessed by a trained examiner or neurologist. The main components tested are relative strength of ten muscles on each side of the body and pinprick and light touch discrimination at sensory dermatomes on each side of the body, left and right. Refer to the ASIA score sheet (Figure 11.6). These individual scores are given to each sensory point: 0, if the sensation is absent; 1, if the sensation is detectable but impaired; and 2, if the sensation is normal. Note that 66 points are possible on one side of the body; a maximum of 112

TABLE 11.1

Patient	Sex	Age	Injury Date	Pre-Op ASIA Exam Date[1]	Days to Pre-OP ASIA[2]	Implant Date[3]	Days to OFS[4]	Explant Date	Injury Level	Cause[5]	Int. Fix.[6]
B W01	M	20	3/17/2001	3/28/2001	11	3/29/2001	12	7/10/2001	T1-T2	8MB	Yes
D V02	F	41	3/24/2001	4/10/2001	17	4/10/2001	17	8/3/2001	C6-C7	MVA	Yes
D P05	M	23	4/22/2011	5/3/2001	11	5/4/2001	12	8/17/2001	T5-T6	ATVA	Yes
SES	F	17	5/19/2001	5/24/2001	5	6/1/2001	13	9/14/2001	T9	ATVA	Yes
MLC03	M	27	10/7/2001	10/22/2001	15	10/23/2001	16	2/5/2002	T4-T5	MVA	Yes
RLZ07	M	21	1/19/2002	1/31/2002	12	2/5/2002	17	5/21/2002	C5	Diving	Yes
D C06	M	43	8/26/2002	9/5/2002	10	9/6/2002	11	12/20/2002	C6-C7	Fall	Yes
L L08	M	27	9/12/2002	9/26/2002	14	9/27/2002	15	1/10/2003	C5	MVA	Yes
D E04	M	23	10/24/2002	11/5/2002	12	11/8/2002	15	2/21/2003	T10	Fall	Yes
BA109	M	22	11/17/2002	11/22/2002	5	11/26/2002	9	3/11/2003	T8-T9	MVA	No
JGR10	M	18	12/26/2002	1/8/2003	13	1/10/2003	15	4/25/2003	C5	Viol.	No
BJM11	F	31	12/10/2005	12/17/2005	7	12/20/2005	10	4/11/2006	C6-C7	MVA	Yes
NAG12	M	19	1/23/2006	2/6/2006	14	2/7/2006	15	5/30/2006	C6-C7	MVA	Yes
NJH13	F	23	3/1/2006	3/16/2006	15	3/16/2006	15	7/1/2006	C6-C7	MVA	Yes
Average		25.4			11.5		13.7				

would be attained by a normal, uninjured person for both sides. The presence or absence of anal sensation is also usually determined. In addition, five important muscles of the upper limb and five found in the lower limb are the basis for the motor score in the ASIA assessment. They are graded by a 0–5 scale, where 0 is total lack of response from the muscle. A score of 5 indicates normal muscle action, including range of motion and relative strength against a resistance provided by the examiner. Thus, a maximum score of 50 for the ten muscles tested on one side of the body would be typical of an uninjured person, and a maximal score of 100 for both sides. It is important to point out that both motor and sensory scores may not mirror each other on each side of the injured person's body, and most often they are not the same.

THE HUMAN OFS STUDIES

Table 11.1 provides some data on patients; note that only nine patients completed the one-year period of study as reported in Shapiro et al. (2005). A total of fourteen are discussed here. I thank Cyberkinetics Corporation (Foxborough, Massachusetts) for the use of some unpublished data on the additional patients and their analysis.

The ASIA test is first given to establish if patients have met the entry requirements. They had to be neurologically complete, scoring 0 below the level of injury. This is performed within two days of the expected implantation date (eleven to eighteen days postinjury). It is very important to point out that this delay in implantation allows all patients to recover whatever functions may spontaneously arise; for example, spinal shock subsides soon after decompressive surgery. Given most injuries were high thoracic/cervical, the more strict definition of "complete," including the absence of anal sphincter tone, was not performed (refer to Table 11.1).

It is well understood that for a population of patients graded acutely—during the first forty-eight hours—significant numbers (often >20%) improve in grade within the first week after surgery. The initial assessment defines a patient as ASIA class A if there is no function to be measured below the injury plane, and ASIA class B if there is some motor or sensory function preserved. Grades of ASIA B, C, and D are called incomplete injuries. In the latter grades, some patients may leave the hospital walking. In our case, any spontaneous recovery would have already occurred, as patients were graded and entered into the study weeks later. Also, consultation with Table 2 in the original report (Shapiro et al. 2005) is very useful. For example, two individuals recovered twenty-three levels of vertebral function in pinprick pain scores and light touch. One of these two people scored 111 out of a possible 112 on the ASIA examination at the one-year follow-up. The other, who gained over twenty vertebral segments of sensory appreciation, scored 78 because some of the dermatomes tested were functional but not found to be completely normal in responsiveness. Recoveries of this magnitude are not to be found in similar patients in the literature. As will be discussed, the majority of patients recovered varying and useful ascending function.

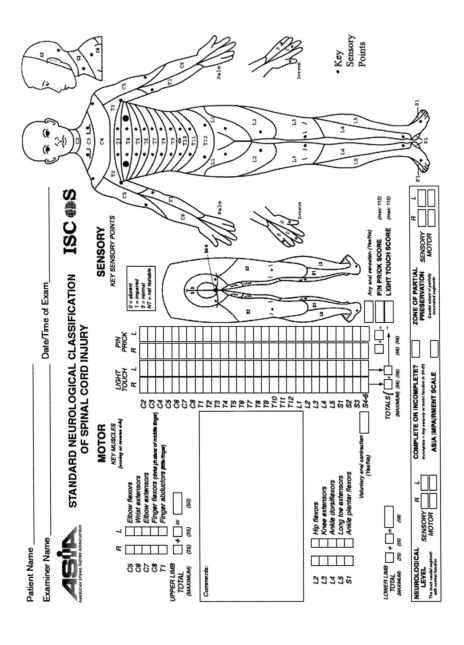

Motor improvement was no better—and no worse—than what has been observed and reported in control groups of other studies.

FUNCTIONAL RECOVERY

Post Hoc Evaluation

Since the publication of that data (Shapiro et al. 2005) significant advancements have become available to permit a more thorough prediction of the frequency of spontaneous recovery in control groups used in SCI clinical investigations (Fawcett et al. 2007; Lammertse et al. 2007).

ICCCP Analysis

Recent advancements in the SCI literature provide a meta-analysis of the recovery from spinal cord injury. The interested reader should refer to *Guidelines for the Conduct of Clinical Trials for Spinal Cord Injury as Developed by the ICCP Panel: Spontaneous Recovery after Spinal Cord Injury and Statistical Power Needed for Therapeutic Clinical Trials* (pages 190–205) and *Guidelines for the Conduct of Clinical Trials for SCI as Developed by the ICCP Panel: Clinical Trial Outcome Measures* (pages 206–221, Fawcett et al. 2007). Spontaneous recovery after surgery was found to be usually restricted to the first spinal segment below the neurological level of the injury, where some preservation of motor function exists (the zone of partial preservation (ZPP)). Recovery of function below this level is extremely uncommon (Fawcett et al. 2007; Lammertse et al. 2007; Steeves et al. 2007). These data supported the common, but unproven, observation that severe and neurologically complete individuals can expect no more than about one vertebral level of improvement. Also of interest to the critical reader are data obtained from the Syngen® placebo group. This randomized, multiple-center, and double-blind trial tested the efficacy of Syngen—a Ganglioside GM 1 sodium salt—compared to a placebo control group (Geisler et al. 2001a, 2001b). A credible comparison would be to compare four-week and six-week scores (Syngen and OFS, respectively) to their one-year data points.

An additional bias against OFS treatment occurs when comparing groups using the ASIA motor scoring system. This difficulty is defined as a ceiling effect. The ASIA motor assay is discontinuous along the body axis as thoracic and abdominal regions are not scored (Figure 11.6). This creates a ceiling where no greater than fifty ASIA motor points can be achieved, unless recovered function extends into the lower limbs. Thus, a T2 – L1 level of motor recovery would not be revealed. To help reconcile this artifact, subsets of placebo patients matched

---

FIGURE 11.6   (*See facing page*) An ASIA score sheet. These individual scores are given to each sensory point: 0, if the sensation is absent; 1, if the sensation is detectable but impaired, and 2, if the sensation is normal. Note that 66 points are possible on one side of the body; a maximum of 112 would be attained by a normal, uninjured person for both sides. In addition, for motor responses, a top score of 100 is possible, totaling scores for each side of the body.

to OFS patients framed the *post hoc* conclusions. I provide a qualitative picture of three cases, including two of the best and worst patient responders from the original OFS group data (Shapiro et al. 2005) (Figures 11.7 to 11.9).

These illustrations were prepared from patient analysis, medical records, and their ASIA scores. Though graded at the worst responder showing few detectable changes, the patient illustrated in Figure 11.7 began to reveal steady and growing lower limb movement and support after one year, allowing the occasional use of a walker, as well as improvements in sexual performance. The significant sensory improvement in the best of the people matched ASIA sensory scores of 111/112 (patient in Figure 11.8) and 78/112 (patient in Figure 11.9; both discussed above).

We do not claim efficacy of OFS relative to motor improvement. Using these new tools available to us, superiority of one group compared to the other in motor scores was not observed. However, this comparison was useful in that it proves there was no detrimental effect of OFS on motor function.

Of most interest were the results of matched comparison of OFS-treated and Syngen placebo patients (unpublished data). The number of OFS-treated patients that improved above the median was highly significant. Most OFS-treated patients scored in the highest of their percentiles in the two sensory modalities tested. For example, when the fourteen OFS-treated patients were matched to a group of fourteen Syngen placebo OFS patients, pinprick scores were above the 50th percentile, and seven were above the 88th percentile. For light touch, eleven patients were above the 50th percentile, and seven were above the 81st percentile. The OFS-treated group sensory scores were more than double or triple those observed in the Syngen placebo group (unpublished data).

A similar dominance of electrically treated patients over the Syngen placebo group was also calculated when applying the guidelines established by Fawcett et al. (2007). OFS cervical, thoracic, and combined groups were statistically improved relative to the expected results from spontaneous recovery (Fawcett et al. 2007).

Finally, only one of the fourteen OFS-treated patients complained of chronic neurogenic pain. Given a nominal frequency of such pain in neurologically complete injuries (~50 to 75%; Nepomuceno et al. 1979; Stover et al. 1995), this would represent a significant loss in sometimes untreatable pain in the OFS-treated patients of ~$p = 0.004$ (Fisher's exact test).

What Is the Status of OFS Therapy for Victims of SCI?

Following the success of a "human use" OFS device in canine paraplegics (Borgens et al. 1999), an FDA application for an investigational device exemption (IDE) was submitted by myself and Professor Scott Shapiro. The request was granted and the first human OFS implantations were begun and reported in 2005. Given the success of this pilot study—and the fact that the investigation was completely free of side effects or surgical issues—a decision was made to pursue commercialization by requesting a humanitarian use device (HUD) classification from the FDA in 2007. This request was based on the low number of severely

Before

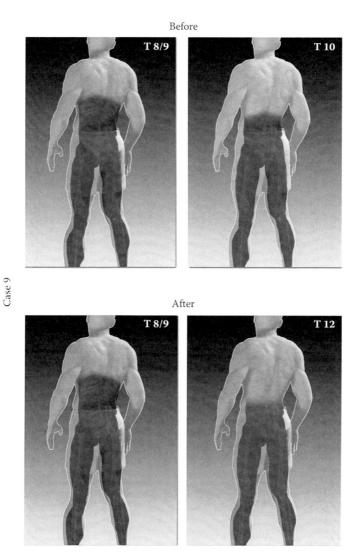

Case 9

After

FIGURE 11.7 Patient 9. Of those cases that responded to OFS, the least recovered patient was number 9. A car accident left him with a severe injury between the eighth and ninth thoracic vertebra. The left panel represents motor function and the right panel represents sensory function. His recovered sensory function, unlike the other patients, was minimal. His motor improvement at the end of one year was also minimal, though he could lift his legs from a sitting position a few inches. This modest motor ability has improved since the end of the trial. Patient 9 can now lift both legs to a position parallel to the ground when he is in his wheelchair. The left leg is stronger, and he can lift it without uncontrollable shaking (spasticity), while the right leg is lifted with more effort and with spasticity. Continued improvement in the right leg would make this patient a candidate for a walker. Not shown is that patient 9 can also curl and wiggle his toes.

Before

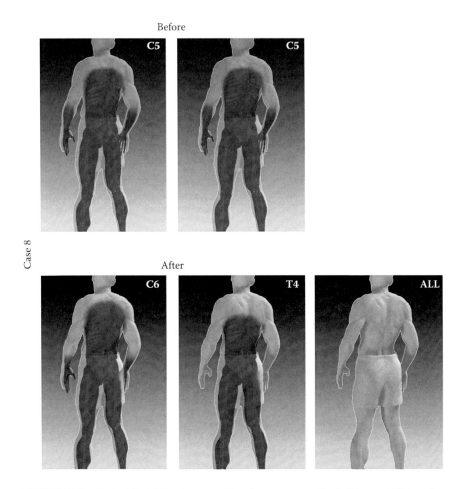

Case 8

After

FIGURE 11.8   Patient 8. A C5 injury resulting from a car accident. His capabilities after OFS treatment would be similar in part to patient 7 (see Figure 11.9). His motor recovery allowed him more independence, not requiring an attendant to move in and out of bed or wheelchairs. Modest recovery in the hand allowed him to transition from a motor-powered to manual wheelchair. Patient 8's sensory recovery to sharp pinprick was modest, recovering about seven vertebral segments to the level of the fourth thoracic vertebra. His recovery to light touch sensation was dramatic (a twenty-three vertebral segment gain), though this ability to sense was somewhat impaired.

injured SCI patients annually (≤7,000), and the fact that there is no modern treatment available to them. The HUD classification was granted by the FDA for OFS. An application for a humanitarian use exemption number for continued human treatment was then applied for (called an HDE number). For more than 2½ years this request has not yet been granted by the FDA, for defensible reasons. This is an unacceptable position by the FDA in the face of the safety record established in the steadily building number of human patients implanted, and the complete lack

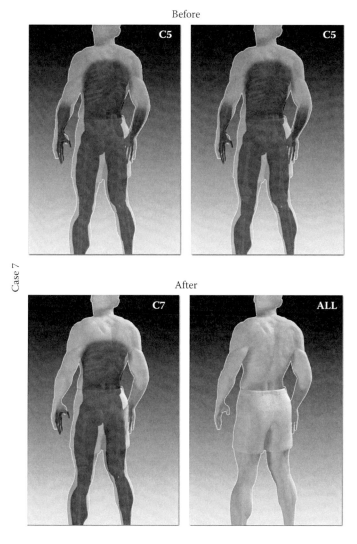

Before

Case 7

After

FIGURE 11.9 Patient 7. A complete injury to the fifth cervical vertebra from a diving accident. His motor function returned in his lower forearm, as well as his hand function, and he is likely able to feed himself. The recovery of sensation was dramatic, scoring as would a normal person. This would make him aware of his lower torso, allowing him to reposition himself to avoid pressure ulcers, reduce the possibility of burns and bone fractures, and have the ability to feel oncoming health problems, such as bladder infections (common among spinal injured patients).

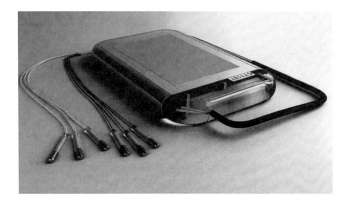

FIGURE 11.10 Titanium-cased manufactured OFS stimulator from a design by Cyberkinetics Corp.

of side effects observed in over three hundred dogs implanted and followed for up to one year postsurgery. In the intervening years of this "noncontroversy" up to twenty thousand catastrophically injured patients could have been treated, as the manufactured version of OFS has been available (Figure 11.10).

We are hopeful that the FDA device panel will see their significant mistakes and misjudgments and reevaluate the error of their judgments, and the human cost in delaying this harmless and effective treatment. They could easily rectify this situation by simply granting the HDE number (after the HUD classification was approved by the agency) so that the work can move ahead.

Finally, the OFS design that has been thus far used in human SCI is more than twenty years old. This is what comes of being locked into a design under examination by the FDA.

Even simple modernizations, such as the use of telemetry, would lead to a demand for yet another human clinical trial at a cost of many millions of dollars, and a delay in years prior to making this treatment generally available to the public. The unwarranted conservatisms and negativisms of the FDA with regard to catastrophic injury to the brain and spinal cord are the single biggest block to modernization of treatment in this subdiscipline of neurosurgery/neurology for which there is no modern remedy. Few corporations will move forward on such costly ventures given the current state of the FDA impediment to progress.

With regard to new embodiments for OFS, we have dropped the use of titanium in our currently manufactured devices (in the bioengineering laboratory of the CPR) bound for human experimental testing. This is because since the units only remain in place within the patient for ~3 months before explanation, there is no justifiable reason to demand titanium-shelled units similar to pacemakers (which remain in place for many years), which also obviates simple and low-cost telemetry. We have already completed such new designs using FDA- approved medical polymers and elastomers complete with a low-cost and dependable telemetry. Initial testing will likely begin implantations in dogs in 2010.

We continue to move on with further modernizations in parallel with pursuing FDA negotiations driving to obtaining our HDE number. All of the above comments relative to the end process of making a technology available to the public should present to the interested reader an opinion of mine, shared by many. Creativity, hard work, and modern design within the investigator's laboratory are the most enjoyable and rapidly progressing phase of this entire process. The interminable stalling by federal oversight is the most vexing, unexplainable, and unjustified part of the "pipeline." The young scientist/entrepreneur should take notice of this and prepare years in advance for such unacceptable intervention. We can only hope for the benefit of patients doomed to lifelong and terrible behavioral loss that this will change.

REFERENCES

Barnard, J. W., and W. Carpenter. 1950. Lack of regeneration in spinal cord of rat. *J Neurophysiol* 13(3):223–28.

Basso, D. M., M. S. Beattie, et al. 1995. A sensitive and reliable locomotor rating scale for open field testing in rats. *J Neurotrauma* 12:1–21.

Blight, A. R., M. E. McGinnis, et al. 1990. Cutaneus trunci muscle reflex of the guinea pig. *J Comp Neurol* 296(4):614–33.

Bohnert, D. M., S. Purvines, et al. 2007. Simultaneous application of two neurotrophic factors after spinal cord injury. *J Neurotrauma* 24(5):846–63.

Borgens, R. B. 1986. The role of natural and applied electric fields in neuronal regeneration and development. *Progr Clin Biol Res* 210:239–50.

Borgens, R. B. 1999. Electrically mediated regeneration and guidance of adult mammalian spinal axons into polymeric channels. *Neuroscience* 91(1):251–64.

Borgens, R. B., A. R. Blight, et al. 1986a. Axonal regeneration in spinal cord injury: A perspective and new technique. *J Comp Neurol* 250(2):157–67.

Borgens, R. B., A. R. Blight, et al. 1986b. Transected dorsal column axons within the guinea pig spinal cord regenerate in the presence of an applied electric field. *J Comp Neurol* 250(168):168–80.

Borgens, R. B., A. R. Blight, et al. 1987. Behavioral recovery induced by applied electric fields after spinal cord hemisection in guinea pig. *Science* 238(366):366–69.

Borgens, R. B., A. R. Blight, et al. 1990. Functional recovery after spinal cord hemisection in guinea pigs: The effects of applied electric fields. *J Comp Neurol* 296(634):634–53.

Borgens, R. B., and D. M. Bohnert. 1997. The responses of mammalian spinal axons to an applied DC voltage gradient. *Exp Neurol* 145(2 Pt 1):376–89.

Borgens, R. B., L. F. Jaffe, et al. 1980. Large and persistent electrical currents enter the transected lamprey spinal cord. *Proc Natl Acad Sci USA* 77(2):1209–13.

Borgens, R. B., E. Roederer, et al. 1981. Enhanced spinal cord regeneration in lamprey by applied electric fields. *Science* 213(611):611–17.

Borgens, R. B., J. P. Toombs, et al. 1993. Effects of applied electric fields on clinical cases of complete paraplegia in dogs. *Restor Neurol Neurosci* 5:305–22.

Borgens, R. B., J. P. Toombs, et al. 1999. An imposed oscillating electrical field improves the recovery of function in neurologically complete paraplegic dogs. *J Neurotrauma* 16(7):639–57.

Bray, G. M., M. P. Villegas-Perez, et al. 1987. The use of peripheral nerve grafts to enhance neuronal survival, promote growth, and permit terminal reconnections in the central nervous system of adult rats. *J Exp Biol* 132:5–19.

Britland, S., and C. McCaig. 1996. Embryonic *Xenopus neurites* integrate and respond to simultaneous electrical and adhesive guidance cues. *Exp Cell Res* 226(1):31–38.

Brown, N. O., M. L. Helphrey, et al. 1977. Thoracolumbar disk disease in the dog: A retrospective analysis of 187 cases. *Jaaha* 13(665):665–72.

Chi, N. H., A. Bignami, et al. 1980. Autologous sciatic nerve grafts to the rat spinal cord: Immunofluorescence studies with neurofilament and gliofilament (GFA) antisera. *Exp Neurol* 68(3):568–80.

Cormie, P., and K. R. Robinson. 2007. Embryonic zebrafish neuronal growth is not affected by an applied electric field *in vitro*. *Neurosci Lett* 411(2):128–32.

Davenport, R. W., and C. D. McCaig. 1993. Hippocampal growth cone responses to focally applied electric fields. *J Neurobiol* 24(1):89–100.

David, S., and A. J. Aguayo. 1981. Axonal elongation into peripheral nervous system "bridges" after central nervous system injury in adult rats. *Science* 214:931–33.

Davies, S. J., M. T. Fitch, et al. 1997. Regeneration of adult axons in white matter tracts of the central nervous system. *Nature* 390(6661):680–83.

Davies, S. J., D. R. Goucher, et al. 1999. Robust regeneration of adult sensory axons in degenerating white matter of the adult rat spinal cord. *J Neurosci* 19(14):5810–22.

Fawcett, J. W., A. Curt, et al. 2007. Guidelines for the conduct of clinical trials for spinal cord injury as developed by the ICCP panel: Spontaneous recovery after spinal cord injury and statistical power needed for therapeutic clinical trials. *Spinal Cord* 45(3):190–205.

Feigin, I., E. H. Geller, et al. 1951. Absence of regeneration in the spinal cord of the young rat. *J Neuropathol Exp Neurol* 10(4):420–25.

Foerester, A. P. 1982. Spontaneous regeneration of cut axons in adult rat brain. *J Comp Neurol* 210(4):335–56.

Geisler, F. H., W. P. Coleman, et al. 2001a. Measurements and recovery patterns in a multicenter study of acute spinal cord injury. *Spine* 26(24 Suppl):S68–86.

Geisler, F. H., W. P. Coleman, et al. 2001b. The Syngen multicenter acute spinal cord injury study. *Spine* 26(24 Suppl):S87–98.

Goldberger, M. E., B. S. Bregman, et al. 1990. Criteria for assessing recovery of function after spinal cord injury: Behavioral methods. *Exp Neurol* 107(2):113–17.

Heinicke, E. A. 1980. Vascular permeability and axonal regeneration in tissues autotransplanted into the brain. *Acta Neuropathol* 49(3):177–85.

Heinicke, E. A., and J. A. Kiernan. 1978. Vascular permeability and axonal regeneration in skin autotransplanted into the brain. *J Anat* 125(Pt 2):409–20.

Hinkle, L., C. D. McCaig, et al. 1981. The direction of growth of differentiating neurones and myeloblasts from frog embryos in an applied electric field. *J Physiol* 314(121):121–35.

Hoerlein, B. F. 1978. The status of the various intervertebral disc surgeries for the dog in 1978. *Jaaha* 14(563):563–70.

Hoerlein, B. F. 1979. Comparative disc disease: Man and dog. *Jaaha* 15(535):535–45.

Horner, P. J., and F. H. Gage. 2000. Regenerating the damaged central nervous system. *Nature* 407(6807):963–70.

Kao, C. C. 1974. Comparison of healing process in transected spinal cords grafted with autogenous brain tissue, sciatic nerve, and nodose ganglion. *Exp Neurol* 44(3):424–39.

Kao, C. C., and L. W. Chang. 1977. The mechanism of spinal cord cavitation following spinal cord transection. *J Neurosurg* 46:197–209.

Kao, C. C., L. W. Chang, et al. 1977. Axonal regeneration across transected mammalian spinal cords: An electron microscopic study of delayed microsurgical nerve grafting. *Exp Neurol* 54(3):591–615.

Keirstead, S. A., M. Rasminsky, et al. 1989. Electrophysiologic responses in hamster superior colliculus evoked by regenerating retinal axons. *Science* 246(4927):255–57.

Lammertse, D., M. H. Tuszynski, et al. 2007. Guidelines for the conduct of clinical trials for spinal cord injury as developed by the ICCP panel: Clinical trial design. *Spinal Cord* 45(3):232–42.

le Gros Clark, W. E. 1942. The problem of neuronal regeneration in the central nervous system. I. The influence of spinal ganglia and nerve fragments grafted in the brain. *J Anat* 77(1):20–418.413.

McCaig, C. D. 1987. Spinal neurite reabsorption and regrowth *in vitro* depend on the polarity of an applied electric field. *Development* 100(1):31–41.

McCaig, C. D., A. M. Rajnicek, et al. 2002. Has electrical growth cone guidance found its potential? *Trends Neurosci* 25(7):354–59.

McCaig, C. D., A. M. Rajnicek, et al. 2005. Controlling cell behavior electrically: Current views and future potential. *Physiol Rev* 85(3):943–78.

Muller, K. J., J. G. Nicholls, et al. 1981. *Neurobiology of the leech.* Cold Spring Harbor, NY: Cold Spring Harbor Laboratory.

Nepomuceno, C., P. R. Fine, et al. 1979. Pain in patients with spinal cord injury. *Arch Phys Med Rehabil* 60(12):605–9.

Pan, L., and R. B. Borgens. 2010. Perpendicular organization of sympathetic neurons within a required physiological voltage. *Exp Neurol* 222(1):161–4.

Patel, N., and M. M. Poo. 1982. Orientation of neurite growth by extracellular electric fields. *J Neurosci* 2(483):483–96.

Rajnicek, A. M., N. A. Gow, et al. 1992. Electric field-induced orientation of rat hippocampal neurones *in vitro*. *Exp Physiol* 77(1):229–32.

Ramon y Cajal, S. 1928. *Degeneration and regeneration of the nervous system.* London: Oxford University Press.

Robinson, K. R., and P. Cormie. 2008. Electric field effects on human spinal injury: Is there a basis in the *in vitro* studies? *Dev Neurobiol* 68(2):274–80.

Roederer, E., N. H. Goldberg, et al. 1983. Modification of retrograde degeneration in transected spinal axons of the lamprey by applied DC current. *J Neurosci* 3(1):153–60.

Rovainen, C. M. 1967. Physiological and anatomical studies on large neurons of central nervous system of the sea lamprey (*Petromyzon marinus*). I. Muller and Mauthner cells. *J Neurophysiol* 30(5):1000–23.

Rovainen, C. M. 1974. Synaptic interactions of reticulospinal neurons and nerve cells in the spinal cord of the sea lamprey. *J Comp Neurol* 154(2):207–23.

Schlaepfer, W. W. 1974. Calcium-induced degeneration of axoplasm in isolated segments of rat peripheral nerve. *Brain Res* 69:203–15.

Shapiro, S., R. Borgens, et al. 2005. Oscillating field stimulation for complete spinal cord injury in humans: A phase 1 trial. *J Neurosurg Spine* 2(1):3–10.

Shi, Y., S. Kim, et al. 2009. Effective repair of traumatically injured spinal cord by nanoscale block copolymer micelles. *Nat Nano* 5(1):80–87.

Steeves, J. D., D. Lammertse, et al. 2007. Guidelines for the conduct of clinical trials for spinal cord injury (SCI) as developed by the ICCP panel: Clinical trial outcome measures. *Spinal Cord* 45(3):206–21.

Stover, S. L., J. A. DeLisa, et al., eds. 1995. *Spinal cord injury: Clinical outcomes from the model systems.* Gaithersburg, MD, Aspen Publishers.

Strautman, A. F., R. J. Cork, et al. 1990. The distribution of free calcium in transected spinal axons and its modulation by applied electrical fields. *J Neurosci* 10(11):3564–75.

Sugar, O., and R. W. Gerard. 1940. Spinal cord regeneration in the rat. *J Neurophysiol* 3(1):1–19.

Theriault, E., and J. Diamond. 1988a. Intrinsic organization of the rat cutaneus trunci motor nucleus. *J Neurophysiol* 60(2):463–77.

Theriault, E., and J. Diamond. 1988b. Nociceptive cutaneous stimuli evoke localized contractions in a skeletal muscle. *J Neurophysiol* 60(2):446–62.

Toombs, J. P., and M. S. Bauer. 1993. Intervertebral disk disease. In *Textbook of Small Animal Surgery*, ed. D. Slatter, 1070–87. Philadelphia: W.B. Saunders Company.

Vidal-Sanz, M., G. M. Bray, et al. 1987. Axonal regeneration and synapse formation in the superior colliculus by retinal ganglion cells in the adult rat. *J Neurosci* 7(9):2894–909.

Wood, M. R., and M. J. Cohen. 1979. Synaptic regeneration in identified neurons of the lamprey spinal cords. *Science* 206(4416):344–47.

Zuberi, M., P. Liu-Snyder, et al. 2008. Large naturally-produced electric currents and voltage traverse damaged mammalian spinal cord. *J Biol Eng* 2(1):17.

12 Bioelectricty of Cancer
Voltage-Gated Ion Channels and Direct-Current Electric Fields

Mustafa B. A. Djamgoz
Division of Cell and Molecular Biology
Neuroscience Solutions to Cancer Research Group
Imperial College London, UK

CONTENTS

INTRODUCTION

Cancer, the leading cause of death in the Western world, results from abnormal expression of genes. Such abnormality can be genetic, involving mutations in proto-oncogenes or tumor suppressor genes, and their heterogeneous accumulation, leading to uncontrolled growth, as in primary tumorigenesis. In the majority of cancers, however, the changes are epigenetic; i.e., it is the abnormal expression levels of otherwise normal genes that lead to cells losing their normal controlled behavior. Epigenetic changes result from hypo/hypermethylation of DNA, leading to dysregulation of gene expression levels or patterns (Hanahan and Weinberg, 2000). Importantly, metastasis—spread of cells from the primary neoplasm ultimately to form secondary tumors—is what causes the majority (>90%) of cancer-related deaths (Nguyen and Massague, 2007; Hunter et al., 2008). Metastasis is a complex, multistage process in which cancer cells (1) disassociate from each other and detach from the extracellular matrix, (2) migrate through the basement membrane and invade local tissue, (3) enter the blood or lymph circulation (in which the survival of the circulating tumor cells is rate limiting), (4) reattach at a distant site and extravasate, and (5) form secondary tumors after proliferation and induction of angiogenesis (Figure 12.1; Fidler, 2002a, 2002b, 2003; Hunter et al., 2004; Bacac and Stamenkovic, 2007). Importantly, primary tumorigenesis and metastasis can be controlled by separate sets of molecular and cellular factors (Hanahan and Weinberg, 2000; Welch et al., 2000, 2004, 2006). Thus, at least eleven metastasis suppressor genes have been identified that do not affect primary tumorigenesis (Berger et al., 2004; Weigelt et al., 2005; Vaidya and Welch, 2007). Furthermore, metastatic potential can be "preprogrammed" and modulated progressively by dynamic response to the local microenvironment (Fidler, 2003; Hunter, 2004; Hunter et al., 2008). The independence (at least, partial) of primary vs. secondary tumorigenesis is consistent with the apparent absence of an identifiable primary tumor in a subset of metastatic disease cases.

Despite this inherent complexity, however, a reductionist approach can be adopted to initially describe the metastatic cascade as an interrelated series of basic but well-coordinated cellular behaviors, including motility, secretion, adhesion/detachment, and gene expression (Bacac and Stamenkovic, 2007). Accordingly, considerable effort has been made to determine the genes → proteins → signaling mechanisms involved in each of these cellular processes (Schwirzke et al., 1999; Weigelt et al., 2005). The ultimate goal of this research is to develop accurate and efficient novel prognostic/diagnostic markers and long-lasting therapies against metastatic disease.

At present, there are significant problems in the clinical management of major cancers, including the most common carcinomas (epithelial cancers), breast cancer (BCa) in women and prostate cancer (PCa) in men (Jemal et al., 2008; Ventura and Merajver, 2008; Gurel et al., 2008; Ramirez et al., 2008). Toward this end, an emerging and promising new field of investigation involves the functional expression of voltage-sensitive/gated ion channels (VGICs). Classically, VGICs are found in so-called excitable cells, such as neurons and muscle cells, where they are responsible for action potential generation and conduction (Hille, 1992). However, it has become apparent that these channels are also present in traditionally nonexcitable cell types, including glia, fibroblasts, and endothelial cells, where their functions are less well understood (Diss et al., 2004). Moreover, increasing evidence suggests that VGICs become abnormally expressed in metastatic carcinomas (Diss et al., 2004; Fiske et al., 2006; Roger et al., 2006; Prevarskaya et al., 2007; Conti, 2007; Fraser and Pardo, 2008; Arcangeli et al., 2009). More generally, ion channels are well known to control the kinds of cellular activities integral to the metastatic cascade (Hille, 1992). Voltage-gated sodium channels (VGSCs) have been found to be strongly associated with metastasis; these channels are expressed at high levels in metastatic carcinomas and control (potentiate) various steps of the metastatic cascade, including adhesion, migration, secretion,

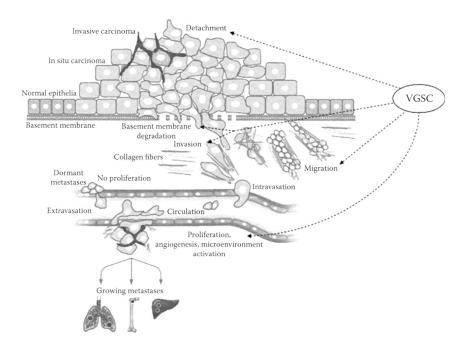

FIGURE 12.1   See color insert. The metastatic cascade. Transformation of normal epithelial cells results in carcinoma *in situ*. Reduced adhesiveness and enhanced migratory behavior of tumor cells progress the disease to an invasive stage. After degradation of the basement membrane, tumor cells invade the surrounding stroma, migrate, and intravasate into the lymph or blood circulation, and surviving cells arrest in the capillaries of a distant organ. Formation of secondary tumors occurs after proliferation, induction of angiogenesis, and microenvironment activation. Voltage-gated sodium channels (VGSCs) have been shown to be involved in controlling various components of the metastatic cascade, as indicated by the arrows. (Modified from Bacac and Stamenkovic, *Annu. Rev. Pathol.*, 3, 221–47, 2007.)

and invasion (Figure 12.1; e.g., Grimes et al., 1995; Roger et al., 2003, 2006; Gillet et al., 2009; Fraser et al., 2005; Brackenbury et al., 2007). Consequently, VGSCs are becoming an exciting new class of targets for determination of metastatic potential or suppression of its progression (Roger et al., 2006; Fiske et al., 2006; Le Guennec et al., 2007; Onkal and Djamgoz, 2009). Ion channel, including VGSC, defects (channelopathies) are already known to be involved in several other pathophysiological conditions, including cardiac arrhythmias, periodic muscle paralysis, epilepsy, migraine, neuropathic pain, and multiple sclerosis (Hoffman, 1995; Waxman, 2001, 2007; Marban, 2002; Meisler et al., 2002; Roberts and Brugada, 2003; Diss et al., 2004; Ashcroft, 2006).

Expression of VGSCs also occurs in cancer cells *in vivo* and correlates with the tissue pathology/metastatic potential (Diss et al., 2005; Fraser et al., 2005; Abdul and Hoosein, 2002, 2006). Importantly, VGICs/VGSCs *in vivo* would be subject to local direct-current electric fields (dcEFs) that can arise by at least two distinct

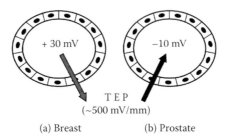

TEP
(~500 mV/mm)

(a) Breast   (b) Prostate

FIGURE 12.2 Typical transepithelial potentials (TEPs) of glandular ducts of human breast (A) and rat prostate (B). The lumen potentials are of opposite signs in the two cases, +30 mV (breast) and –10 mV (prostate), thus generating dcEFs (arrows) of opposite directions. However, since the galvanotaxes of breast and prostate cells are anodal and cathodal, respectively, the effect of the dcEF is to facilitate movement into the lumen in both cases. Original data from (A) Faupel et al. (1997) and (B) Szatkowski et al. (2000). (Adapted from McCaig et al., *J. Cell Sci.*, 122, 4267–76, 2009.)

mechanisms. First, in the central nervous system, the dense packing of glial cells (which can outnumber neurones by 10:1) and the associated blood-brain barrier can generate local standing field potentials (SFPs) of a few mV (Jefferys, 1995). However, under particular pathophysiological conditions (e.g., hypoxia, spreading depression), such SFPs can reach 10 to 15 mV (Somjen, 2001). Somewhat surprisingly, the cleft of chemical synapses is thought to be associated with dcEFs of ~10^4 mV/mm (Eccles and Jaeger, 1958; Sylantyev et al., 2008). Second, in epithelia, dcEFs of ~500 mV/mm are inherent to the cells' polarized membrane structure and chemical segregation of apical (lumen) and basal (blood) compartments, the direction of the field depending upon the gland (Figure 12.2; Faupel et al., 1997; Szatkowski et al., 2000).

In this chapter, we aim to give an overview of the bioelectricity of cancer, dealing with both basic and applied aspects. Bioelectricity is a wide field. In relation to cancer/carcinomas, the most relevant subtopics are membrane potential, voltage-gated ion channels, transporters, cell-surface charge, dcEFs, body electrolytes (tissue and fluid), and bioimpedance (Figure 12.3). A typical membrane potential (~$|70|$ mV) is equivalent to a voltage gradient of ~$|10^7|$ V/m, and a typical VGIC can permeate ions at a rate of ~10^4 ions/ms (in single file). It is to be expected, therefore, that such bioelectrical characteristics would have a profound effect upon cellular behavior in health and disease, including cancer. The specific aims of this chapter are (1) to review the ion channels and cancer field, and (2) to incorporate dcEFs and galvanotaxis into the pathophysiology of metastatic disease. These aspects are extended to measurements of cell-surface charge, body electrolytes, and tissue impedance. Throughout, emphasis is upon VGSC expression in PCa and BCa. Other aspects of VGIC/VGSC involvement in cancer have been reviewed before (Diss et al., 2004; Fiske et al., 2006; Roger et al., 2006; Prevarskaya et al., 2007, 2010; Le Guennec et al., 2007; Onkal and Djamgoz, 2009).

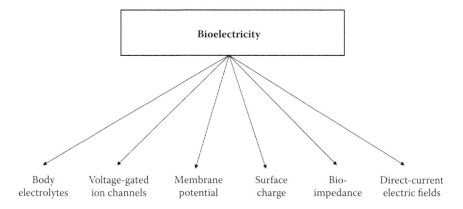

FIGURE 12.3 Aspects of bioelectricity relevant to cancer. Six different bioelectric phenomena, covered in this chapter, are illustrated. Additional aspects, not covered here, would include electrogenic transporters (exchangers and pumps), alternating electric fields, and even G-protein-coupled receptors. (From Mahaut-Smith et al., *Trends Pharmacol. Sci.*, 29, 421–29, 2008.)

EXPRESSION OF VOLTAGE-GATED ION CHANNELS IN CANCER CELLS AND TISSUES: INVASIVENESS VS. PROLIFERATION, AND CELEX HYPOTHESIS

According to the National Cancer Institute (United States), 6,955 genes have been identified from the literature as having an association with cancer (https://cabig.nci.nih.gov/inventory/data-resources/cancer-gene-index). Those genes that directly contribute to the cancer process give rise to a variety of subcellular changes, including an increase in cell-surface negativity and altered permeability (Figure 12.4). Electrophysiological comparison of strongly vs. weakly metastatic cells, by whole-cell patch-clamp recording, revealed that progression from weak → strong metastatic potential involved (1) upregulation of functional VGSC activity, and (2) downregulation of voltage-gated outward (mainly K^+) currents (Figure 12.5). In addition, the resting membrane potential depolarized (Fraser et al., 2005). Such differences were seen clearly in cells of PCa (Grimes et al., 1995; Laniado et al., 1997) and BCa (Fraser et al., 2005). There is also comparable evidence from cancers of the lung (small-cell, non-small-cell, and mesothelioma), ovary, and cervix, and parallels occur *in vivo*.

Taken together, the concomitant changes in inward and outward currents would imply that membranes of strongly metastatic cells are electrically excitable, and indeed, all-or-none regenerative activity (action potentials) can be elicited in such cells by injection of small depolarizing stimuli (Figure 12.6). It is conceivable, therefore, that such excitability is one of the mechanistic factors underlying the hyperactivity of strongly metastatic cells. We refer to this phenomenon as the cellular excitability (CELEX) hypothesis of metastatic disease. This hypothesis

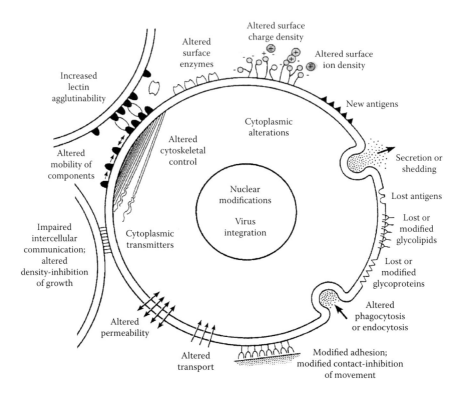

FIGURE 12.4 See color insert. Subcellular changes in cancer. The figure illustrates these for a hypothetical cell. A variety of changes occur that affect the cell surface, cytoplasm, and intercellular junctions. Of particular relevance to this article are the alterations in surface charge density, surface ion density, permeability, and transport. (Modified from Nicolson, *Biochim. Biophys. Acta*, 458, 1–72, 1976.)

states that metastatic cell membranes are geared up for excitation, and this potentiates the hyperactive behavior of aggressive cancer cells.

As regards the molecular nature of these ion channels, a series of experiments revealed the culprit VGSC to be Nav1.5 in BCa (Fraser et al., 2005) and Nav1.7 in PCa (Diss et al., 2001). Importantly, in both cases, the channels were expressed as apparent neonatal splice variants, consistent with the dedifferentiated nature of cancer cells/tissues and (re)expression of normally developmentally regulated (the so-called oncofeotal) genes in cancer (e.g., Biran et al., 1994; Wepsic, 1983; Monk and Holding, 2001). An extension of such a consideration would be consistent with cancer being a disease of stem cells, a concept currently undergoing intense discussion (e.g., Takebe and Ivy, 2010).

A comparative study on the Dunning model of rat PCa gave an interesting insight into the dynamic role of voltage-gated K⁺ channel (VGPC) activity in proliferation vs. invasiveness of weakly vs. strongly metastatic AT-2 and Mat-LyLu cells, respectively (Fraser et al., 2003a). The predominant VGPC in both these

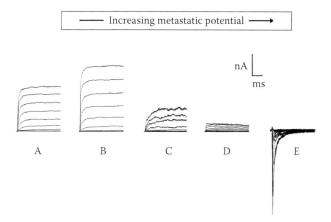

FIGURE 12.5 Voltage-activated whole-cell membrane currents recorded from human breast epithelial cell lines of varying metastatic potential. Recordings from a normal human breast epithelial cell line (MCF-10A) (A) and breast cancer cells with increasing metastatic potential are shown: (B) MDA-MB-453, (C) MCF-7, (D) MDA-MB-435, and (E) MDA-MB-231 cells. The currents were generated by pulsing the membrane potential from a holding voltage of −100 mV, in 5 mV steps, from −60 to +60 mV for 200 ms. Voltage pulses were applied with a repeat interval of 20 s. For clarity, every second current trace generated is displayed. With increasing metastatic potential, voltage-gated outward (K^+) currents are reduced, and inward (Na^+) currents appear only in the strongly metastatic MDA-MB-231 cells (E). Scale bars: 4 nA/50 ms (A,B), 1 nA/50 ms (C,D), and 200 pA/10 ms (E). (Modified from Onkal and Djamgoz, *Eur. J. Pharmacol.*, 625, 206–19, 2009; original data mostly from Fraser et al., *Clin. Cancer Res.*, 11, 5381–89, 2005.)

cells is Kv1.3 (the so-called lymphocyte K^+ channel), which is expressed at a significantly lower level in Mat-LyLu cells (Fraser et al., 2003a). Blocking Kv1.3 activity in the AT-2 cells using margatoxin suppressed proliferation, consistent with the role of VGPC activity in many other cancer and noncancer cells, including lymphocytes (e.g., Blackiston et al., 2009). In contrast, margatoxin had no effect on proliferation of Mat-LyLu cells. It appeared, therefore, that functional VGPC expression was downregulated in strongly metastatic cells so as to promote VGSC activity and excitability in the invasive mode where proliferation was not the critical prerequisite. Accordingly, we can propose basic models of (1) VGSC-dependent control of metastatic cell behavior (MCB) and (2) concerted VGPC and VGSC expression/activities in metastatic progression (Figure 12.7). In this model, proliferation in the primary tumor is controlled mainly by VGPC activity. However, for invasiveness, VGPC expression/activity is reduced while VGSC expression/activity is upregulated, enabling the cells to become excitable and invasive (Figure 12.7B). It is highly likely that this relationship is dynamic and concerted interactions of varying complexity occur during the various stages of the metastatic cascade.

We should note that a variety of other VGICs, particularly EAG1 and HERG, and TRP channels have also been associated with the cancer process

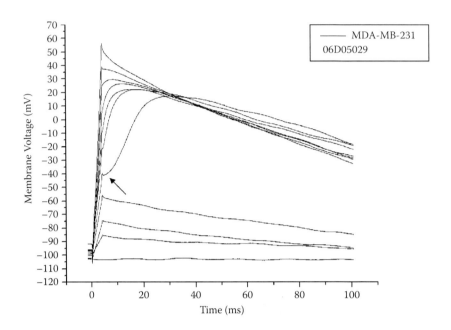

FIGURE 12.6 Current-clamp recording of action potentials from MDA-MB-231 cells. The cell was stimulated from the holding level by currents in the range 0 to 340 pA (20 pA steps, 2 s intervals). Only every other trace is shown for clarity. At the eighth stimulation (indicated by the arrow), the cell fires into an action potential that is repeated for subsequent stimuli in an all-or-none fashion. In this recording, the action potentials appear broad due to these strongly metastatic cells having little outward (e.g., K^+) currents for repolarization. (Previously unpublished recording by Rustem Onkal.)

(Prevarsakaya et al., 2010). Also, some ligand-gated ion channels have been found to be involved in cancer (e.g., Lang and Bastian, 2007), and the balance of excitatory vs. inhibitory neurotransmission has been suggested to influence cancer progression (Schuller, 2008). Mechanosensitive ion channels have received surprisingly little attention, although one would expect extensive mechanosensitive interactions to occur extensively during the metastatic process, especially during extra/intravasation, thought to be a critical stage in the metastatic cascade (Wyckoff et al., 2000). However, these aspects are outside the scope of the present review, as are the role of ion channels in related issues like gene expression and proliferative disease.

FUNCTIONAL CONTRIBUTION OF ION CHANNEL ACTIVITY TO THE CANCER PROCESS

There is substantial evidence, obtained using various independent techniques, for VGSC-mediated control (enhancement) of a range of MCBs.

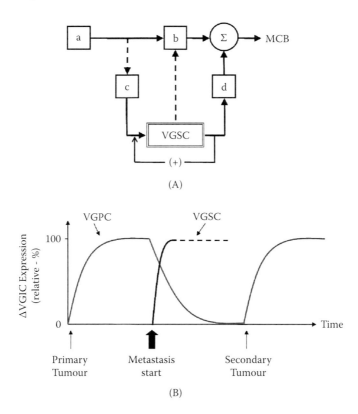

FIGURE 12.7 Models of VGSC-dependent control of metastatic cell behavior (MCB). (A) A primary mechanism (a) may drive MCBs through intermediary signal(s) (b). VGSC upregulation/activity occurs through an intermediary c driven by a, or by an independent mechanism. VGSC-dependent and VGSC-independent mechanisms summate (at Σ) to potentiate metastasis. (B) Schematic diagram showing dynamic behavior of voltage-gated K$^+$ channel (VGPC) and voltage-gated Na$^+$ channel (VGSC) during cancer progression. Initially, during primary tumorigenesis, proliferative activity is driven mainly by VGPCs. The onset of metastasis is accompanied by downregulation of VGPC activity concomitant with VGSC upregulation, together inducing excitability in the cell membrane. Secondary proliferation involves resurgence of VGPC expression/activity.

EFFECTS OF DOWNREGULATING VGSC EXPRESSION/ACTIVITY

The natural toxin tetrodotoxin (TTX) has been used extensively as a highly selective blocker of VGSCs and has been shown to suppress various *in vitro* MCBs in human and rodent cell lines of BCa, PCa, and lung cancer. These VGSC-enhanced cellular behaviors include invasion (Grimes et al., 1995; Laniado et al., 1997; Smith et al., 1998; Roger et al., 2003; Bennett et al., 2004; Brackenbury et al., 2007; Gillet et al., 2009), colony formation in Matrigel-composed three-dimensional matrices (Gillet et al., 2009), cell morphology and process extension

(Fraser et al., 1999), galvanotaxis (Djamgoz et al., 2001; Fraser et al., 2005), lateral motility (Fraser et al., 2003b), transverse (Transwell) migration (Roger et al., 2003; Fraser et al., 2005; Brackenbury and Djamgoz, 2006), reduced adhesion (Palmer et al., 2008), gene expression (Mycielska et al., 2005), endocytic membrane activity (Mycielska et al., 2003; Onganer and Djamgoz, 2005; Nakajima et al., 2009), vesicular patterning (Krasowska et al., 2004), and nitric oxide production (Williams and Djamgoz et al., 2005). Similar data were obtained using pharmacological agents (including clinical drugs) blocking VGSC activity as their main mode of action. Such agents include lidocaine (a local anaesthetic) and phenytoin (an anticonvulsant), which inhibited the VGSC-dependent enhancement of endocytic membrane activity in small-cell lung cancer cell lines (Onganer and Djamgoz, 2005). Phenytoin also significantly reduced the lateral motility of Mat-LyLu rat PCa cells (Fraser et al., 2003b; Sikes et al., 2003). Moreover, phenytoin and carbamezapine (another anticonvulsant) inhibited secretion of prostate-specific antigen (PSA) and interleukin-6 from human PCa cells (Abdul and Hoosein, 2001). Interestingly, even dietary agents blocking VGSC activity (e.g., omega-3 polyunsaturated fatty acids, eicosapentaenoic acid, and docosahexaenoic acid) suppressed VGSC-dependent MCBs of human and rat PCa and BCa cells (Isbilen et al., 2006; Nakajima et al., 2009). Comparable experiments showed that TTX had no effect on weakly metastatic cell lines, consistent with the absence of functional VGSCs in these cells. Interestingly, proliferation of metastatic PCa and BCa cells was not affected by TTX (Fraser et al., 2003b, 2005), implying that VGSCs are involved mainly in metastatic progression. This is another clear example of primary and secondary tumorigenesis being governed by separate sets of molecular determinants (Welch et al., 2004, 2006).

In an alternative approach, downregulating specifically functional neonatal Nav1.5 (SCN5A) expression, using RNAi, suppressed VGSC-dependent migration and invasion in MDA-MBA-231 human BCa cells *in vitro* (Brackenbury et al., 2007; Gillet et al., 2009). Similarly, in human metastatic PCa PC-3 cells, siRNA targeting the predominant VGSCs, Nav1.6 (SCN8A), and Nav1.7 (SCN9A) inhibited endocytosis and invasiveness (Nakajima et al., 2009). Nakajima et al. (2009) also reported that silencing VGSC activity by RNAi had no effect on PC-3 cell proliferation, as shown previously by Fraser et al. (2003b, 2005). These results further support the notion that the mechanisms controlling primary and secondary tumorigenesis are different. Another aspect of VGSC expression that emerged from the siRNA work on BCa is that silencing nNav1.5 protein level by only ~30% was sufficient to eliminate its role in promoting invasiveness, i.e., in all-or-none fashion (Brackenbury et al., 2007). This raised the possibility that functional VGSC expression acts like a switch promoting metastatic behavior to higher gear (Figure 12.7B).

EFFECTS OF UPREGULATING VGSC EXPRESSION/ACTIVITY

Conversely, VGSC openers promoted MCBs; veratridine enhanced galvanotaxis, while aconitine and ATX II increased motility of Mat-LyLu cells (Djamgoz et al.,

2001; Fraser et al., 2003b). However, these effects were minimal, consistent with the view that VGSCs, when functionally expressed, already make a maximal contribution to enhancement of MCBs, i.e., in all-or-none switching fashion, as also noted above.

Transient overexpression of Nav1.4 in a weakly metastatic human PCa cell line (LNCaP) significantly increased invasion, which was reversed by TTX (Bennett et al., 2004). Hence, Bennett et al. (2004) suggested that VGSC activity is "necessary and sufficient" for the invasive behavior of PCa cells. An important implication of this study was that the identity of the VGSC subtype expressed in cancer cells may not strictly be critical. Accordingly, although the culprit VGSC expressed in metastatic human PCa cells is Nav1.7, overexpression of Nav1.4 alone was adequate to enhance invasiveness. This raises the important question of whether the subtype of VGSC expressed is important to invasiveness or whether any VGSC (e.g., simply to allow Na^+ influx into cells) would suffice. In the case of the latter possibility, *in vivo* expression of the different VGSC subtypes in different cancers would just be a matter of tissue origin (given the basic tissue-specific distribution of VGSC subtypes), rather than a strict functional (pathophysiological) prerequisite. Further studies are required to address this important question.

REGULATION OF ION CHANNEL EXPRESSION: ASSOCIATION WITH MAINSTREAM CANCER MECHANISMS

Since functional ion channel expression differs between cancer and corresponding normal cells, and such differences also occur in parallel *in vivo*, it would follow that the channel expression is likely to be an integral part of the cancer process. This view is supported by a number of mainstream cancer mechanisms controlling ion channel expression/activity, and the subsequent impact upon MCBs. Thus, ion channel expression/activity has been shown generally to be significantly controlled by steroid hormones (e.g., Liu et al., 2008; Zhang et al., 2009) and growth factors (e.g., Rose et al., 2004; Goldfarb et al., 2007; Kim et al., 2010). Even some drugs (e.g., tamoxifen) that affect steroid hormone action also modulate ion channel activity (Borg et al., 2002; Tsang et al., 2004; Bolanz et al., 2009). In fact, it is already apparent from cells' resting states that there is a direct association with hormone sensitivity and VGSC expression. Thus, PC-3 (PCa) and MBA-MBA-231 (BCa) cells are both devoid of classic receptors for androgen and estrogen, respectively, and both express functional VGSCs. The converse is also true for corresponding hormone-sensitive cells that do not express functional VGSCs. Interestingly, steroid hormones can also have nontranscriptional (nongenomic) roles (Leung et al., 2007; Prossnitz et al., 2008). Such an effect was recently demonstrated for the potentiating action of estrogen on VGSC activity in MBA-MD-231 cells, possibly involving GPR30 as a cell-surface receptor and decreased adhesion as a functional output (Fraser et al., 2010).

When cancer cells become hormone insensitive, they frequently become dependent on growth factors (Santen et al., 2005, 2008), which are well-known modulators of VGSC expression/function. In particular, epidermal growth factor (EGF) increased VGSC activity and enhanced migration of Mat-LyLu cells (Ding et al., 2008). Similarly, in the strongly metastatic human PC-3M cells, EGF enhanced migration, endocytosis, and invasion via mainly functional Nav1.7 upregulation (Uysal-Onganer and Djamgoz, 2007). Although nerve growth factor (NGF) also upregulated functional VGSC density in plasma membrane of Mat-Ly-Lu cells, via PKA activity, there was no subsequent effect on cellular migration (Brackenbury and Djamgoz, 2007). These results suggest that there are multiple growth factors regulating VGSCs. This is a potentially important notion since there appears to be a multiplicity of expression of growth factor systems in cancer cells, resulting in a new system becoming prominent following the suppression of another. If so, blocking a common downstream mechanism, such as a VGSC, could be an expedient way of controlling growth factor–dependent metastasis. It is also possible that VGSCs are compartmentalized within cancer cells such that different VGSC pools control different MCBs.

VGSC β-subunits are also associated with mainstream cancer mechanisms (Brackenbury et al., 2008). Thus, upregulation of β3 in mouse fibroblasts and human cancer cell lines was dependent on p53, a major tumor suppressor (Adachi et al., 2004). In fact, VGSCβ3 is conspicuous by its absence in PCa (Diss et al., 2007) and BCa (Chioni et al., 2009). This is consistent with cancer cells generally being resistant to apoptosis. VGSCβ1 was shown to function as a cell adhesion molecule in BCa (Chioni et al., 2010). Importantly, it is highly likely that the interaction between cell adhesion and VGSC activity is reciprocal, whereby VGSC α-subunit expression is upregulated transcriptionally when BCa cells are in a state of weak adhesion, e.g., following β1 downregulation (Brackenbury et al., 2008; Chioni et al., 2009). As regards hormonal control, VGSCβ1 expression was found to be downregulated following treatment with an androgen analogue (Diss et al., 2007).

Interestingly, VGSC expression may also be under activity-dependent regulation (ADR) (Figure 12.8). In Mat-LyLu cells, ADR kept Nav1.7 expression upregulated via a positive feedback mechanism involving both increased SCN9A transcription and PKA-mediated VGSC protein trafficking to plasma membrane (Brackenbury and Djamgoz, 2006). There is a comparable situation in human MDA-MBA-231 (BCa) cells (Chioni et al., 2010). Indeed, a variety of protein-protein interactions may be involved in VGSC regulation (Diss et al., 2004; Herfst et al., 2004; Shao et al., 2009). Such positive feedback, which is unusual in ADR of VGSCs, would ensure that maximal functional VGSC expression would be maintained in the plasma membrane. Importantly, on the other hand, it would imply that blocking VGSC activity would suppress VGSC activity *and* expression in the long term. Such a characteristic would be therapeutically expedient and add to the criteria for VGSCs being an ideal target for inducing regression of metastasis.

In overall conclusion, therefore, VGSC expression in carcinomas is not an isolated phenomenon but a part of the overall cancer process in close association with the mainstream mechanisms of cancer (metastasis).

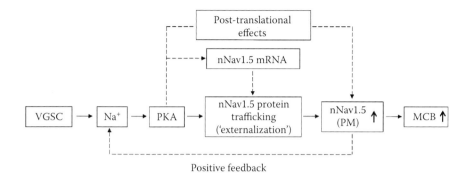

FIGURE 12.8 A model and possible mechanisms of activity-dependent regulation of VGSC expression/activity and control of metastatic cell behavior (MCB). The main pathway is denoted by solid lines and arrows. Activation of PKA by VGSC (nNav1.5)-induced Na+ influx would alter the balance of intracellular VGSC protein trafficking in favor of externalization to plasma membrane (PM), in turn enhancing MCB. This pathway has intrinsic ability to self-sustain by positive feedback. Additional, minor pathways (above broken lines and arrows) include the following: (1) posttranslational effects involving regulatory proteins (e.g., β-subunits) and direct phosphorylation of VGSC protein in PM, and (2) *de novo* mRNA synthesis for possible subsequent translation to protein. (Modified from Chioni et al., *Int. J. Biochem. Cell. Biol.*, 42, 346–58, 2010; also Brackenbury and Djamgoz, *J. Physiol.*, 573, 343–56, 2006.)

RESPONSE OF CANCER CELLS TO DIRECT-CURRENT ELECTRIC FIELDS

Galvanotaxis—a directional response to small direct current electric fields (dcEFs)—is an inherent property of cells (McCaig et al., 2009). Mycielska and Djamgoz (2004) advanced the view that, in its most basic form, galvanotaxis would be cathodal. Indeed, many cells respond to dcEFs by moving toward the cathode. dcEFs would also be expected to play a significant role in metastatic carcinomas since the tumor cells emerging from the epithelia would be subject to the transcellular voltage difference. The intriguing question was whether galvanotaxis would be qualitatively or quantitatively different between strongly vs. weakly/nonmetastatic cells. This question was addressed originally by Djamgoz et al. (2001) using the Dunning rodent model of PCa, comparing strongly vs. weakly metastatic Mat-LyLu and AT-2 cells, respectively. In this study, using dcEFs of 10 to 400 mV/mm, Mat-LyLu cells moved briskly toward the cathode, while AT-2 cells were much slower and tended to move toward the anode (Figure 12.9). Importantly, the galvanotactic response of the Mat-LuLu cells was suppressed significantly by TTX, which had no effect on the weakly metastatic cell line, consistent with the expression pattern of the channel in these cells. Interestingly, compared with PCa, galvanotaxis of analogous, strongly metastatic human BCa cells was in the opposite direction, toward the anode (Fraser et al., 2005; Pu et al., 2007). Again, nonmetastatic MCF-7 (BCa) cells showed much less reaction to dcEF, and this was

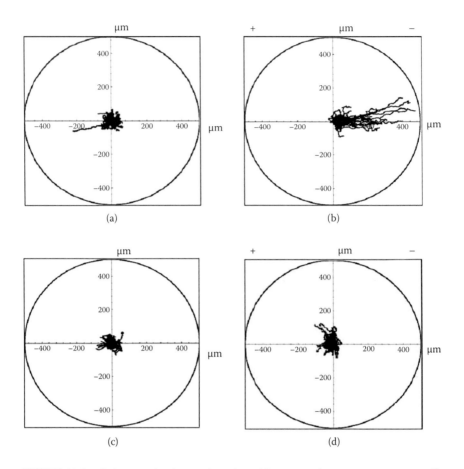

FIGURE 12.9 Galvanotaxis of strongly and weakly metastatic rat prostate cancer cells: Mat-LyLu (A,B) and AT-2 cells (C,D), respectively. Composite trajectories of fifty cells of each, migrating in the absence (A,C) and presence (B,D) of an electric field (300 mV/mm), are shown as circular diagrams. In each diagram, the initial point for each trajectory was placed at the center of the circle. The x axis corresponds to the direction of the electric field. The cathode (– pole) is always placed at the right-hand side of the diagram. Treatment with TTX (1 μM) suppressed the response of Mat-LyLu cells but had no effect on AT-2 cells (not shown). (Modified from Djamgoz et al., *J. Cell Sci.*, 114, 2697–705, 2001.)

cathodal (Fraser et al., 2005). At present, it is not clear why metastatic PCa and BCa cells move in opposite directions in comparable dcEFs. Apart from species difference, one possibility is that this is due to some inherent property of the cells related to the fact that the lumen potentials in the respective glands are of opposite polarities, positive in breast, negative in prostate (Figure 12.2). Using a specially constructed sealed recording chamber, in the form of a microfluidic cell culture chip, similar results were obtained from human lung cancer cell lines of differing invasive/metastatic potential (Huang et al., 2009). Thus, strongly metastatic

CL1-5 cells were strongly galvanotactic toward the anode, while the parental, nonmetastatic CL1-0 cells were unresponsive.

Two other cell types related to the cancer process also demonstrate galvanotactic responses. First, endothelial cells possess ion channels, including VGSCs (Chang et al., 1996; Andrikopoulos et al., 2010), and are associated with endogenous dcEFs in the form of transendothelial potentials or tissue SFPs (Revest et al., 1993, 1994; Graves et al., 1975). Endothelial cells are indeed galvanotactic (Chang et al., 1996; Li and Kolega, 2002; Zhao et al., 2004) (Chapter 7). Further, it is possible that galvanotaxis is involved in angiogenesis, since application of dcEFs to human umbilical vein endothelial cells stimulated production of vascular endothelial growth factor, known to play a significant role in the angiogenic process (e.g., Hicklin and Ellis, 2005). Second, leukocytes also possess a range of ion channels, including VGSCs (Gallin, 1991). Circulating leukocytes would be exposed to endothelial dcEFs, as well as, possibly, the cancer cells' surface charges. Lymphocytes have also been shown to be galvanotactic (Lin et al., 2008), and infiltration into tumors has prognostic significance (e.g., Nzula et al., 2003) (Chapter 8).

As regards possible biochemical/molecular mechanisms controlling galvanotaxis, Djamgoz et al. (2001) focused on VGSC activity for two reasons: (1) the response was fast (response to field reversal being complete within ~30 s), so it was likely to involve a small molecule such as an ion, and (2) only the strongly metastatic cells expressing functional VGSCs were galvanotactic, (e.g., Grimes et al., 1995; Fraser et al., 2005). Treating the cells with TTX significantly suppressed the galvanotaxis of Mat-LyLu and MDA-MBA-231 cells; there was no effect on the corresponding weakly/nonmetastatic cell lines (Djamgoz et al., 2001; Fraser et al., 2005). Regional changes in intracellular Ca^{2+} concentration were suggested to be downstream of VGSC activity in generating directional movement (Mycielska and Djamgoz, 2004). On the other hand, Pu et al. (2007) suggested that galvanotaxis of BCa cells involved asymmetrical distribution of EGF receptors (EGFRs) in the cell membrane and EGFR signaling. Interestingly, EGF is known to upregulate functional VGSC expression in several types of cell, including cancer cells (e.g., Toledo-Aral et al., 1995; Uysal-Onganer and Djamgoz, 2007). It is possible, therefore, that a common mechanism operates involving both EGFR and VGSC signaling.

A further bioelectric characteristic that may contribute to galvanotaxis is surface charge, which can give rise to a "charge cloud" (charged residues or free ions) on the plasma membrane and also control localization and function of proteins (Goldenberg and Steinberg, 2010). Increased surface negativity is associated with malignancy (Abercrombie and Ambrose, 1962). Surface charge can respond briskly to dcEF, and digesting sialic acid moieties (the most abundant charged molecules on mammalian cell surfaces) with neuraminidase can significantly suppress galvanotaxis (McLaughlin and Poo, 1981).

Galvanotaxis is likely to be involved in metastatic disease for several reasons. First, carcinoma cells detaching and locally migrating from epithelium will come under the influence of the transcellular potential (~500 mV/mm), which is comparable to the experimental values of dcEFs used to induce galvanotaxis. Also, during extra/intravasation, tumor cells will be subject to the transendothelial potential, which could further modulate the metastatic progression. Finally, the inflammatory response will be affected by the galvanotactic property of infiltrating lymphocytes (Lin et al., 2008; Chapter 8). Importantly, the latter is likely to be a dynamic process since ion channel expression in cells of the immune system can be stage dependent (Zsiros et al., 2009).

TISSUE ELECTROLYTES AND BIOIMPEDANCE

The molecular switches that occur in cancer cells result in changes in the cells' biochemical contents, leading to modification of membrane conductance and tissue electrolytes. In particular, according to the CELEX hypothesis, the enhanced VGSC activity would lead to an increase in the intracellular Na^+ concentration. This has been confirmed at tissue level in human BCa patients. Thus, using noninvasive ^{23}Na–magnetic resonance imaging (MRI), Ouwerkerk et al. (2007) showed that the sodium content of breast tissue is higher, in the following order: locally advanced BCa > poorly differentiated ductal carcinoma > benign (Figure 12.10). Tumor ^{23}Na-MRI has also been suggested to be useful for monitoring effectiveness of therapy (e.g., Schepkin et al., 2006a, 2006b). A complementary technique is magnetic resonance spectroscopic imaging (MRSI), which uses radio frequency transmission reception to map relative concentrations of metabolic markers of disease, e.g., citrate for PCa (Mycielska et al., 2009; Near et al., 2009).

Electrolyte analyses have also been carried out on human breast cyst fluid. Thus, Sisman et al. (2009) showed that extracellular fluid from cysts associated

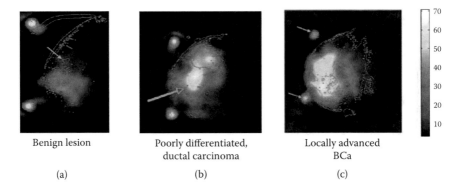

Benign lesion	Poorly differentiated, ductal carcinoma	Locally advanced BCa
(a)	(b)	(c)

FIGURE 12.10   See color insert. ^{23}Na-MRI images from three different breast cancer patients. (A) A benign lesion (proliferative fibrocystosis), indicated by the arrow, aligned from gadolinium-enhanced image. (B) Infiltrating poorly differentiated ductal carcinoma (outlined in blue). Below (indicated by a green arrow) is a region with edema. (C) A large locally advanced breast cancer (outline in middle blue). The arrows indicate positioning landmarks. The intensity scale on the far right indicates approximate sodium concentrations (relative). (Modified from Ouwerkerk et al., *Breast Cancer Res. Treat.*, 106, 151–60, 2007.)

with higher risk of developing BCa had lower Na$^+$, higher K$^+$, and a lower Na$^+$:K$^+$ concentration ratio than those of lower risk. These changes would appear to mirror those seen in tissue. Another electrolyte of pathological significance is calcium (Saidak et al., 2009). Breast calcification, which can readily be picked up on mammograms, has been associated with increased risk of BCa (Tse et al., 2008). On the other hand, calcium has been suggested to be beneficial to colon cancer, and further work is required to determine the role and clinical potential of calcium, including its imaging, in different cancers (Chakravarti et al., 2009).

The modified tissue, body fluid electrolytes, or membrane conductances could lead to changes in tissue bioimpedance in cancer that have been measured by applying sinusoidal currents, using either sets of several surface electrodes or individual probes, depending on the organ being studied. So far, this technique has been applied mainly to cancers of the breast (Cuzick et al., 1998), cervix (Abdul et al., 2006; Brown et al., 2000), and in a preliminary study, tongue (Ching et al., 2010). A methodology for BCa has already been approved by the American Federal Drug Administration as an adjunct to mammography (www.fda.gov/cdrf/pma/pmaapr99.html). In bioimpedance measurements, several parameters (e.g., α/β-dispersion, impedance, real part and imaginary part of impedance, and phase angle) need to be measured at a range of stimulation frequencies (Hz to MHz), and ideally, multivariate analyses should be used before clear differences can be shown for cancer vs. normal tissue. Malignant tumors typically have around twofold lower impedances than surrounding normal tissue. A major advantage of bioimpedance measurements is the possibility of detecting tumors as soon as some biochemical change has occurred, before any metastasis.

FUTURE PERSPECTIVES AND CLINICAL POTENTIAL

The main message of this chapter is that bioelectricity can play a significant role in the cancer process, as in cell physiology generally and other pathophysiologies. Thus, any suggestion that electrical phenomena are somehow restricted to classically excitable cells/tissues (neurones and muscles) is preconceived and essentially wrong. Accordingly, the ion channels and cancer field is a worthwhile, multidisciplinary area of research where much interesting work remains to be done. On the whole, for example, we need to (1) elucidate the mechanisms up- and downstream of ion channels' expression/activity in cancer cells, (2) functionally relate ion channels, transporters, etc., to mainstream cancer mechanisms, and (3) increase the *in vivo* evidence for the role of bioelectricity in cancer. For these, it is necessary to use well-defined models (*in vitro* and *in vivo*). As regards the clinical potential of cancer bioelectricity, several possibilities can be considered. First, VGSC expression is likely to occur early in metastasis, so it could serve as an efficient biomarker for functional diagnosis of metastatic potential in cancer. Such a procedure can be based upon mRNA (RT-PCR) or protein (immunohistochemistry) (Fraser et al., 2005). More generally, it may be possible to quantitatively map the dcEF or associated electrolyte changes. Second, the expression of normally developmentally regulated ion channel (e.g., VGSC) splice variants in cancer offers novel possibilities for selective targeting and killing of tumors using antibody-based techniques (Onkal and Djamgoz, 2009). In addition, use of small-molecule inhibitors of VGSCs, toxins, and siRNA may be possible. Finally, since tumor cells generate specific galvanotactic responses to dcEFs, an intriguing clinical possibility would be drawing metastatic cells out of organs and body fluids by manipulating them with exogenous fields. Indeed, electric stimulation with dcEFs or mild/low-frequency fields can produce a variety of benefits *in vivo* (e.g., Ainsworth et al., 2006; Nelson et al., 2003; Mobbs et al., 2004; Morino et al., 2008; Kai et al., 2009). A number of studies have indicated that dcEFs can also produce anticancer effects (galvanotherapy), although the precise underlying mechanisms are not yet known (Miklavcic et al., 1997; Jarm et al., 2003; Cameron et al., 1980; Smith et al., 1978; Xin et al., 1994; Vogl et al., 2007).

ACKNOWLEDGMENTS AND DEDICATION

We are grateful to the Pro Cancer Research Fund (PCRF; originally the Prostate Cancer Research Fund) and all its many volunteers for supporting our work on the neuroscience approach to cancer, especially as it went through its critical, formative stages. I thank Refika Güzel for many useful discussions and help in preparing the manuscript. I dedicate this chapter to the memory of my father (Ali Şevki Camgőz), who taught me how to operate with my hands but died of metastatic fibrosarcoma at a time (1965) when it took too long for the cancer to be diagnosed.

REFERENCES

Abdul M and Hoosein N. 2001. Inhibition by anticonvulsants of prostate-specific antigen and interleukin-6 secretion by human prostate cancer cells. *Anticancer Res* 21:2045–48.

Abdul M and Hoosein N. 2002. Voltage-gated sodium ion channels in prostate cancer: Expression and activity. *Anticancer Res* 22:1727–30.

Abdul M and Hoosein N. 2006. Reduced Kv1.3 potassium channel expression in human prostate cancer. *J Membr Biol* 214:99–102.

Abdul S, Brown BH, Milnes P, Tidy JA. 2006. The use of electrical impedance spectroscopy in the detection of cervical intraepithelial neoplasia. *Int J Gynecol Cancer* 16:1823–32.

Abercrombie M and Ambrose EJ. 1962. The surface properties of cancer cells: A review. *Cancer Res* 22:525–548.

Adachi K, Toyota M, Sasaki Y, Yamashita T, Ishida S, Ohe-Toyota M, Maruyama R, Hinoda Y, Saito T, Imai K, Kudo R, Tokino T. 2004. Identification of SCN3B as a novel p53-inducible proapoptotic gene. *Oncogene* 23:7791–98.

Ainsworth L, Budelier K, Clinesmith M, Fiedler A, Landstrom R, et al. 2006 Transcutaneous electrical nerve stimulation (TENS) reduces chronic hyperalgesia induced by muscle inflammation. *Pain* 120:182–87.

Andrikopoulos P, Fraser SP, Patterson L, Ahmad Z, Burcu H, Ottaviani D, Diss JKJ, Box C, Eccles SA, Djamgoz MBA. 2010. Angiogenic function of voltage-gated Na+ channels in human endothelial cells: Implications for VEGF signalling. MS submitted for publication.

Arcangeli A, Crociani O, Lastraioli E, Masi A, Pillozzi S, Becchetti A. 2009. Targeting ion channels in cancer: A novel frontier in antineoplastic therapy. *Curr Med Chem* 16:66–93.

Ashcroft FM. 2006. From molecule to malady. *Nature* 440:440–47.

Bacac M and Stamenkovic I. 2007. Metastatic cancer cell. *Annu Rev Pathol* 3:221–47.

Bennett ES, Smith BA, Harper JM. 2004. Voltage-gated Na+ channels confer invasive properties on human prostate cancer cells. *Pflugers Arch* 447:908–14.

Berger JC, Vander-Griend D, Stadler WM, Rinker-Schaeffer C. 2004. Metastasis suppressor genes: Signal transduction, cross-talk, and the potential for modulating the behavior of metastatic cells. *Anticancer Drugs* 15:559–68.

Biran H, Ariel I, de Groot N, Shani A, Hochberg A. 1994. Human imprinted genes as oncodevelopmental markers. *Tumour Biol* 15:123–34.

Blackiston DJ, McLaughlin KA, Levin M. 2009. Bioelectric controls of cell proliferation: Ion channels, membrane voltage, and the cell cycle. *Cell Cycle* 8:3519–28.

Bolanz KA, Kovacs GG, Landowski CP, Hediger MA. 2009. Tamoxifen inhibits TRPV6 activity via estrogen receptor-independent pathways in TRPV6-expressing MCF-7 breast cancer cells. *Mol Cancer Res* 7:2000–10.

Borg JJ, Yuill KH, Hancox JC, Spencer IC, Kozlowski RZ. 2002. Inhibitory effects of the antiestrogen agent clomiphene on cardiac sarcolemmal anionic and cationic currents. *J Pharmacol Exp Ther* 303:282–92.

Brackenbury WJ, Chioni AM, Diss JKJ, Djamgoz MBA. 2007. The neonatal splice variant of Nav1.5 potentiates *in vitro* invasive behavior of MDA-MBA-231 human breast cancer cells. *Breast Cancer Res Treat* 101:149–60.

Brackenbury WJ and Djamgoz MBA. 2006. Activity-dependent regulation of voltage-gated Na+ channel expression in Mat-LyLu rat prostate cancer cell line. *J Physiol* 573:343–56.

Brackenbury WJ and Djamgoz MBA. 2007. Nerve growth factor enhances voltage-gated Na+ channel activity and Transwell migration in Mat-LyLu rat prostate cancer cell line. *J Cell Physiol* 210:602–8.

Brackenbury WJ, Djamgoz MBA, Isom LL. 2008. An emerging role for voltage-gated Na+ channels in cellular migration: Regulation of central nervous system development and potentiation of invasive cancers. *Neuroscientist* 14:571–83.

Brown BH, Tidy JA, Boston K, Blackett AD, Smallwood RH, Sharp F. 2000. Relation between tissue structure and imposed electrical current flow in cervical neoplasia. *Lancet* 355:892–95.

Cameron IL, Smith NK, Pool TB, Sparks RL. 1980. Intracellular concentration of sodium and other elements as related to mitogenesis and oncogenesis *in vivo*. *Cancer Res* 40:1493–500.

Chakravarti B, Dwivedi SK, Mithal A, Chattopadhyay N. 2009. Calcium-sensing receptor in cancer: Good cop or bad cop? *Endocrine* 35:271–84.

Chang PC, Sulik GI, Soong HK, Parkinson WC. 1996. Galvanotropic and galvanotaxic responses of corneal endothelial cells. *J Formos Med Assoc* 95:623–27.

Ching CT, Sun TP, Huang SH, Hsiao CS, Chang CH, Huang SY, Chen YJ, Cheng CS, Shieh HL, Chen CY. 2010. A preliminary study of the use of bioimpedance in the screening of squamous tongue cancer. *Int J Nanomed* 5:213–20.

Chioni AM, Brackenbury WJ, Calhoun JD, Isom LL, Djamgoz MBA. 2009. A novel adhesion molecule in human breast cancer cells: Voltage-gated Na⁺ channel beta1 subunit. *Int J Biochem Cell Biol* 41:1216–27.

Chioni AM, Shao D, Grose R, Djamgoz MBA. 2010. Protein kinase A and regulation of neonatal Nav1.5 expression in human breast cancer cells: Activity-dependent positive feedback and cellular migration. *Int J Biochem Cell Biol* 42:346–58.

Conti M. 2007. Targeting ion channels for new strategies in cancer diagnosis and therapy. *Curr Clin Pharmacol* 2:135–44.

Cuzick J, Holland R, Barth V, Davies R, Faupel M, Fentiman I, Frischbier HJ, LaMarque JL, Merson M, Sacchini V, Vanel D, Veronesi U. 1998. Electropotential measurements as a new diagnostic modality for breast cancer. *Lancet* 352:359–63.

Ding Y, Brackenbury WJ, Onganer PU, Montano X, Porter LM, Bates LF, Djamgoz MBA. 2008. Epidermal growth factor upregulates motility of Mat-LyLu rat prostate cancer cells partially via voltage-gated Na⁺ channel activity. *J Cell Physiol* 215:77–81.

Diss JK, Archer SN, Hirano J, Fraser SP, Djamgoz MBA. 2001. Expression profiles of voltage-gated Na⁺ channel alpha-subunit genes in rat and human prostate cancer cell lines. *Prostate* 48:165–78.

Diss JK, Fraser SP, Djamgoz MBA. 2004. Voltage-gated Na⁺ channels: Multiplicity of expression, plasticity, functional implications and pathophysiological aspects. *Eur Biophys J* 33:180–93.

Diss JKJ, Fraser SP, Walker MM, Patel A, Latchman DS, Djamgoz MBA. 2007. Beta-subunits of voltage-gated sodium channels in human prostate cancer: Quantitative *in vitro* and *in vivo* analyses of mRNA expression. *Prostate Cancer Prostatic Dis* 11:325–33.

Diss JKJ, Stewart D, Pani F, Foster CS, Walker MM, Patel A, Djamgoz MBA. 2005. A potential novel marker for human prostate cancer: Voltage-gated sodium channel expression *in vivo*. *Prostate Cancer Prostatic Dis* 8:266–73.

Djamgoz MBA, Mycielska M, Madeja Z, Fraser SP, Korohoda W. 2001. Directional movement of rat prostate cancer cells in direct-current electric field: Involvement of voltage-gated Na⁺ channel activity. *J Cell Sci* 114:2697–705.

Eccles JC and Jaeger JC. 1958. The relationship between the mode of operation and the dimensions of the junctional regions at synapses and motor end-organs. *Proc R Soc Lond B Biol Sci* 148:38–56.

Faupel M, Vanel D, Barth V, Davies R, Fentiman IS, Holland R, Lamarque JL, Sacchini V, Schreer I. 1997. Electropotential evaluation as a new technique for diagnosing breast lesions. *Eur J Radiol* 24:33–38.

Fidler IJ. 2002a. Critical determinants of metastasis. *Semin Cancer Biol* 12:89–96.

Fidler IJ. 2002b. The organ microenvironment and cancer metastasis. *Differentiation* 70:498–505.

Fidler IJ. 2003. The pathogenesis of cancer metastasis: The 'seed and soil' hypothesis revisited. *Nat Rev Cancer* 3:453–58.

Fiske JL, Fomin VP, Brown ML, Duncan RL, Sikes RA. 2006. Voltage-sensitive ion channels and cancer. *Cancer Metastasis Rev* 25:493–500.

Fraser SP, Ding Y, Liu A, Foster CS, Djamgoz MBA. 1999. Tetrodotoxin suppresses morphological enhancement of the metastatic Mat-LyLu rat prostate cancer cell line. *Cell Tissue Res* 295:505–12.

Fraser SP, Diss JKJ, Chioni AM, Mycielska ME, Pan H, Yamaci RF, Pani F, Siwy Z, Krakowska M, Grzywna Z, Brackenbury WJ, Theodorou D, Koyutűrk M, Kaya H, Battaloğlu E, Tamburo De Bella M, Slade MJ, Tolhurst R, Palmieri C, Jiang J,

Latchman DS, Coombes RC, Djamgoz MBA. 2005. Voltage-gated sodium channel expression and potentiation of human breast cancer metastasis. *Clin Cancer Res* 11:5381–89.

Fraser SP, Grimes JA, Diss JK, Stewart D, Dolly JO, Djamgoz MBA. 2003a. Predominant expression of Kv1.3 voltage-gated K+ channel subunit in rat prostate cancer cell lines: Electrophysiological, pharmacological and molecular characterisation. *Pflugers Arch* 446:559–71.

Fraser SP, Ozerlat-Gunduz I, Onkal R, Diss JK, Latchman DS, Djamgoz MBA. 2010. Estrogen and non-genomic upregulation of voltage-gated Na+ channel activity in MDA-MBA-231 human breast cancer cells: Role in adhesion. *J Cell Physiol* 224:527–39.

Fraser SP and Pardo LA. 2008. Ion channels: Functional expression and therapeutic potential in cancer. Colloquium on ion channels and cancer. *EMBO Rep* 9:512–15.

Fraser SP, Salvador V, Manning EA, Mizal J, Altun S, Raza M, Berridge RJ, Djamgoz MBA. 2003b. Contribution of functional voltage-gated Na+ channel expression to cell behaviors involved in the metastatic cascade in rat prostate cancer. I. Lateral motility. *J Cell Physiol* 195:479–87.

Gallin EK. 1991. Ion channels in leukocytes. *Physiol Rev* 71:775–811.

Gillet L, Roger S, Besson P, Lecaille F, Gore J, Bougnoux P, Lalmanach G, Le Guennec JY. 2009. Voltage-gated sodium channel activity promotes cysteine cathepsin-dependent invasiveness and colony growth of human cancer cells. *J Biol Chem* 284:8680–91.

Goldenberg NM and Steinberg BE. 2010. Surface charge: A key determinant of protein localization and function. *Cancer Res* 70:1277–80.

Goldfarb M, Schoorlemmer J, Williams A, Diwakar S, Wang Q, Huang X, Giza J, Tchetchik D, Kelley K, Vega A, Matthews G, Rossi P, Ornitz DM, D'Angelo E. 2007. Fibroblast growth factor homologous factors control neuronal excitability through modulation of voltage-gated sodium channels. *Neuron* 55:449–63.

Graves CN, Sanders SS, Shoemaker RL, Rehm WS. 1975. Diffusion resistance of endothelium and stroma of bullfrog cornea determined by potential response to K+. *Biochim Biophys Acta* 389:550–56.

Grimes JA, Fraser SP, Stephens GJ, Downing JEG, Laniado ME, Foster CS, Abel PD, Djamgoz MBA. 1995. Differential expression of voltage-activated Na+ currents in two prostatic tumor cell lines: Contribution to invasiveness *in vitro*. *FEBS Lett* 369:290–94.

Gurel B, Iwata T, Koh CM, Yegnasubramanian S, Nelson WG, De Marzo AM. 2008. Molecular alterations in prostate cancer as diagnostic, prognostic, and therapeutic targets. *Adv Anat Pathol* 15:319–31.

Hanahan D and Weinberg RA. 2000. The hallmarks of cancer. *Cell* 100:57–70.

Herfst LJ, Rook MBA, Jongsma HJ. 2004. Trafficking and functional expression of cardiac Na+ channels. *J Mol Cell Cardiol* 36:185–93.

Hicklin DJ and Ellis LM. 2005. Role of the vascular endothelial growth factor pathway in tumor growth and angiogenesis. *J Clin Oncol* 23:1011–127.

Hille B. 1992. *Ionic channels of excitable membranes*. Sunderland, MA: Sinauer Associates.

Hoffman EP. 1995. Voltage-gated ion channelopathies: Inherited disorders caused by abnormal sodium, chloride, and calcium regulation in skeletal muscle. *Annu Rev Med* 46:431–41.

Huang CW, Cheng JY, Yen MH, Young TH. 2009. Electrotaxis of lung cancer cells in a multiple-electric-field chip. *Biosens Bioelectron* 24:3510–16.

Hunter KW. 2004. Host genetics and tumor metastasis. *Br J Cancer* 90:752–55.

Hunter KW, Crawford NP, Alsarraj J. 2008. Mechanisms of metastasis. *Breast Cancer Res* 10(1):S2.

Jarm T, Cemazar M, Steinberg F, Streffer C, Sersa G, Miklavcic D. 2003. Perturbation of blood flow as a mechanism of anti-tumor action of direct current electrotherapy. *Physiol Meas* 24:75–90.

Jefferys JG. 1995. Nonsynaptic modulation of neuronal activity in the brain: Electric currents and extracellular ions. *Physiol Rev* 75:689–723.

Jemal A, Siegel R, Ward E, Hao Y, Xu J, Murray T, Thun MJ. 2008. Cancer statistics, 2008. *CA Cancer J Clin* 58:71–96.

Kai H, Suico MA, Morino S, Kondo T, Oba M, Noguchi M, Shuto T, Araki E. 2009. A novel combination of mild electrical stimulation and hyperthermia: General concepts and applications. *Int J Hyperthermia* 25:655–60.

Kim SY, Shin DH, Kim SY, Koo B-S, Bae C-D, Park J, Jeon S. 2010. Chloride channel conductance is required for NGF-induced neurite outgrowth in PC12 cells *Neurochem Int* 56:663–69.

Krasowska M, Grzywna ZJ, Mycielska ME, Djamgoz MBA. 2004. Patterning of endocytic vesicles and its control by voltage-gated Na^+ channel activity in rat prostate cancer cells: Fractal analyses. *Eur Biophys J* 33:535–42.

Lang K and Bastian P. 2007. Neurotransmitter effects on tumor cells and leukocytes. *Prog Exp Tumor Res* 39:99–121.

Laniado ME, Lalani EN, Fraser SP, Grimes JA, Bhangal G, Djamgoz MBA, Abel PD. 1997. Expression and functional analysis of voltage-activated Na^+ channels in human prostate cancer cell lines and their contribution to invasion *in vitro*. *Am J Pathol* 150:1213–21.

Le Guennec JY, Ouadid-Ahidouch H, Soriani O, Besson P, Ahidouch A, Vandier C. 2007. Voltage-gated ion channels, new targets in anti-cancer research. *Recent Pat Anticancer Drug Discov* 2:189–202.

Leung SW, Teoh H, Keung W, Man RY. 2007. Non-genomic vascular actions of female sex hormones: Physiological implications and signalling pathways. *Clin Exp Pharmacol Physiol* 34:822–26.

Li X and Kolega J. 2002. Effects of direct current electric fields on cell migration and actin filament distribution in bovine vascular endothelial cells. *J Vasc Res* 39:391–404.

Lin F, Baldessari F, Gyenge CC, Sato T, Chambers RD, Santiago JG, Butcher EC. 2008. Lymphocyte electrotaxis *in vitro* and *in vivo*. *J Immunol* 181:2465–71.

Liu H, Wu MM, Zakon HH. 2008. A novel Na^+ channel splice form contributes to the regulation of an androgen-dependent social signal. *J Neurosci* 28:9173–82.

Mahaut-Smith MP, Martinez-Pinna J, Gurung IS. 2008. A role for membrane potential in regulating GPCRs? *Trends Pharmacol Sci* 29:421–29.

Marban E. 2002. Cardiac channelopathies. *Nature* 415:213–18.

McCaig CD, Song B, Rajniek AM. 2009. Electric dimensions in cell science. *J Cell Sci* 122:4267–76.

McLaughlin S and Poo MM. 1981. The role of electro-osmosis in the electric-field-induced movement of charged macromolecules on the surfaces of cells. *Biophys J* 34:85–93.

Meisler MH, Kearney JA, Sprunger LK, MacDonald BT, Buchner DA, Escayg A. 2002. Mutations of voltage-gated sodium channels in movement disorders and epilepsy. *Novart Fdn Symp* 241:72–86.

Miklavcic D, Jarm T, Cemazar M, Sersa G, An DJ. Belehradek J, Mir LM. 1997. Tumor treatment by direct electric current. Tumor perfusion changes. *Bioelectrochem Bioenerg* 43:253–56.

Mobbs RJ, Nair S, Blum P. 2007. Peripheral nerve stimulation for the treatment of chronic pain. *J Clin Neurosci* 14:216–21.

Monk M and Holding C. 2001. Human embryonic genes re-expressed in cancer cells. *Oncogene* 20:8085–91.

Mycielska ME and Djamgoz MBA. 2004. Cellular mechanisms of direct-current electric field effects: Galvanotaxis and metastatic disease. *J Cell Sci* 117:1631–39.

Mycielska ME, Fraser SP, Szatkowski M, Djamgoz MBA. 2003. Contribution of functional voltage-gated Na$^+$ channel expression to cell behaviors involved in the metastatic cascade in rat prostate cancer. II. Secretory membrane activity. *J Cell Physiol* 195:461–69.

Mycielska ME, Palmer CP, Brackenbury WJ, Djamgoz MBA. 2005. Expression of Na+-dependent citrate transport in a strongly metastatic human prostate cancer PC-3M cell line: Regulation by voltage-gated Na$^+$ channel activity. *J Physiol* 563:393–408.

Mycielska ME, Patel A, Rizaner N, Mazurek MP, Keun H, Patel A, Ganapathy V, Djamgoz MBA. 2009. Citrate transport and metabolism in mammalian cells: Prostate epithelial cells and prostate cancer. *BioEssays* 31:10–20.

Nakajima T, Kubota N, Tsutsumi T, Oguri A, Imuta H, Jo T, Oonuma H, Soma M, Meguro K, Takano H, Nagase T, Nagata T. 2009. Eicosapentaenoic acid inhibits voltage-gated sodium channels and invasiveness in prostate cancer cells. *Br J Pharmacol* 56:420–31.

Near J, Romagnoli C, Curtis AT, Klassen LM, Izawa J, Chin J, Bartha R. 2009. High-field MRSI of the prostate using a transmit/receive endorectal coil and gradient modulated adiabatic localization. *J Magn Reson Imaging* 30:335–43.

Nelson FR, Brighton CT, Ryaby J, Simon BJ, Nielson JH, et al. 2003. Use of physical forces in bone healing. *J Am Acad Orthop Surg* 11:344–54.

Nguyen DX and Massague J. 2007. Genetic determinants of cancer metastasis. *Nat Rev Genet* 8:341–52.

Nicolson GL. 1976. Trans-membrane control of the receptors on normal and tumor cells. II. Surface changes associated with transformation and malignancy. *Biochim Biophys Acta* 458:1–72.

Nzula S, Going JJ, Stott DI. 2003. Antigen-driven clonal proliferation, somatic hypermutation, and selection of B lymphocytes infiltrating human ductal breast carcinomas. *Cancer Res* 63:3275–80.

Onganer PU and Djamgoz MBA. 2005. Small-cell lung cancer (human): Potentiation of endocytic membrane activity by voltage-gated Na$^+$ channel expression *in vitro*. *J Membr Biol* 204:67–75.

Onkal R and Djamgoz MBA. 2009. Molecular pharmacology of voltage-gated sodium channel expression in metastatic disease: Clinical potential of neonatal Nav1.5 in breast cancer. *Eur J Pharmacol* 625:206–19.

Ouwerkerk R, Jacobs MA, Macura KJ, Wolff AC, Stearns V, Mezban SD, Khouri NF, Bluemke DA, Bottomley PA. 2007. Elevated tissue sodium concentration in malignant breast lesions detected with non-invasive (23)Na MRI. *Breast Cancer Res Treat* 106:151–60.

Palmer CP, Mycielska ME, Burcu H, Osman K, Collins T, Beckerman R, Perrett R, Johnson H, Aydar E, Djamgoz MBA. 2008. Single cell adhesion measuring apparatus (SCAMA): Application to cancer cell lines of different metastatic potential and voltage-gated Na$^+$ channel expression. *Eur Biophys J* 37:359–68.

Prevarskaya N, Skryma R, Bidaux G, Flourakis M, Shuba Y. 2007. Ion channels in death and differentiation of prostate cancer cells. *Cell Death Differ* 14:1295–304.

Prevarskaya N, Skryma R, Shuba Y. 2010. Ion channels and the hallmarks of cancer. *Trends Mol Med* 16:107–21.

Prossnitz ER, Arterburn JB, Smith HO, Oprea TI, Sklar LA, Hathaway HJ. 2008. Estrogen signaling through the transmembrane G protein-coupled receptor GPR30. *Annu Rev Physiol* 70:165–90.

Pu J, McCaig CD, Cao L, Zhao Z, Segall JE, Zhao M. 2007. EGF receptor signalling is essential for electric-field-directed migration of breast cancer cells. *J Cell Sci* 120:3395–403.

Ramirez ML, Nelson EC, Evans CP. 2008. Beyond prostate-specific antigen: Alternate serum markers. *Prostate Cancer Prostatic Dis* 11:216–29.

Revest, PA, Jones HC, Abbott NJ. 1993. The transendothelial DC potential of rat blood-brain barrier vessels *in situ. Adv Exp Med Biol* 331:71–74.

Revest PA, Jones HC, Abbott NJ. 1994. Transendothelial electrical potential across pial vessels in anaesthetised rats: A study of ion permeability and transport at the blood-brain barrier. *Brain Res* 652:76–82.

Roberts R and Brugada R. 2003. Genetics and arrhythmias. *Annu Rev Med* 54:257–67.

Roger S, Besson P, Le Guennec JY. 2003. Involvement of a novel fast inward sodium current in the invasion capacity of a breast cancer cell line. *Biochim Biophys Acta* 1616:107–11.

Roger S, Potier M, Vandier C, Besson P, Le Guennec JY. 2006. Voltage-gated sodium channels: New targets in cancer therapy? *Curr Pharm Des* 12:3681–95.

Roger S, Rollin J, Barascu A, Besson P, Raynal PI, Iochmann S, Lei M, Bougnoux P, Gruel Y, Le Guennec JY. 2007. Voltage-gated sodium channels potentiate the invasive capacities of human non-small-cell lung cancer cell lines. *Int J Biochem Cell Biol* 39:774–86.

Rose CR, Blum R, Kafitz KW, Kovalchuk Y, Konnerth A. 2004. From modulator to mediator: Rapid effects of BDNF on ion channels. *BioEssays* 26:1185–94.

Saidak Z, Mentaverri R, Brown EM. 2009. The role of the calcium-sensing receptor in the development and progression of cancer. *Endocr Rev* 30:178–95.

Santen RJ, Song RX, Masamura S, Yue W, Fan P, Sogon T, Hayashi S, Nakachi K, Eguchi H. 2008. Adaptation to estradiol deprivation causes up-regulation of growth factor pathways and hypersensitivity to estradiol in breast cancer cells. *Adv Exp Med Biol* 630:19–34.

Santen RJ, Song RX, Zhang Z, Kumar R, Jeng MH, Masamura A, Lawrence J Jr, Berstein L, Yue W. 2005. Long-term estradiol deprivation in breast cancer cells up-regulates growth factor signaling and enhances estrogen sensitivity. *Endocr Relat Cancer* 12:S61–73.

Schepkin VD, Chenevert TL, Kuszpit K, Lee KC, Meyer CR, Johnson TD, Rehemtulla A, Ross BD. 2006a. Sodium and proton diffusion MRI as biomarkers for early therapeutic response in subcutaneous tumors. *Magn Reson Imaging* 24:273–78.

Schepkin VD, Lee KC, Kuszpit K, Muthuswami M, Johnson TD, Chenevert TL, Rehemtulla A, Ross BD. 2006b. Proton and sodium MRI assessment of emerging tumor chemotherapeutic resistance. *NMR Biomed* 19:1035–42.

Schuller HM. 2008. Neurotransmission and cancer: Implications for prevention and therapy. *Anticancer Drugs* 19:655–71.

Schwirzke M, Schiemann S, Gnirke AU, Weidle UH. 1999. New genes potentially involved in breast cancer metastasis. *Anticancer Res* 19:1801–14.

Shao D, Okuse K, Djamgoz MBA. 2009. Protein-protein interactions involving voltage-gated sodium channels: Post-translational regulation, intracellular trafficking and functional expression. *Int J Biochem Cell Biol* 41:1471–81.

Sikes RA, Walls AM, Brennen WN, Anderson JD, Choudhury-Mukherjee I, Schenck HA, Brown ML. 2003. Therapeutic approaches targeting prostate cancer progression using novel voltage-gated ion channel blockers. *Clin Prostate Cancer* 2:181–87.

Sisman AR, Sis B, Canda T, Onvural B. 2009. Electrolytes and trace elements in human breast cyst fluid. *Biol Trace Elem Res* 128:18–30.

Smith NR, Sparks RL, Pool TB, Cameron IL. 1978. Differences in the intracellular concentration of elements in normal and cancerous liver cells as determined by x-ray microanalysis. *Cancer Res* 38:1952–59.

Smith P, Rhodes NP, Shortland AP, Fraser SP, Djamgoz MBA, Ke Y, Foster CS. 1998. Sodium channel protein expression enhances the invasiveness of rat and human prostate cancer cells. *FEBS Lett* 423:19–24.

Somjen GG. 2001. Mechanisms of spreading depression and hypoxic spreading depression-like depolarization. *Physiol Rev* 81:1065–96.

Sylantyev S, Savtchenko LP, Niu YP, Ivanov AI, Jensen TP, Kullmann DM, Xiao MY, Rusakov DA. 2008. Electric fields due to synaptic currents sharpen excitatory transmission. *Science* 319:1845–49.

Szatkowski M, Mycielska M, Knowles R, Kho A-L, Djamgoz MBA. 2000. Electrophysiological recordings from the rat prostate gland *in vitro*. Identified single cell and trans-epithelial (lumen) potentials. *BJU Int* 86:1–8.

Takebe N and Ivy SP. 2010. Controversies in cancer stem cells: Targeting embryonic signaling pathways. *Clin Cancer Res* 16:3106–12.

Toledo-Aral JJ, Brehm P, Halegoua S, Mandel G. 1995. A single pulse of nerve growth factor triggers long-term neuronal excitability through sodium channel gene induction. *Neuron* 14:607–11.

Tsang SY, Yao XQ, Wong CM, Chan FL, Chen ZY, Huang Y. 2004. Differential regulation of K⁺ and Ca²⁺ channel gene expression by chronic treatment with estrogen and tamoxifen in rat aorta. *Eur J Pharmacol* 483:155–62.

Tse GM, Tan PH, Pang AL, Tang AP, Cheung HS. 2008. Calcification in breast lesions: Pathologists' perspective. *J Clin Pathol* 61:145–51.

Uysal-Onganer P and Djamgoz MBA. 2007. Epidermal growth factor potentiates *in vitro* metastatic behavior of human prostate cancer PC-3M cells: Involvement of voltage-gated sodium channel. *Mol Cancer* 6:76.

Vaidya KS and Welch DR. 2007. Metastasis suppressors and their roles in breast carcinoma. *J Mammary Gland Biol Neoplasia* 12:175–90.

Ventura AC and Merajver SD. 2008. Genetic determinants of aggressive breast cancer. *Annu Rev Med* 59:199–212.

Vogl TJ, Mayer HP, Zangos S, Selby JB Jr, Ackermann H, Mayer FB. 2007. Prostate cancer: MR imaging-guided galvanotherapy—Technical development and first clinical results. *Radiology* 245:895–902.

Waxman SG. 2001. Acquired channelopathies in nerve injury and MS. *Neurology* 56:1621–27.

Waxman SG. 2007. Channel, neuronal and clinical function in sodium channelopathies: From genotype to phenotype. *Nat Neurosci* 10:405–9.

Weigelt B, Hu Z, He X, Livasy C, Carey LA, Ewend MG, Glas AM, Perou CM, Van't Veer LJ. 2005. Molecular portraits and 70-gene prognosis signature are preserved throughout the metastatic process of breast cancer. *Cancer Res* 65:9155–58.

Welch DR. 2004. Microarrays bring new insights into understanding of breast cancer metastasis to bone. *Breast Cancer Res* 6:61–64.

Welch DR. 2006. Do we need to redefine a cancer metastasis and staging definitions? *Breast Dis* 26:3–12.

Welch DR, Steeg PS, Rinker-Schaeffer CW. 2000. Molecular biology of breast cancer metastasis. Genetic regulation of human breast carcinoma metastasis. *Breast Cancer Res* 2:408–16.

Wepsic HT. 1983. Overview of oncofetal antigens in cancer. *Ann Clin Lab Sci* 13:261–66.

Williams EL and Djamgoz MBA. 2005. Nitric oxide and metastatic cell behavior. *BioEssays* 27:1228–38.

Wyckoff JB, Jones JG, Condeelis JC, Segall JE. 2000. A critical step in metastasis: *In vivo* analysis of intravasation at the primary tumor. *Cancer Res* 60:2504–11.

Xin YL. 1994. Advances in the treatment of malignant tumors by electrochemical therapy (ECT). *Eur J Surg Suppl* 574:31–35.

Zhang C, Bosch MA, Rick EA, Kelly MJ, Rønnekleiv OK. 2009. 17Beta-estradiol regulation of T-type calcium channels in gonadotropin-releasing hormone neurons. *J Neurosci* 29:10552–62.

Zhao M, Bai H, Wang E, Forrester JV, McCaig CD. 2004. Electrical stimulation directly induces pre-angiogenic responses in vascular endothelial cells by signalling through VEGF receptors. *J Cell Sci* 117:397–405.

Zsiros E, Kis-Toth K, Hajdu P, Gaspar R, Bielanska J, Felipe A, Rajnavolgyi E, Panyi G. 2009. Developmental switch of the expression of ion channels in human dendritic cells. *J Immunol* 183:4483–92.

Index

293

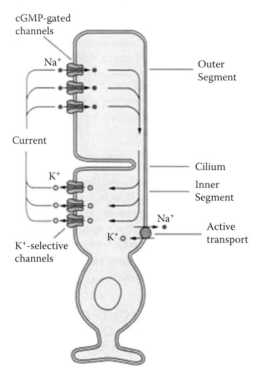

FIGURE 1.1 Diagram of a single retinal rod cell illustrating the segregation of ion channels that leads to the generation of a dark current. Na⁺ channels found only in the outer segment are gated by cGMP and pass the positive inward current there. K⁺ channels are localized in the inner segment and pass the outward current. Photon absorbance by rhodopsin in the outer segment triggers a transduction reaction that results in the reduction of cGMP and leads to the reduction of the inward Na⁺ current.

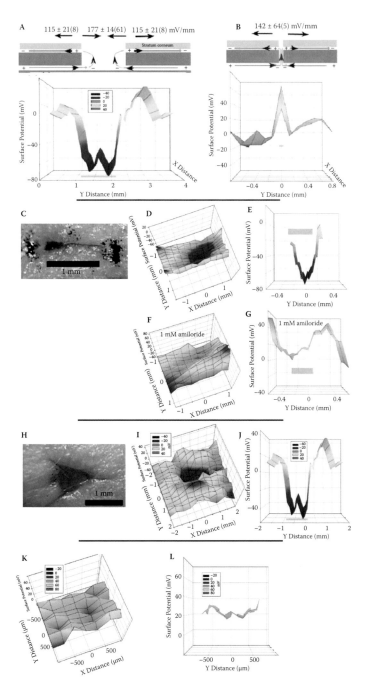

FIGURE 1.4 Summary of results observed on mouse skin wounds. Pink bars mark the wound location on the scan. (For complete description, please see page 10.)

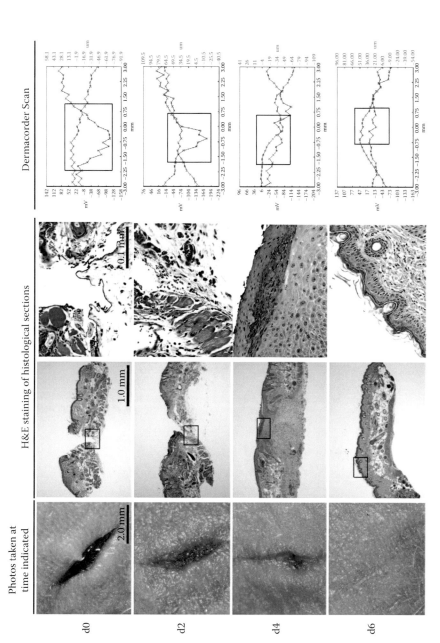

FIGURE 1.5 Dermacorder scan of four different skin wounds. Left column shows reflected light photo of wound on the day indicated to the left of the photo. Two center columns are stained histological sections taken at 40× and 400× magnification. The rectangular outline indicates the region of the image that is further magnified on the right center column. The far right shows the Dermacorder scan of the wound on the left. The blue line represents the surface potential and the red line represents the surface topology. The rectangle in the far right column outlines the region of the scan taken when the probe was over the wound.

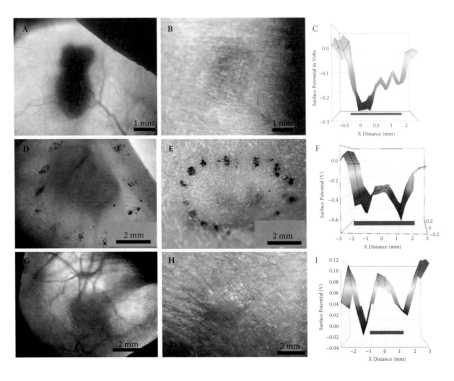

COLOR FIGURE 1.6 Skin tumors and bacterial infections influence the surface potential on the epidermis above them. (For complete description, please see page 13.)

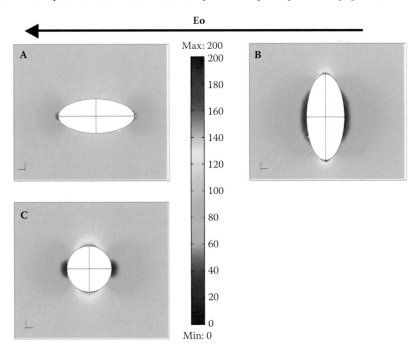

COLOR FIGURE 2.1 The total electric field distribution around a hemiellipsoidal cell placed in a uniform, 100 V/m electric field directed along the x axis. The color range is from 0 (deep blue) to 200 V/m (red) for an applied field of 100 V/m (green).

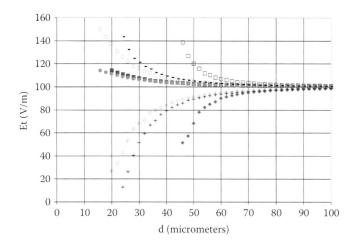

COLOR FIGURE 2.2 The total electric field at various distances from the center of the cell described in Figure 2.1. The solid symbols represent the field components for the long axis parallel to the field, and the open symbols represent the field components for the long axis perpendicular to the field. The blue diamonds represent the field values along the x axis; the red squares, along the y axis; the green circles, along the z axis (distance above the substrate). The + signs represent the values along the x axis for the hemispherical cell; the – signs, along either the y or z directions for that cell.

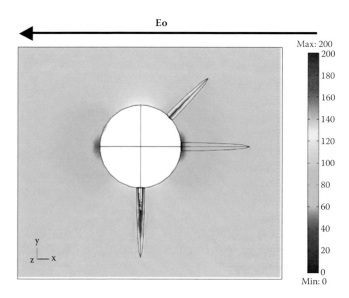

COLOR FIGURE 2.3 The total electric field distribution around a stylized model of a neurite placed in a uniform, 100 V/m electric field directed along the x axis. The field is evaluated in the xy plane at a height z = 3 μm above the substrate. The color range is from 0 (deep blue) to 200 V/m (red) for an applied field of 100 V/m (green).

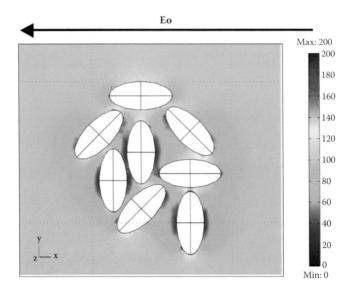

COLOR FIGURE 2.4 The total electric field distribution within a closely spaced group of hemiellipsoidal cells placed in a uniform, 100 V/m electric field directed along the x axis. The field is evaluated in the xy plane at a height z = 2 μm above the substrate. The color range is from 0 (deep blue) to 200 V/m (red) for an applied field of 100 V/m (green).

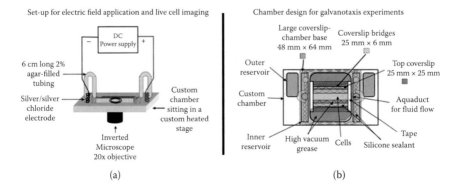

COLOR FIGURE 2.5 Apparatus used in a typical galvanotaxis experiment. The electrical connections used to apply the field are illustrated in (a). The design of the experimental cell chamber is shown in (b). (From Pullar, *J. Wound Technol. Iss.*, 6, 20–24, 2009.)

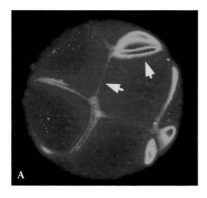

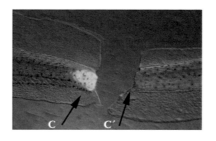

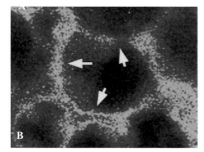

COLOR FIGURE 3.1 Ascertaining transmembrane potential *in vivo* using voltage-sensitive dyes. Fluorescence of the voltage-sensitive dyes DiBAC and DiSBAC (Krotz et al., 2004; Wolff et al., 2003) reveals transmembrane potential in early frog embryo blastomeres (A) as well as COS cells in monolayer culture (B). Both kinds of cells exhibit significant variations of membrane voltage level around the cell surface, indicating that a single V_{mem} number for a given cell drastically underestimates the amount of information that can be encoded in the plasma membrane's physiological state and potentially communicated to neighboring cells. Blastemas of regenerating (C) and nonregenerating (C') tadpole tails differ significantly in their membrane voltage. (Images in A and B courtesy of D. S. Adams.)

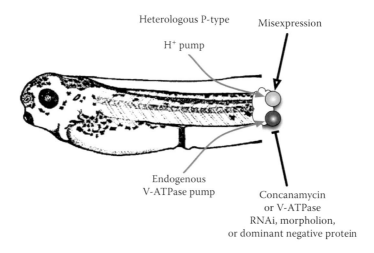

COLOR FIGURE 3.2 Strategy for molecular rescue experiment to test role of ion transporter in patterning. In tadpoles, the V-ATPase H⁺ pump complex is required to regenerate the tail after amputation (Adams et al., 2007). A rescue experiment, in which endogenous V-ATPase is inhibited by concanamycin (pharmacological blocker) or morpholino/RNAi (genetically), can be performed by misexpressing a heterologous (yeast) proton pump, PMA1 (Bowman et al., 1997; Masuda and Montero-Lomeli, 2000), which bears no sequence or structure similarity to the V-ATPase. The resulting restoration of regenerative ability (Adams et al., 2007) proves that it is the bioelectrical signal (proton pumping), not some cryptic other role of V-ATPase proteins, that is responsible for the induction of tail regeneration. Different ion transporter proteins with similar physiological functions can be used to test loss and gain of function for individual ion fluxes in many contexts, with molecular specificity.

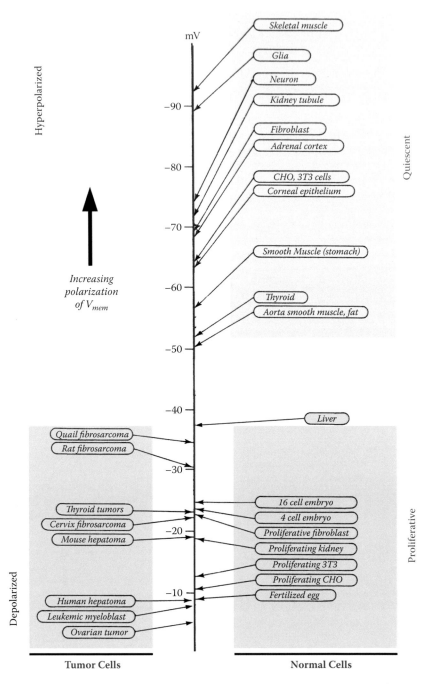

COLOR FIGURE 3.3 Transmembrane potential as a determinant of proliferation and plasticity. Membrane voltage is correlated to, and indeed controlled by, proliferative potential and differentiation state. This sample of data (from Binggeli and Weinstein, 1986) illustrates the observation that tumor and embryonic (proliferative) cells tend to be highly depolarized; in contrast, terminally differentiated quiescent cells tend to be strongly polarized. Data suggest that this relationship is functional and not merely an epiphenomenon.

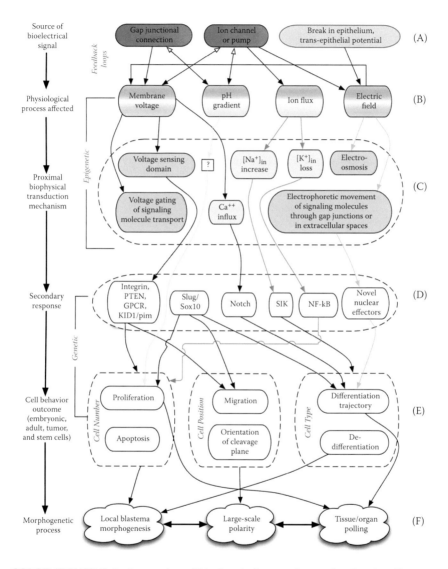

COLOR FIGURE 3.4 Integration of bioelectrical events into molecular signaling cascades. The arrows indicate sample cases where the whole pathway has been traced for bioelectrical control of patterning. (From Levin, *Semin. Cell. Dev. Biol.*, 20, 548, 2009.)

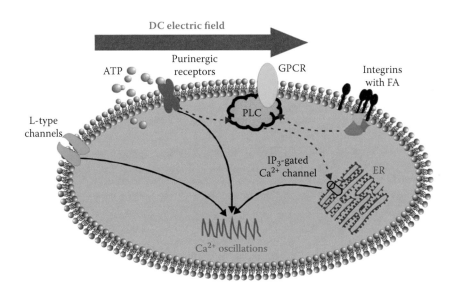

COLOR FIGURE 4.2 Potential electrocoupling mechanisms of hMSC calcium dynamics modulated by electrical stimulation. The coupling mechanisms of dc and low-frequency oscillatory fields are likely confined to the cell surface, and might include multiple types of calcium channels (voltage-gated calcium channels (VGCCs) and stretch-activated cation channels (SACCs)), membrane receptors (e.g., G-protein-coupled receptors), and integrins. As a result, an electrical stimulus leads to the PLC-mediated signaling through the IP_3 pathway, intracellular ATP secretion, and ATP-sensitive purinergic receptor activation. All these pathways likely contribute to $[Ca^{2+}]_i$ oscillation modulation by an external electric field.

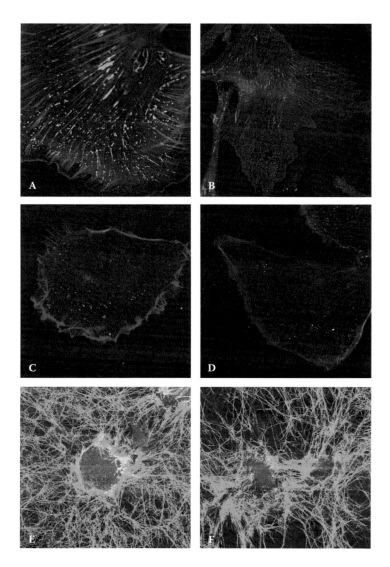

COLOR FIGURE 4.3 Actin cytoskeleton remodeling and cell adhesion modulation by dc electric field exposure. (A–D) Immunofluorescent images of a thin filamentous actin meshwork (red) and vinculin (green) showed that hMSCs contain thick actin stress fibers and multiple large adhesion contacts (A). In contrast, mature osteoblasts showed fewer and smaller focal adhesions (C). Actins and focal adhesions were partially disassembled in both hMSCs (B) and osteoblasts (D) after exposure to a 2 V/cm dc field for 60 min in serum-free saline. (E,F) Stem cells in three-dimensional collagen gel (1 mg/ml) were visualized using second harmonic generation signals from collagen (green) and cell tracker dye fluorescence (red). Collagen fiber bundles are involved in tight stem cell adhesion in the three-dimensional scaffold (E), and only partially relax in response to a 10 V/cm electrical stimulus without significant cell reorientation (F).

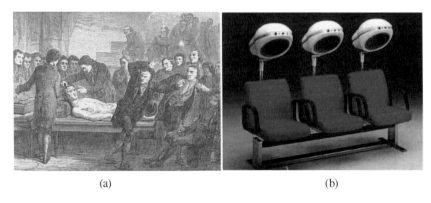

(a)   (b)

COLOR FIGURE 5.1 (a) A demonstration of galvanism, the raising of a corpse using electricity. The print is called "The Galvanisation of Matthew Clydesdale" (see http://scienceonstreets.phys.strath.ac.uk/galv05play.html). (b) Electrical "therapies" for baldness.

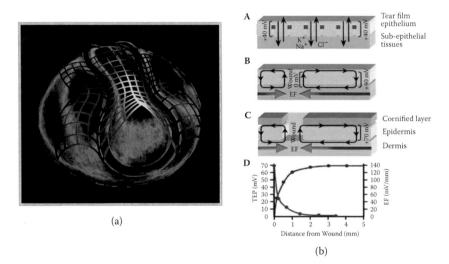

(a)   (b)

COLOR FIGURE 5.2 (a) The skin potential across amphibian embryos varies spatially and temporally. A pseudocolored map of the skin p.d. is superimposed on the dorsal surface of an amphibian embryo during neurulation. The rostral end of the embryo faces outward. Higher TEP values are shown rostrally (yellow) than caudally, and higher TEP values are evident more medially (yellow) than laterally (pink/purple). (Taken from Shi and Borgens, *Dev. Dyn.*, 202, 101–14, 1995.) (b) Model of transporting corneal epithelium. (A) A single epithelial layer is shown (there are around six in cornea) with each cell separated by tight junctional electrical seals (brown squares). Inward transport of Na^+ and K^- and active efflux of Cl^- create a transepithelial potential (TEP) difference of around +40 mV, inside positive. (B,C) Wounding the epithelial layer in cornea (B) and skin (C) short-circuits the TEP, which falls to 0 mV at the wound, but remains high distal from the wound edge. This induces a flow of ionic current (black arrowheads), and a steady voltage gradient is established with the cathode at the wound (EF red and green arrows). (D) Direct measurements of mammalian skin TEP (red line) as a function of distance from the wound edge. The EF (green line) resulting from the TEP gradient is 140 mV/mm very near the wound edge. The color gradients of the EF arrows in C and D indicate that the EF is strongest very near the wound, and that the potential gradient (hence EF) is steeper in the mammalian skin wound than in the corneal wound, where the TEP is smaller. Therefore, the gradient of TEP per unit distance would be expected to be smaller in the cornea. (Redrawn from Vanable, in *Electric Fields in Vertebrate Repair*, ed. Borgens et al., Alan R. Liss, New York, 1989, pp. 171–224, Figure 3. With permission.)

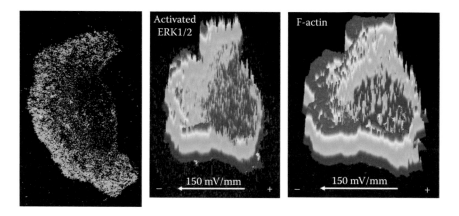

COLOR FIGURE 5.3 Cathodal accumulation of EGF receptor (left), second messenger molecules (middle), and the F-actin cytoskeleton (right) may underpin cathodally directed cell migration. In all three frames the cathode is at the left and the EF applied was 150mV/mm for three hours. (From Zhao, Pu, Forrester, and McCaig, unpublished.)

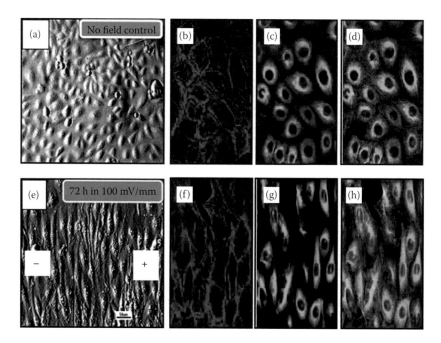

COLOR FIGURE 7.5 Perpendicular orientation and elongation of endothelial cells in a small physiological EF. (a) HUVEC cultured in the absence of an EF showed a typical cobblestone morphology and random orientation. (b–d) No-field controls showed no obvious alignment and cell elongation. (e) Cells exposed to a small applied EF showed dramatic elongation and perpendicular orientation in the EF (seventy-two hours, 100 mV/mm). (f–h) Most actin filaments (red) and microtubules (green) became aligned along the long axes of the cells (twelve hours at 150 mV/mm). (a,e) Images taken with Hoffman modulation optics. (d,h) Merged images scale bar = 50 µm. (Modified from Zhao et al., *J. Cell Sci.*, 117(Pt 3), 397–405, 2004.)

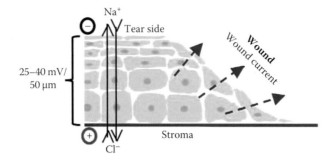

COLOR FIGURE 7.7 Mechanisms of endogenous electric fields at a corneal epithelium wound. Active pumping of Na⁺ from the apical (tear) side of the cornea across the epithelium and stroma to the basal side (aqueous humor), and of Cl⁻ from the basal to apical side, establishes a transepithelial potential difference of about 25 mV across 50 μm, with the inside positive. Damage to the epithelium disrupts tight junctions that maintain high resistance. This generates a short circuit at the wound site, which drains high concentrations of ions toward the wound from surrounding tissue. This generates a net flow of positive charge out of the wound into the tear solution, i.e., an endogenous wound electric current (blue dashed arrows). This wound electric current can be measured, and pharmacological manipulation by ion substitution or drug treatment to alter ion flow should enhance or decrease the capacity of the batteries, and therefore the wound electric current. (Redrawn from Reid et al., *FASEB J.*, 19, 379–86, 2005.)

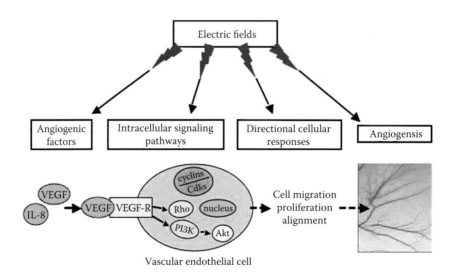

COLOR FIGURE 7.8 Schematic diagram shows the electrical control of angiogenesis. Electrical stimulation may affect many molecular and cellular components of angiogenesis. Application of EFs upregulates production of pro-angiogenic factors, notably VEGF, in endothelial cells and many other types of cells. Important intracellular signaling pathways PI3K and ERK may be activated, and in a directional manner in direct current EFs, which results in directional organization of the cytoskeleton and directional cell migration and alignment. When multiple cells organize into vessels, the organized directional behaviors may orientate the formation of blood vessels.

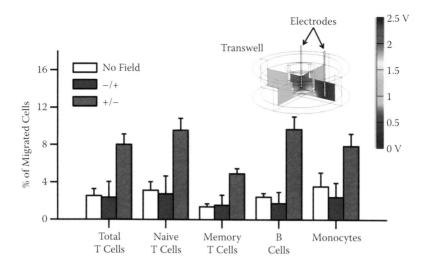

COLOR FIGURE 8.1 Migration of human blood lymphocytes and monocytes in an applied electric field. (For complete description, please see page 179.)

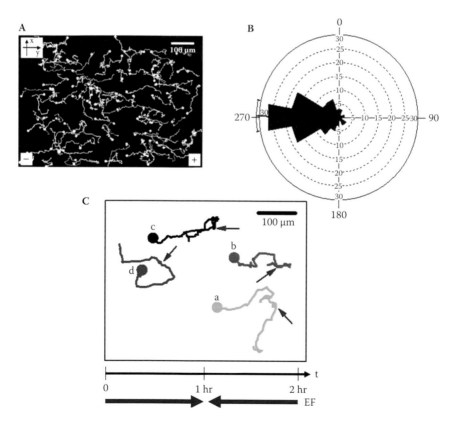

COLOR FIGURE 8.2 Directional migration of memory T cells in an applied electric field *in vitro*. (For complete description, please see page 180.)

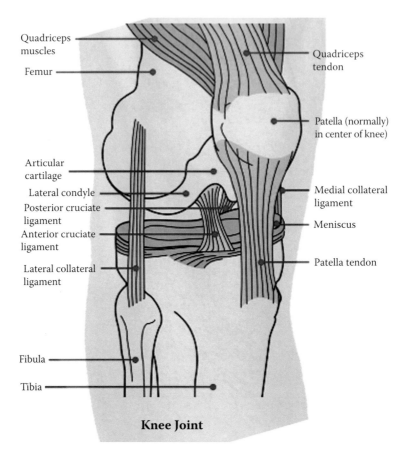

COLOR FIGURE 9.1 Anatomy of the human knee.

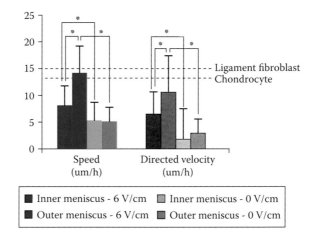

COLOR FIGURE 9.2 Cell speed and directed velocity of inner and outer meniscus cells. Asterisks represent conditions in which mean values show honestly significant differences (HSD) between groups. Data for each group were analyzed by one-way ANOVA, followed by Tukey's post hoc test of the mean for each group to determine whether differences between them exceeded the computed HSD. Dashed lines represent average speed of ligament fibroblasts and chondrocytes at an applied DC electric field strength of 6 V/cm.

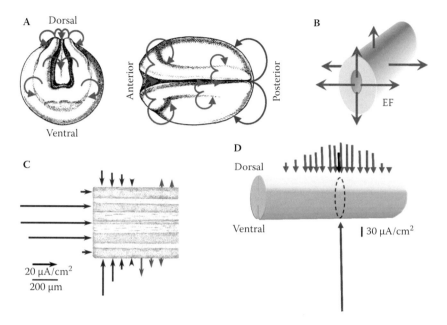

COLOR FIGURE 10.1 Ionic currents in the developing and regenerating nervous systems. (A,B) A stage 17 *Xenopus* embryo (Nieuwkoop and Faber, 1994). The neural ridges have not yet fused completely, particularly at the anterior region where the brain will develop. The red arrows show ionic current loops around the developing neural tube. By convention biological ionic currents are drawn to show the direction of flow of positive ions. (A) The drawing on the left is a face-on view of the anterior end of the embryo, and the drawing on the right is a dorsal view. The neural ridges and the central neural groove can be seen dorsally. The blue arrow represents the outward blastopore current. (B) At a later developmental stage (> stage 21), after the neural tube has sealed and has moved away from the overlying skin. Na+ ions are transported away from the lumen. (C) Ionic current pattern near the severed lamprey spinal cord (Borgens et al., 1980). Blue arrows are inward current and red arrows are outward current. (D) Current pattern measured in an *ex vivo* preparation of guinea pig spinal cord (Zuberi et al., 2008). Red arrows are inward currents measured on the dorsal side of the cord near a crush injury (black oval). The black arrow indicates the median size of the maximum dorsal current. The maximum current near the injury is higher at the ventral side of the cord (blue arrow).

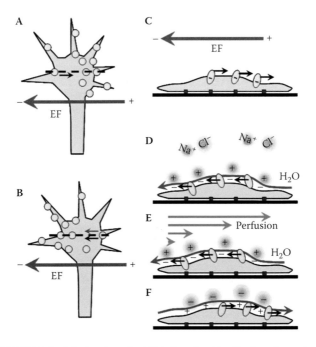

COLOR FIGURE 10.3 Mechanisms for EF-induced redistribution of membrane proteins in growth cones. (A,B) Membrane receptors (yellow) are moved within the plane of the plasma membrane by lateral electrophoresis (A) or electro-osmosis (B). The black arrow indicates the direction of receptor movement, and the blue arrow in B indicates the direction of electro-osmotic fluid flow over the cell surface. (C–F) View through a cross section of the growth cones in A and B at the level of the dashed lines. Growth cone to substratum contacts are indicated by squares. (C) Lateral electrophoresis. The receptors, which have a net negative charge on the extracellular domain, are drawn toward the anode. (D) Electro-osmosis. Na^+ ions accumulate near the membrane to counter its negative charge. The Na^+ ions are drawn toward the cathode of the EF, dragging their shells of hydration with them passively. The result is fluid flow (blue arrow) very near the cell surface in the direction of the counter-ion flow (in this case, toward the cathode). This is sufficient to move membrane proteins in the direction of fluid flow. (E) Perfusion of the bulk medium in a direction opposite to electro-osmotic fluid flow does not affect electro-osmotic membrane protein redistribution because the bulk fluid flow does not extend sufficiently near the cell surface. (F) Changing the charge of the cell surface reverses the direction of electro-osmotic fluid flow and the direction of receptor accumulation in an EF.

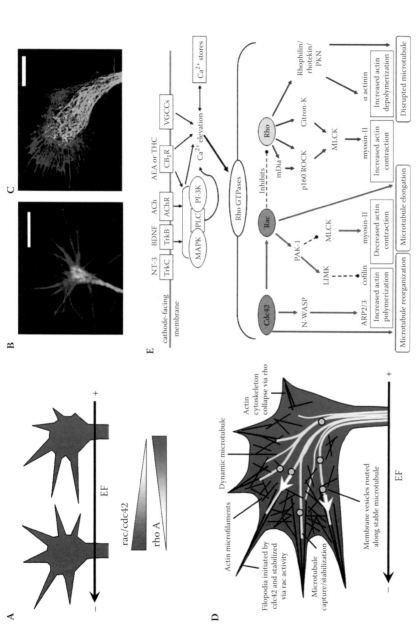

COLOR FIGURE 10.4 Hypothetical mechanism for cathodal turning of *Xenopus* growth cones. (A) The EF establishes a gradient of rac/cdc42, activities relative to rhoA, such that rac/cdc42 is higher cathodally and rho A is higher anodally. This gradient is amplified by their mutual antagonism. (B,C) Rho GTPase signaling affects local cytoskeletal organization in response to the EF. Red staining shows actin microfilaments and green shows microtubules. Scale bars are 10 μm. (B) Control growth cone with no EF applied. (C) Growth cone exposed to 150 mV/mm for five hours, cathode is at left. This cell was treated with a peptide that inhibits cdc42 signaling, reducing the extent of filopodia at the periphery. The growth cone has turned toward the cathode and the anode-facing side has collapsed. (D) Cytoskeletal events underpinning cathodal turning. (E) Molecular signals at the cathode-facing membrane. (Modified from Rajnicek et al., 2006a, 2006b.)

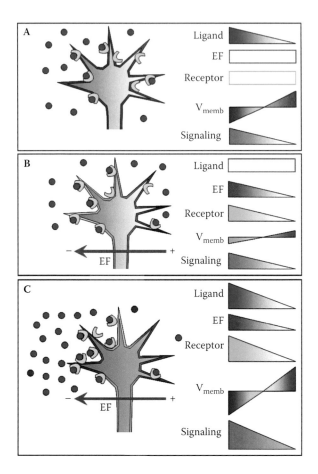

COLOR FIGURE 10.5 Hypothetical mechanism for interaction of chemical and electrical gradients in growth cone guidance. (A) A growth cone in a gradient of a chemoattractive ligand (high at left). Ligand (blue dots) binds to appropriate receptors (green) and the membrane potential (V_{memb}) is affected; the membrane becomes depolarized by ~15 mV on the side of the growth cone facing the ligand source (left). Since there is more ligand at the left side of the growth cone, "attractive" signaling is stimulated in the cytoplasm and the growth cone turns toward the left, up the ligand gradient. (B) A growth cone bathed in both a uniform ligand distribution and an extracellular voltage gradient (EF) (cathode at left). The EF on its own does not affect V_{memb} significantly at physiological EFs, but receptors for the ligand have been redistributed toward the cathode by the EF (see Figure 10.3). Even though ligand is uniformly distributed, attractive cytoplasmic signaling is biased cathodally due to locally enhanced ligand-receptor interactions. Therefore, the cathode is analogous to a chemoattractant. (C) Interaction of electrical and chemical signals could amplify the cytoplasmic signaling gradient. The EF could establish, stabilize, or amplify a gradient of a charged chemoattractant while also redistributing its membrane receptors cathodally. Receptors may cluster (e.g., AChRs) once they reach a critical density, forming stable aggregates resistant to back-diffusion. This would increase receptor-ligand binding frequency and with consequent depolarization of the cathode-facing membrane. This polarized V_{memb} change would boost the modest effects on V_{memb} by the EF itself (depolarized cathodally). This additive effect would amplify the cathode-facing attractive cytoplasmic signaling, perhaps sufficiently to open voltage-gated channels asymmetrically. The growth cone would therefore turn to the left (up the concentration gradient and toward the cathode). Although this model describes an attractive cue, the anode is analogous to a repulsive cue (including membrane hyperpolarization), so both sets of events could happen simultaneously in spatially restricted sides of the growth cone, further amplifying the signaling gradient.

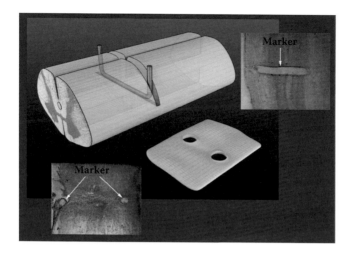

COLOR FIGURE 11.1 The placement of a U-shaped spinal cord marking device. The device is removed just prior to histology. The illustration and actual histological images show the fudicial holes left in the tissue. In the top right, the elongated hole reveals the base of the devices.

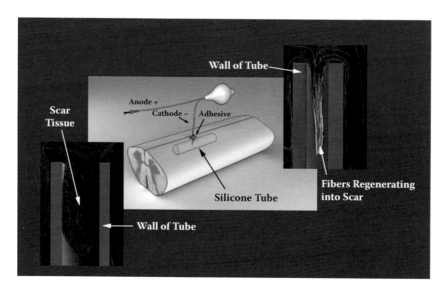

COLOR FIGURE 11.2 A cathode inserted within the hollow tube implanted within the dorsal spinal cord. The two insets show the plug of scar tissue in control tubes and the robust growth of axons into electrically active tubes. (From Borgens, *Neuroscience*, 91, 251–64, 1999.)

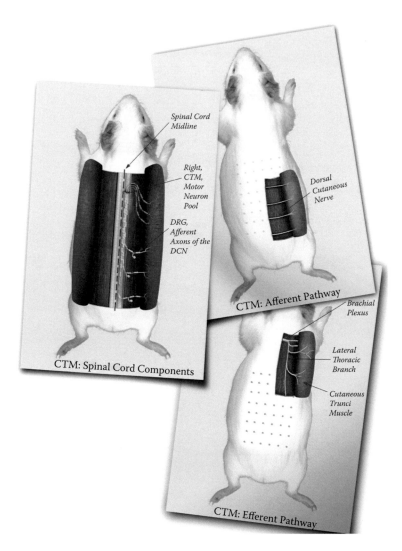

COLOR FIGURE 11.3 The anatomical components of the guinea pig CTM reflex. The tattooed dots on the back skin provide quantitative information about movement of the skin after direct and local stimulation. The process of stimulation before and after injury is videotaped from above, and stop frame analysis of the movement of the dots when the skin contracts achieves the following data: (1) the boundaries of function and nonfunctional receptive fields, (2) the vector of skin movement relative to a local stimulation, (3) the speed of contraction, (4) the latency of contraction, and other such quantitative evaluation of this behavior. (From Blight et al., *Journal of Comparative Neurology*, 296(4), 614–33, 1990.)

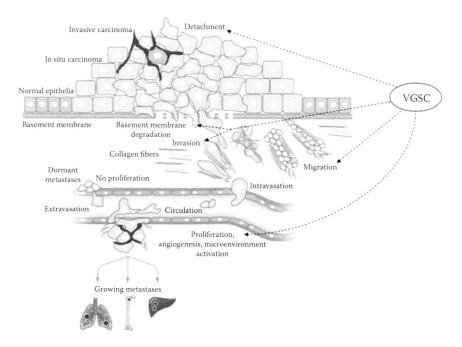

FIGURE 12.1 The metastatic cascade. Transformation of normal epithelial cells results in carcinoma *in situ*. Reduced adhesiveness and enhanced migratory behavior of tumor cells progress the disease to an invasive stage. After degradation of the basement membrane, tumor cells invade the surrounding stroma, migrate, and intravasate into the lymph or blood circulation, and surviving cells arrest in the capillaries of a distant organ. Formation of secondary tumors occurs after proliferation, induction of angiogenesis, and microenvironment activation. Voltage-gated sodium channels (VGSCs) have been shown to be involved in controlling various components of the metastatic cascade, as indicated by the arrows. (Modified from Bacac and Stamenkovic, *Annu. Rev. Pathol.*, 3, 221–47, 2007.)

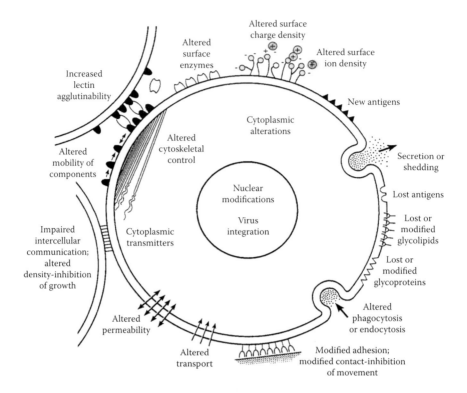

FIGURE 12.4 Subcellular changes in cancer. The figure illustrates these for a hypothetical cell. A variety of changes occur that affect the cell surface, cytoplasm, and intercellular junctions. Of particular relevance to this article are the alterations in surface charge density, surface ion density, permeability, and transport. (Modified from Nicolson, *Biochim. Biophys. Acta*, 458, 1–72, 1976.)

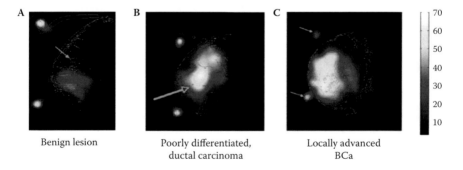

A	B	C
Benign lesion	Poorly differentiated, ductal carcinoma	Locally advanced BCa

FIGURE 12.10 ^{23}Na-MRI images from three different breast cancer patients. (A) A benign lesion (proliferative fibrocystosis), indicated by the arrow, aligned from gadolinium-enhanced image. (B) Infiltrating poorly differentiated ductal carcinoma (outlined in blue). Below (indicated by a green arrow) is a region with edema. (C) A large locally advanced breast cancer (outline in middle blue). The arrows indicate positioning landmarks. The intensity scale on the far right indicates approximate sodium concentrations (relative). (Modified from Ouwerkerk et al., *Breast Cancer Res. Treat.*, 106, 151–60, 2007.)

Printed and bound by CPI Group (UK) Ltd, Croydon, CR0 4YY

18/10/2024

01776270-0001